# Sustainable Products in the Circular Economy

This book explores how the Circular Economy influences product design in today's business and society. Drawing on contributions from a wide range of expert thinkers, this book assesses the existing approaches, strategies, and tools which facilitate socially and environmentally responsible production and consumption systems. It then goes on to highlight the ways in which the Circular Economy conceptual framework could be implemented effectively at both micro-level (product policy) and macro-level (sustainable consumption) in order to alter the industrial landscape and increase its interconnectedness with materials and scarce resources. Highlighting the pros and cons of transitioning to this new model, the book also cautions that it will only be made possible via significant behavioural change at both industry and consumer levels. *Sustainable Products in the Circular Economy* will be of great interest to students and scholars of sustainable manufacturing, sustainable consumption, corporate social responsibility, and business ethics. It will also be relevant to industry professionals whose work dovetails with these areas.

**Magdalena Wojnarowska** is Assistant Professor at the Department of Technology and Ecology of Products at Cracow University of Economics, Poland.

**Marek Ćwiklicki** is Professor of Business Studies and Public Management and Head of Chair for Management of Public Organisations at Cracow University of Economics, Poland.

**Carlo Ingrao** is currently Assistant Professor in Commodity Science at the Department of Economics of University of Foggia, Italy.

**Routledge – SCORAI Studies in Sustainable Consumption**
Series Editors:
**Halina Szejnwald Brown**
*Professor Emerita at Clark University, USA.*
**Philip J. Vergragt**
*Emeritus Professor at TU Delft, The Netherlands; Research Professor at Clark University, USA.*
**Lucie Middlemiss**
*Associate Professor and Co-Director of the Sustainability Research Institute, Leeds University, UK.*
**Daniel Fischer**
*Associate Professor for Consumer Communication and Sustainability, Wageningen Research and University, The Netherlands.*

This series aims to advance conceptual and empirical contributions to this new and important field of study. For more information about The Sustainable Consumption Research and Action Initiative (SCORAI) and its activities please visit www.scorai.org.

For more information about this series, please visit: www.routledge.com/Routledge-SCORAI-Studies-in-Sustainable-Consumption/book-series/RSSC

# Sustainable Products in the Circular Economy

Impact on Business and Society

**Edited by Magdalena Wojnarowska,
Marek Ćwiklicki and Carlo Ingrao**

Routledge
Taylor & Francis Group

earthscan
from Routledge

LONDON AND NEW YORK

First published 2022
by Routledge
4 Park Square, Milton Park, Abingdon, Oxon OX14 4RN

and by Routledge
605 Third Avenue, New York, NY 10158

*Routledge is an imprint of the Taylor & Francis Group, an informa business*

© 2022 selection and editorial matter, Magdalena Wojnarowska, Marek Ćwiklicki and Carlo Ingrao; individual chapters, the contributors

The right of Magdalena Wojnarowska, Marek Ćwiklicki and Carlo Ingrao to be identified as the authors of the editorial material, and of the authors for their individual chapters, has been asserted in accordance with sections 77 and 78 of the Copyright, Designs and Patents Act 1988.

*British Library Cataloguing-in-Publication Data*
A catalogue record for this book is available from the British Library

*Library of Congress Cataloging-in-Publication Data*
A catalog record for this book has been requested

ISBN: 978-1-032-01701-3 (hbk)
ISBN: 978-1-032-01719-8 (pbk)
ISBN: 978-1-003-17978-8 (ebk)

DOI: 10.4324/9781003179788

Typeset in Times New Roman
by Apex CoVantage, LLC

# Contents

# Figures

# Tables

# Preface

In the current era of transformation, the conceptual framework of Circular Economy (CE) is increasingly becoming meaningful and relevant in terms of the future and competitiveness of enterprises, which are the after-effects of the Fourth Industrial Revolution. Programmes of integrating the concepts of the CE with industrial activities encompass alterations in manufacturing processes intended to mitigate their impact upon natural environment, the development of new ecological products, and redesign of existing business models. Alterations in manufacturing and distribution paradigms may be considered in terms of diverse aspects; however, they require participation and engagement of various groups of stakeholders. Thus, the primary aim of the intended book is to answer the following research question: what is the role of sustainable product in the CE?

The CE is, in fact, different from the linear economy, because it is essentially based upon slowing and closing resource loops: two features that can be considered to be complementary, rather than alternatives, to each other. In particular, slowing happens when long-life goods and product-life extension solutions are designed. Therefore, the utilisation time of products is extended and/or intensified, thereby contributing to the slowing down of resource flows, whereas closing happens when the loop between end-of-life and production is closed, with the consequence that post-use products are recirculated within the life cycle as zero-burden resources to produce secondary raw materials (Moraga et al., 2019). The CE plays a number of key roles in enabling industrial economics to pursue Sustainable Development. It is, indeed, acknowledged to be capable of harmonising ambitions favouring not only economic growth but also environmental protection, thereby opposing the conventional high-impact linear model of the economy (Hysa, Kruja, Rehman, & Laurenti, 2020).

In the twentieth century, environmental problems were often viewed just on a local scale due to the impact associated with product life cycles. However, today it is becoming clearer that these problems are much more complex and concern all phases of those cycles, from the extraction of material to waste or waste product management (Berkhout & Smith, 1999). Both product design and manufacturing processes have been developed already in the past to meet the needs of producing high-quality products at minimal cost in order to promote the company's competitiveness. Thus, recycling and recovery were taken into account, but they had to compete on purely economic terms with the use of primary raw materials and disposal (O'Brien, 1999).

Transition towards CEs will be feasible only after the operating model of all market players has been rearranged. Gradual transition from the linear to the CE is a strategic goal at the European level, which is dependent upon coordinated efforts of both authorities and society, since such transformation results in significant changes to the market. Therefore, the success of CE is reliant on active involvement of authorities in the creation of relevant legal frameworks, taking into account that CE will become a reality only when the intensification of research and development has been reflected in subsequent implementation of both technologies and relevant strategies.

As a consequence of the aforementioned explanations, the core content of the proposed book centres on the characteristics of resources management compliant with the CE principles. The adopted general assumption is that such a transformation determines the change from the current model of the linear economy into the CE at the level of consumers and enterprises.

This book addresses the impact of sustainable manufacturing in compliance with key principles of the CE. It focuses on exploring how CE influences product design in today's business and society. Respective chapters assess existing approaches, strategies, and tools, thus facilitating creation, promotion, and support of socially and environmentally responsible production and consumption systems. Services are incorporated as an inherent part of product design.

The book was conceived to fill a research gap in the extant scholarship. An analysis of both the advantages and disadvantages of applying the CE in the context of product policy with respect to sustainable manufacturing is missing in current literature. The integration of CE and sustainable product policy becomes crucial, inter alia, due to technological and social progress, as well as the adverse impact of current economic models. This lacuna translates into the requirement for behavioural change in the entire environment. Therefore, this book aims to fill the identified gap by integrating sustainable product design with CE principles, thereby providing a real chance to alter the industrial landscape and its interconnectedness with materials and scarce resources.

The content of this book comprises 16 chapters prepared by 35 authors specialising in areas related to sustainable products, including design, manufacturing, marketing, business models, and consumption.

In the first chapter, an explanation of sustainable products is provided based on manufacturing strategies, which have been launched in Europe, focusing in particular on Italy and Poland. Chapter 2 presents guidelines for eco-design, which require consideration before the commencement of integrated product manufacturing. Next, the third chapter demonstrates the use of Life Cycle Assessment. It shows how calculation assumptions impact the final result. Chapter 4 presents the most popular environmental certification schemes that can be applied in line with CE principles. Chapter 5 refers to environmental labelling as a communication tool for CE solutions. It thus serves to potentially inform e-customers. The sixth chapter depicts relations between sustainable manufacturing and the CE based on an example from the building sector. Chapter 7 identifies the enablers and barriers to transit towards a circular business model. Chapter 8 develops previous topics by focusing on the costs and benefits of such change. Chapter 9 is about the role of digitalisation in fostering the implementation of sustainable manufacturing. Chapter 10 shows how sustainability is being introduced in inter-firm networks. In the following Chapter 11, the exemplary case of a recycling company is discussed in terms of using deliberation as a tool for communication with stakeholders. The issue of cooperation among companies is also a topic of the next chapter. Chapter 12 presents the theme of how to integrate CE principles with supply chains. Chapter 13 focuses on consumer behaviour by addressing how young consumers perceive sustainable consumption. The same topic is undertaken in Chapter 14; however, in this case, from an economic perspective. Chapter 15 refers to the role of universities in the diffusion of CE in society. The last chapter (Chapter 16) raises the issue of the resilience of the CE as a concept.

**Funding:** The project was financed by the Ministry of Science and Higher Education within the "Regional Initiative of Excellence" Program for 2019–2022. Project no.: 021/RID/2018/19. Total financing: 11,897,131.40 PLN.

## References

Berkhout, F., & Smith, D. (1999). Products and the environment: An integrated approach to policy. *European Environment, 9*(5), 174–185.

Hysa, E., Kruja, A., Rehman, N. U., & Laurenti, R. (2020). Circular economy innovation and environmental sustainability impact on economic growth: An integrated model for sustainable development. *Sustainability, 12*(12), 4831. Doi:10.3390/su12124831

Moraga, G., Huysveld, S., Mathieux, F., Blengini, G. A., Alaerts, L., Van Acker, K., . . . Dewulf, J. (2019). Circular economy indicators: What do they measure? *Resources, Conservation and Recycling, 146*, 452–461. Doi:10.1016/j.resconrec.2019.03.045

O'Brien, C. (1999). Sustainable production – a new paradigm for a new millennium. *International Journal of Production Economics*, 1–7, 60–61. Doi:10.1016/S0925-5273(98)00126-1

# Abbreviations

| | |
|---|---|
| AHP | Analytic Hierarchy Process |
| AIJ | Aggregation of individual judgements |
| B2B | Business-to-business |
| CBM | Circular business models |
| CC | Cleaning company |
| CE | Circular Economy |
| CR | Consistency ratio |
| CSC | Circular supply chains |
| EC | European Commission |
| EDIMS | EcoDesign Integration Method for SMEs |
| EEFP | Environmental Evaluation Tool for Food Packaging |
| EOD | Environmental Objectives Deployment |
| ERPA | Environmentally Responsible Product Assessment Matrix |
| EU | European Union |
| FGIO | Online focus group interview |
| FND | Function Network Diagram |
| FSC | Forest Stewardship Council |
| GTBE | Glycol tertiary butyl ether |
| HTC | HydroThermal Carbonisation |
| IoT | Internet of Things |
| IS | Industrial symbiosis |
| ISO | International Organization for Standardization |
| LCA | Life Cycle Assessment |
| LCAM | Life Cycle Asset Management |
| LOHAS | Lifestyle of Health and Sustainability |
| R&D | Research and Development |
| RDF | Refuse-Derived Fuels |
| RES | Renewable energy sources |
| SBD | Sustainable Behaviour Design |
| SD | Sustainable Development |
| SSC | Sustainable supply chains |
| SSCM | Sustainable supply chain management |
| TBL | Triple bottom line |
| WSC | Water and sewage company |

# 1 Characteristics of sustainable products

*Magdalena Wojnarowska, Mariusz Sołtysik, and Carlo Ingrao*

## Introduction

The economy of the last 150 years has been based upon a one-way track model (take, make, use, and dispose) that was characterised by the extraction of resources for production and consumption and by no plans for reutilising waste or actively regenerating the economy (Venkata Mohan, Modestra, Amulya, Butti, & Velvizhi, 2016). Over time, that linear model of the economy has been shown to be responsible for a number of problems, mainly related to the fact that:

- Virgin materials are extracted faster than the capacity for their replenishment;
- Post-use products are often land filled or are treated in incineration plants, with the consequence that valuable and scarce natural resources are extracted anew – and so the original resources are lost for the manufacturing of new products;
- The unsafe way in which waste is managed, which is often characteristic of the linear economy, leads to hazardous substances that leach into soil, water, and air and thus generates alarming conditions of environmental pollution;
- The manufacturing and the transportation of products are responsible for extensive energy usage and environmental pollution.

Hence, linear economies can be considered to be totally unsustainable from each of the environmental, economic, and social dimensions of sustainability (Ingrao, Arcidiacono, Siracusa, Niero, & Traverso, 2018; Korhonen, Nuur, Feldmann, & Birkie, 2018).

In this context, the Circular Economy (CE) may represent a valid alternative, as it would help to maintain products, components, and materials at their highest level of utility and value (Ingrao, Arcidiacono, Siracusa, Niero, & Traverso, 2021; Webster, 2020).

A sustainable CE involves the design and promotion of products that last and that can be reused, repaired, and remanufactured before being recycled. This aspect is a priority when implementing CE models, as it best retains the functional value of products, rather than just recovering the material or energy content and continuously making products anew. The CE, in fact, outlines how to reuse,

DOI: 10.4324/9781003179788-1

repair, and recycle items, thereby reducing waste and, overall, increasing sustainable manufacturing and consumption. In addition to this, the approach contributes to saving energy and helping to avoid irreversible environmental damage due to the extraction and usage of resources at a rate that exceeds the aforementioned capacity of the earth to renew them (EU, 2019).

The CE is, in fact, different from the linear economy, because it is essentially based upon slowing and closing resource loops: two features that can be considered to be complementary rather than alternatives to each other. In particular, slowing happens when long-life goods and product-life extension solutions are designed. Therefore, the utilisation time of products is extended and/or intensified, thereby contributing to the slowing down of resource flows. Whereas closing occurs when the loop between end-of-life and production is closed, with the consequence that post-use products are recirculated within the life cycle as zero-burden resources to produce secondary raw materials (Moraga et al., 2019).

In an eco-design-based CE, consumable products such as food, drinks, cosmetics, and detergents should be produced with the minimum impact on resources and should be consumed generating as little waste as possible. While meeting these requirements, actions should be taken to minimise the emissions into the environment and impacts on the climate across the whole life cycle (EU, 2019). According to the EU (EU, 2019), this would result in less usage of resources, less waste, more jobs in repair and recycling sectors, and monetary savings, while maintaining the services provided by products.

Products that are obtained from such CE-based production systems can be considered to be more sustainable than the conventional ones (EU, 2019). So, in the light of the aforementioned points, there is evidence of the potential of CE to contribute to enhancing the sustainability of products and services from a life-cycle perspective. This can be considered as one main reason why CE has been receiving a lot of attention from researchers, decision- and policy-makers, and managers (Hysa, Kruja, Rehman, & Laurenti, 2020). As a matter of fact, Geissdoerfer, Savaget, Bocken, and Hultink (2017) highlighted that the CE is a condition for sustainability, as it acts like a regenerative system that minimises material and energy inputs as well as emissions and wastes. It is, however, desirable that CE-oriented measures are tested using tools like Life Cycle Sustainability Assessment (LCSA) and Life Cycle Assessment (LCA) already in the design phase, so that the CE can be truly effective in making material and energy commodities holistically sustainable from a life-cycle perspective (Ingrao et al., 2021).

So, it is in the light of this understanding that CE and sustainability are intricately connected and feed off each other. It is this interaction leading to the manufacturing of sustainable products that this chapter wants to explore with a focus on the Italian and Polish strategies in this area, in line with the overall objective of the book project that this chapter is a part of.

After an in-depth analysis of the state of the art in the CE, the chapter includes a section dedicated to exploring the ways the application of CE measures can enable products to be made sustainably.

At the end, in the second part of this chapter, CE strategies for sustainable product manufacturing are explored at the European level with a focus on those implemented by Poland and Italy.

## Circular Economy: analysis of the state of the arts

The interest towards CE has led to a divergence in views on the methods of assessment and measurement of implementation of the CE and an overwhelming number of different definitions that currently dominate the specialised literature on the subject. This results in a lack of conceptual clarity and of any accepted definition of the CE, as has also been documented by several studies in the literature.

All of these definitions relevantly address the different facets of the CE, with the consequence of generating discrepancies.

However, what those definitions seem to have in common is the vision of the CE as a sustainable economic model where economic growth is decoupled from material consumption through the reduction and recirculation of natural resources (Corona et al., 2019; Ingrao et al., 2021). In the CE, goods at the end of their life cycle as well as the waste generated during the manufacturing and use/maintenance of those goods are in fact reutilised as zero-burden resources. The latter are utilised as material inputs in recycling processes for the production of secondary raw materials that, then, are manufactured into value-added commodities (Ingrao et al., 2018, 2021).

Recent reviews of the literature seeking to identify the key conceptual elements of the CE and their relationships to other concepts, like Sustainable Development, point to the CE as an alternative model of production and consumption and even a growth strategy that allows resource use to be decoupled from economic growth, thus contributing to Sustainable Development (Geissdoerfer et al., 2017). Therefore, both Sustainable Development (SD) and the CE have now become key concepts for creating a sustainable, low-carbon, resource-efficient, and competitive economy. The relationship between SD and the CE is confirmed by a research done by González-Ruiz, Botero-Botero, and Duque-Grisales (2018), who indicated eco-innovation, eco-design, and waste management as the main trends in CE research, as well as the relations of the CE to Sustainable Development. Cecchin, Salomone, Deutz, Raggi, and Cutaia (2021) and others add that the concept of the CE proposes a rebuilding of the production and consumption system into a regenerative system by closing the entry and exit cycles of the economy, which could help in the transition to a sustainable future. Thus, the concept of the CE follows an evolutionary path similar to that of SD, but at a much faster pace (Cecchin et al., 2021). It should be emphasised that integrating Sustainable Development and the CE with industrial activities should include changes in production processes with a view to minimising their impact on the environment. This involves the development of new ecological products and even the redesign of the business model, which has several environmental and socioeconomic benefits (Kallis, 2011). Despite numerous studies on the relationship between the CE and SD, as noted by Millar, McLaughlin, and Börger (2019), it is still unclear how the CE promotes economic growth while protecting the environment and ensuring intra- and intergenerational social equality (Millar et al., 2019). Due to numerous doubts raised by authors in the literature on the subject, one can also find more critical voices regarding the CE, which questioned the potential attributed to the CE (Hobson, 2013; Lazarevic & Valve, 2017). The 2011 UNEP Report "Decoupling natural resource use and environmental impacts from economic growth" also reveals that related Sustainable Development concepts and approaches, such as industrial ecology (IE), eco-efficiency, and cleaner production (CP), have contributed to achieving relative but not absolute decoupling from production (UNEP, 2011). Also, according to Kiser (2016), economic growth clearly contradicts the concept of resource efficiency in the supply chain, because the goal of selling more materials and using fewer resources is an environmental paradox (Kiser, 2016). In addition to this, other authors have also questioned the thermodynamic parameters of the CE and emphasise the need to consider environmental impacts and resource consumption when implementing a CE strategy to avoid overestimating their benefits, which is not often done in practice (Bianchini, Rossi, & Pellegrini, 2019; Korhonen, Honkasalo, & Seppälä, 2018). Research by Zink and Geyer (2017) shows how separation can be weakened by the rebound effect (Zink & Geyer, 2017). The social consequences of implementing the CE, an often overlooked aspect in research to date, also need to be addressed (Murray, Skene, & Haynes, 2017; Sauvé, Bernard, & Sloan, 2016; Schulz, Hjaltadóttir, & Hild, 2019).

## Sustainable products

Increasing pressure to adopt a more sustainable approach to both product design and manufacturing is one of the key challenges facing industries in the twenty-first century. This situation is moreover influenced by the growing total number of products, the increasing diversity of products and their functions, new types of products being created as a result of innovation, global product turnover, and increasing product complexity (Thorpe, 2015). According to Garg (2015), the manufacturing sector accounts for almost half of the world's total energy consumption, which has doubled over the past 60 years. These are reasons why manufacturers are not only under enormous pressure to be competitive on the one hand through increased productivity, but, on the other hand, under enormous pressure to deliver more sustainable products (due to an increased awareness of environmental responsibility) as well.

Previous research combining the concept of Sustainable Development with products, however, focussed mainly on an ecological product, that is one that is beneficial for the environment (Bhardwaj, Garg, Ram, Gajpal, & Zheng, 2020; Biswas & Roy, 2015; Nuryakin & Maryati, 2020; Qiu, Jie, Wang, & Zhao, 2020; Sdrolia & Zarotiadis, 2019; Tezer & Bodur, 2021). According to Sdrolia (Sdrolia & Zarotiadis, 2019), there are around 50 definitions of green products. On the basis of these definitions, it can be concluded that ecological products aim to protect or improve the condition of the environment by saving energy and/ or resources and limiting or eliminating the use of toxic agents, pollution, and waste (Ottman, Stafford, & Hartman, 2006). Undoubtedly, product research in the context of their environmental impact has made significant progress in explaining how companies can develop greener products that should allow companies to be successful in this area, although this is not always the case (Hofenk, van Birgelen, Bloemer, & Semeijn, 2019).

It is emphasised that products manufactured in the production process interact directly and indirectly with the society (employees, business owners, community, and customers) throughout their life cycle. Therefore, it is necessary to optimise not only the environmental impacts, but also the economic and social ones in an integrated, holistic approach to sustainability (Lin, Belis, & Kuo, 2019).

The concept of developing sustainable products is, in fact, evolving as a key element in cleaner production and in the CE. In response to the shift in environmental policy and legislation (through initiatives such as Integrated Product Policy and Extended Producer Responsibility for packaging cars and electronics), there is an increasing legal, market, and financial pressure on the manufacturing industry to develop sustainable products (Maxwell & van der Vorst, 2003). Since 2001, the European Commission has been putting emphasis on promoting its Integrated Product Policy (IPP) which, as defined by the European Commission, aims at supporting the development of environmental product innovations to achieve a broad reduction in all environmental impacts throughout a product's life cycle (Commission of the European Communities, 2001). It will be important to harness the Green Markets Policy Toolbox through greening on both the demand (consumption) and supply (product development) sides. The IPP is in line with the growing trend in environmentally advanced European countries, towards a product-oriented environmental policy (Charter, 2001).

Research into the definition of sustainable products shows a lack of understanding of the fact that our planet itself is not a sustainable system. Only by adopting this assumption can a sustainable product be defined as:

> a product, which will give as little impact on the environment as possible during its life cycle. The life cycle in this simple definition includes extraction of raw material, production, use and final recycling (or deposition). The material in the product as well as the material (or element) used for producing energy is also included here.
>
> (Ljungberg, 2007)

Whereas Shuaib et al. define a sustainable product through the prism of Sustainable Development as: "Sustainable products are those that provide environmental, societal, and economic benefits while protecting public health, welfare, and the environment over their full commercial cycle". The authors of this definition also point out that the design and production of sustainable products must be based on a comprehensive approach that simultaneously takes into account the economic, environmental, and social aspects of the TBL. To achieve this, you need to focus on all phases of the product life cycle. Such a holistic approach also often requires the adoption of the 6Rs (reduction, reuse, recycling, recovery, redesign, and remanufacturing) which must be applied throughout the product life cycle to achieve a circular material flow (Shuaib et al., 2014).

According to (Ljungberg, 2007; Zhou, Yin, & Hu, 2009), in order to develop sustainable products, it is required to follow rules such as:

- Reducing the consumption of materials and energy in a product, including services, throughout its useful life;
- Reducing the emissions, dispersion, and toxin formation throughout the product life cycle;
- Increasing the amount of recyclable materials;
- Utilising renewable resources for production;
- Extending the useful life of the product;
- Minimising environmental impact throughout the product life cycle;
- Replacing products with services;
- Utilising "reverse logistics";
- Increasing the performance of the product during the use phase;
- Using materials with low environmental pollution;
- Limiting the use of rare materials;
- The choice of clean materials for the production process;
- Avoiding the generation of hazardous and toxic materials;
- The use of materials with low energy consumption.

In fact, it is extremely difficult to meet all sustainability demands throughout the entire life cycle (Anex & Lifset, 2014). Therefore, in practice, different types of sustainable products emphasise different aspects in relation to different stages of the life cycle. Assuming that a sustainable product is one that meets the challenges of Sustainable Development, that is generates ecological, economic, and social benefits, contributing to the protection of public health, welfare, and the environment throughout the entire life cycle, it can be concluded that it is a form of excellence, an ideal that manufacturers can constantly strive for, perfecting selected aspects of the product (Sanyé-Mengual et al., 2014).

A detailed analysis of selected instruments related to a sustainable product is presented in the following chapters, including:

- Eco-design;
- LCA;
- Management systems;
- Environmental labelling;
- Market contacts and product phases in the marketplace;
- Legislation and precautions;
- Cultural aspects;
- Fashion and trends.

## A review of CE-oriented strategies on the European, Italian, and Polish scales for sustainable product manufacturing

The information reported here has been taken from the report titled "Circular economy strategies and roadmaps in Europe: Identifying synergies and the potential for cooperation and alliance building" developed by the European Economic and Social Committee (Salvatori, Holstein, & Böhme, 2019). The report reviewed 33 CE strategy documents in support of the European Circular Economy Stakeholder Platform (ECESP).

Strategies were documented to be more effective when the CE was addressed comprehensively and broad partnerships were included in the spirit of the five elements of Sustainable Development (i.e. planet, people, partnership, peace, and prosperity). The inclusiveness of partnerships takes into account the number and type of the different players in strategies and the ways and opportunities for interactions between them. In this regard, the report highlighted that CE strategies have different degrees of inclusiveness in terms of transversal tools and policies, sectors approached, and partners involved. Involvement is considered through specific objectives that depend upon the country making the strategy and its priorities through governance structures or through a combination of the two.

All strategies developed at the European level have aimed to further the transition to a sustainable CE model through a strategic plan that clearly defines objectives and desired outcomes and to include milestones at the end of key-step developments. The transition is addressed comprehensively by considering all of the main stages in value supply chains, namely production, consumption, waste management, secondary raw materials, and innovation and investments. To pick up on what has been said earlier in this regard, Salvatori et al. (2019) emphasised that comprehensiveness is a key added value in the reviewed strategies and so should be taken into consideration in the subsequent, new CE strategies and/or should be maintained in the strategies currently in existence that, however, will be improved in the future. In addition to this, all the reviewed strategy documents were found by Salvatori et al. (2019) to:

- Provide overarching frameworks for in-progress initiatives in different sectors, by different actors, and at different steps in the value chain or development;
- Provide a common objective for each provided activity;
- Describe ways and approaches for transitioning towards a sustainable CE model by defining tools and roles to make the transition clear and transparent for the stakeholders; and so
- Effectively contribute to inspiring other actors to get involved in the transition.

Differences were recorded by Salvatori et al. (2019) based upon the territorial context, as territories can have different opportunities and challenges in making the move to the CE, such as density, industrial clusters, and natural resources.

Strategies were found to follow the approach of closing the material loops in specific supply chains and, alternatively, that of focussing upon integrated, horizontal approaches.

In the light of this, Salvatori et al. (2019) categorised the strategies reviewed, making a distinction between integrated strategies, restricted strategies, and broader strategies with a clear set of priorities.

Integrated strategies, like the ones of Italy, Poland, and other EU countries, represent around 30% of all strategies reviewed. They are politically driven, generally top-down, and focussed on larger geographic scopes. They are typical of territories where the concept of the CE is relatively new to the public debate, as their aim is more to steer public opinion than to provide tools for implementing a full-fledged CE model.

Restricted strategies do not address a wide range of sectors but rather are restricted to only one sector. By contrast, broader strategies represent the major group of strategies in numeric terms, with 19 documents out of a total of 33, and are to be found at all territorial levels and at different levels of CE development.

Among the economic sectors that were analysed in the strategies reviewed, those which recurred the most frequently are manufacturing, construction, waste processing, and the production of foods and feeds.

In addition to this, horizontal themes were addressed to introduce new innovative concepts and practices that contribute to the enhancement of circularity in the aforementioned sectors. The themes which recurred most were found by Salvatori et al. (2019) to be repairing, reusing, and refurbishing; public procurement; design and eco-design; and urban planning and development.

Manufacturing is taken into consideration as it presents some of highest potential for circularisation due to the large quantities of materials consumed and of waste generated. The aim is to ensure that waste is "designed out" of products and that product and process design is done in ways that enable the recycling, recovery, and remanufacture of materials.

Another sector that was found by Salvatori et al. (2019) to be extensively considered in CE strategies is construction, mainly because it is the largest consumer of resources and generates huge amounts of heterogeneous waste. In this regard, ensuring circularity of material flows is one of the key features of the CE that can contribute to enhancing the sustainability of the waste management system.

In the conclusion to their report, Salvatori et al. (2019) highlighted that the reviewed strategies provided a wide range of approaches in many different sectors that can most benefit from application of the CE. Overall, they found those strategies to touch upon the key aspects of the CE and to provide a very good understanding of the challenges and ways forward.

The review report did, however, highlight that there is an urgent need for strategies to develop approaches that are inclusive not just with regard to the value chain but also to the range of partners that Salvatori et al. (2019) recommend should be the widest possible.

Like many other member states, Italy and Poland have also developed a strategy for introducing the concept of a CE to the domestic economy. The CE strategies that two governments have put together appear to have the features recommended by Salvatori et al. (2019); they are reviewed in the following two sections.

## *A focus on the Italian strategy*

In 2017, the Italian government released the report titled "Towards a model of Circular Economy for Italy" (MiSE-MATTM, 2017), with the aim of providing a general framework on the principles of the CE and of defining the Italian strategic position on such an issue. This section is dedicated to reviewing and building upon the content of that report. The report is part of the process of implementation of the wider strategies that Italy has made for Sustainable Development by the Italian government and specifically contributes to defining objectives like efficient resource consumption and sustainable production and consumption.

In this regard, the "National Action Plan on Sustainable Production and Consumption", set out by Italian Law 221/2015, represents the essential point of departure and is also one of the effective tools available for implementing the national CE-oriented policies and strategies. Six macro-areas of intervention were addressed by the aforementioned Action Plan, as they were identified to form the base of the Italian production system and, also, identified to be highly burdening from an environmental point of view; they were as follows: small- and medium-sized enterprises (SMEs), production chains and districts, agriculture and food production, built environments, tourism, organised distribution, and sustainable consumption and behaviour.

This CE strategy framework was developed around the important challenge for Italy to adequately and effectively respond to the complex environmental and socio-economic dynamics, while maintaining the competitiveness of its production system. In this regard, the document highlighted the importance of making policies that are oriented towards sustainable innovation and, at the same time, increasing the competitiveness of the Italian production and manufacturing system. Therefore, Italy is required to initiate a paradigm shift that is based upon rethinking and redesigning the ways to consume and do business, taking it as the opportunity to develop new business models that maximise the value of "*Made in Italy*" and the role of SMEs.

In this regard, the report highlights that transitioning towards CE means culturally and structurally triggering a radical change that provides a profound revision of the Italian patterns of consumption to abandon the conventional, unsustainable linear model of the economy and establish a well-rooted trend in innovation.

An important part of the report was the one dedicated to an analysis of the Italian context, highlighting the importance of transforming necessities into opportunities. Key necessities were identified as improving both the efficiency and sustainability of resource consumption and waste management: the latter was identified by the Italian government as being central to the process of transitioning to a sustainable CE.

Opportunities are based upon designing products in a way that, when they reach the end of their useful life, they are treated as zero-burden resources to feed into downstream production cycles. From this perspective, Italy is a technologically advanced country, with a strong background in innovation and sustainability, which must necessarily move to adopt the current European vision of transition towards sustainable CE models by using opportunities to create and promote concrete initiatives.

As is known to be the case, the CE brings many environmental and social benefits and allows natural capital to be preserved by reducing pressure on resources and on land by reducing its use for the disposal of waste in landfills. This was highlighted in the report to be an issue of considerable importance for a country like Italy where the natural factor is actually one of the main levers of economic development, as shown by the growing demand for sustainable and cultural tourism prior to the COVID-19 pandemic.

In the report, it was highlighted that, from an economic point of view, building a CE means stimulating the creativity of Italian entrepreneurs as a function of the economic value embedded in the reuse and recovery of materials that, in this way, never become waste. In this regard, the report puts due emphasis upon the need to rethink the concept of "waste" and states that one of the key challenges that the transition poses to all stakeholders, from politicians to citizens, is to consider what is now waste as an element from which value can be extracted, *"a brick for a new production cycle"*.

In this regard, Italian SMEs are being called upon to invest effectively in research and development, with the aim of rethinking and changing their production models and of consolidating their presence in the global value chain. In addition to this, implementing and spreading the CE throughout the country would help to transform the current well-known problems of the Italian production and manufacturing system into opportunities for sustainable forms of innovation, improvement, and growth. According to the report, waste recycling and recovery in line with the principles of the CE to produce value-added material and energy commodities can help to make countries like Italy, which are poor in raw materials, less dependent upon foreign procurements, with lesser vulnerability to the volatility of market prices. The reduction of the dependence on foreign countries should be, however, coupled with the rationalisation of the production and consumption systems in order to optimise the costs of production activities with benefits for both businesses and citizens. This would increase the Italian competitiveness on the international scale, thanks to the higher quality at lower prices, and, to achieve that, the market for secondary raw materials needs to be developed and consolidated, as also recommended by Potting et al. (2017).

At the end, the importance for all CE-based actions to be measurable by indicators was highlighted in the report, as being essential to give substance to those actions to be pursued towards greater transparency for the market and for consumers, as well. From this perspective, the Italian government is making quite a lot of effort to identify suitable indicators that enable the circularity of the economy and the efficient usage of resources to be measured and monitored at the macro- and micro-level.

*A focus on the Polish strategy*

The Polish CE strategy was adopted by a resolution of the Council of Ministers in September 2019. Over 200 social and economic partners as well as representatives of central and local government administration participated in the development of the Circular Economy (CE) strategy for Poland. It should be emphasised that in the work on the strategy, existing experience related to the implementation of other CE-related concepts was used, such as Sustainable Development, green economy, or cleaner production. This procedure was deliberate, with the aim of achieving greater coherence of measures in the field of CE with measures in other areas of socio-economic development in Poland. As a result, the Circular Economy Road Map is one of the projects of the Strategy for Responsible Development, fitting into the overall vision of the country's development (Kuzincow, 2018).

Basically, it can be said that the Circular Economy Road Map, like that of Italy, is a document that contains a set of tools aimed at creating conditions for the implementation of a new economic model in Poland. The proposed activities relate primarily to analytical, conceptual, informational, promotional, and coordination tasks. The Polish Circular Economy Road Map is based upon the CE model, commonly used in the EU, and developed by the Ellen MacArthur Foundation, that is it assumes the existence of two biological and technical cycles (Webster, 2020).

The Polish Circular Economy Road Map consists of five chapters.

*Chapter I* "Sustainable Industrial Production" emphasises the important role of industry in the Polish economy and new opportunities for its development. It is noted that in Poland there is a great potential for improvement in the field of industrial waste, in particular from mining and quarrying activities, industrial processing, as well as energy generation and supply. Running a production activity that generates less and less waste as well as the management of as much industrial waste as possible from this activity in other production processes and in other sectors of the economy may significantly contribute to increasing the profitability of production in Poland and reducing its negative impact on the environment.

*Chapter II* "Sustainable Consumption" discusses actions aimed at consumers as part of the transformation towards a CE. Among them, attention is paid to ensuring the availability of information on repair and spare parts, better enforcement of warranties, eliminating false claims about environmental impact, and determining the maximum shelf life of a product without harming the consumer and the environment.

*Chapter III* "Bioeconomy" concerns the management of renewable resources (the biological cycle of the CE), which seems to have unexploited potential in Poland. The Circular Economy road map focuses, on the one hand, on general activities aimed at creating conditions for the development of the bioeconomy in Poland and, on the other hand, on activities related to the development of bioeconomy in selected areas, that is in the area of creating local value chains in industry and in the energy sector.

*Chapter IV* "New Business Models" indicates the possibilities of reorganising the ways of functioning of various market participants based on the idea of the CE. In this part of the Circular Economy Road Map, it mainly refers to business models of enterprises, understood as the sum of resources and activities that simultaneously serve to provide value for the customer and to "close the loop".

At the end, Chapter V concerns the implementation, monitoring, and financing of the CE. It should be emphasised that the concept of CE is firmly established in the country's strategic documents, including the SRD, the draft Productivity Strategy and the draft National Environmental Policy. As the basis for the country's development policy, these documents are, and will continue to be in the future, a reference point for directing the support system in the area of CE, including in particular the Cohesion Policy and the Common Agricultural Policy (Rada Ministrów, 2019).

In line with European practice, the goal of the Circular Economy Road Map is to indicate horizontal measures that would concern the largest possible segment of socio-economic life. It also prioritises areas whose development will make it possible to take advantage of the opportunities facing Poland and, at the same time, will address the currently existing or expected threats. Poland's priorities within the CE include innovation, strengthening cooperation between industry and the science sector, and the implementation of innovative solutions in the economy as a result; creating a European market for secondary raw materials in which their circulation is easier; ensuring high-quality secondary raw materials that result from sustainable production and consumption; and development of the service sector (Smol, Kulczycka, Czaplicka-Kotas, & Włóka, 2019).

## Conclusion

One of the key factors determining Sustainable Development is the ability to associate the laws of ecology and economy in decision-making processes (Fernandes, Limont, & Bonino, 2020). It is essential that this process takes place at all institutional levels, both at the level of policy of states and enterprises and at that of households. Therefore, balancing economic goals with environmental and social goals is a big challenge not only for modern producers and consumers, but also for governments, social organisations, and other economic actors (Eisenmenger et al., 2020).

Environmental protection requirements have a significant impact on enterprises, including due to the applicable legal regulations that regulate it (Jose et al., 2020). However, environmental protection is perceived as a source of additional costs, because, for example, enterprises have to budget for the growing costs of using the environment and outlays for environmental protection in their budgets. Therefore, modern company management should perceive environmental protection as an integral part of the management process (Haldar, 2019).

To this end, it is necessary to change the current linear model of the economy into a sustainable circular one. The objective is to achieve the highest possible level of recovery and recycling of waste and then its re-management in production.

Creating a CE model requires meeting certain conditions while promoting a policy based on renewable resources in natural processes.

The key task will also be to develop products focussed on the production of products and services that are safe for the environment. It is possible to implement this principle by giving products ecological features already at the design stage. It is therefore important to design in a way which allows the transfer of waste with certain properties back to the production process or for its use by other entities. At the design stage, it is also recommended to use one of the models of operation within the CE, that is the ReSOLVE model or the R strategy. The ReSOLVE model is implemented through six paths of action, that is regenerate, share, optimise, loop, virtualise, and exchange. The R strategy is to reduce the consumption of resources and materials throughout the entire life cycle. It allows for the formulation of a CE strategy while maintaining the basic function of the product. It is therefore important to stimulate innovation in the field of environmentally safer products, not only through the development of cleaner technologies but also cleaner products through the dissemination of a life-cycle approach.

It is also worth returning to re-examine new phenomena in the sphere of consumption. Not only the state but also other market participants, including consumers (Tunn, Bocken, Van Den Hende, & Schoormans, 2018), must undertake activities with the aim, *inter alia*, of promoting environmental protection. That is why it is so important to raise environmental awareness and shape a modern image of effective economic processes based on ethical and ecological components (Nikolaou, Tsalis, & Evangelinos, 2018). The condition for the functioning of the CE model is reliable knowledge resulting from the high environmental awareness of all market participants. One of the key challenges, therefore, is to develop a system that not only educates consumers about the environmental impact of products throughout their life cycle, but also, at the same time, gives producers the opportunity to inform consumers about the benefits of their products. One such solution is eco-labelling, which is considered to be one of the key tools of consumer education in the field of environmentally friendly products (Bertrandias & Bernard, 2017; Buelow & Lewis, 2010; Martino, Nanere, & Dsouza, 2019). One of the undoubted benefits of buying organic products is that it reduces the negative impact of humanity on the environment and thus helps us achieve the main goals of sustainable production and consumption. It depends, however, on the increased environmental awareness of consumers. Consequently, environmental education is a key communication and information tool, and its aim should be to make the consumer able to consciously interpret the eco-label and make the right product choices based on it. If this condition is not met, an overabundance of information from advertising and marketing campaigns will lead to target audience members misinterpreting messages from senders. However, if they are to have a positive effect, the eco-label must be scientifically standardised and the environmental awareness of consumers raised.

Another challenge is to implement an effective strategy of replacing traditional products with sustainable products. A key element of the strategy should be to set the prices of sustainable products at the right level. The customer will be interested in such a product if he or she experiences a direct financial benefit. Preferential prices for this type of products can be achieved, for example, through differentiated taxation (applying a reduced VAT tax on biodegradable products). Therefore, it is important to create and support markets for more environmentally friendly products using, for example, preferential pricing and tax policies, a well-functioning environmental labelling system, and an administratively strong system of standardisation.

# References

Anex, R., & Lifset, R. (2014). Life cycle assessment – a guide to approaches, experiences and information sources. *Journal of Industrial Ecology, 18*.

Bertrandias, L., & Bernard, Y. (2017). The environmental labelling rollout of consumer goods by public authorities: Analysis of and lessons learned from the French case. *Journal of Cleaner Production, 161*, 688–697. doi:10.1016/j.jclepro.2017.05.179

Bhardwaj, A. K., Garg, A., Ram, S., Gajpal, Y., & Zheng, C. (2020). Research trends in green product for environment: A bibliometric perspective. *International Journal of Environmental Research and Public Health*. Doi:10.3390/ijerph17228469

Bianchini, A., Rossi, J., & Pellegrini, M. (2019). Overcoming the main barriers of Circular Economy implementation through a new visualization tool for circular business models. *Sustainability*. doi:10.3390/su11236614

Biswas, A., & Roy, M. (2015). Green products: An exploratory study on the consumer behaviour in emerging economies of the East. *Journal of Cleaner Production*. doi:10.1016/j.jclepro.2014.09.075

Buelow, S., & Lewis, H. (2010). The role of labels in directing consumer packaging waste. *Management of Environmental Quality: An International Journal, 21*(1477–7835), 198–213. doi:10.1108/14777831011025544

Cecchin, A., Salomone, R., Deutz, P., Raggi, A., & Cutaia, L. (2021). What is in a name? The rising star of the Circular Economy as a resource-related concept for sustainable development. *Circular Economy and Sustainability*. doi:10.1007/s43615-021-00021-4

Charter, M. (2001). Integrated product policy (IPP) and eco-product development (EPD). *Proceedings Second International Symposium on Environmentally Conscious Design and Inverse Manufacturing*, 672–677. doi:10.1109/ECODIM.2001.992445

Commission of the European Communities. (2001). *Green paper on integrated product policy*. Brussels: Author.

Corona, B., Shen, L., Reike, D., Rosales Carreón, J., & Worrell, E. (2019). Towards sustainable development through the Circular Economy – a review and critical assessment on current circularity metrics. *Resources, Conservation and Recycling, 151*, 104498. doi:10.1016/j.resconrec.2019.104498

Eisenmenger, N., Pichler, M., Krenmayr, N., Noll, D., Plank, B., Schalmann, E., Theres, M., & Simone, W. (2020). The sustainable development goals prioritize economic growth over sustainable resource use: A critical reflection on the SDGs from a socio-ecological perspective. *Sustainability Science, Sachs, 2012*. doi:10.1007/s11625-020-00813-x

European Commission (EU). (2019). *Sustainable products in a Circular Economy – towards an EU product policy framework contributing to the circular economy*. SWD(2019) 92 final. Retrieved from https://ec.europa.eu/transparency/documents-register/detail?ref=SWD(2019)91&lang=en

Fernandes, V., Limont, M., & Bonino, W. (2020, January). Sustainable development assessment from a capitals perspective: Analytical structure and indicator selection criteria. *Journal of Environmental Management, 260*, 110147. Doi:10.1016/j.jenvman. 2020.110147

Garg, A. (2015). Green marketing for sustainable development: An industry perspective. *Sustainable Development, 23*(5), 301–316. Doi:10.1002/sd.1592

Geissdoerfer, M., Savaget, P., Bocken, N. M. P., & Hultink, E. J. (2017). The Circular Economy – a new sustainability paradigm? *Journal of Cleaner Production, 143*, 757–768. Doi:10.1016/j.jclepro.2016.12.048

González-Ruiz, J. D., Botero-Botero, S., & Duque-Grisales, E. (2018). Financial eco-innovation as a mechanism for fostering the development of sustainable infrastructure systems. *Sustainability (Switzerland)*. Doi:10.3390/su10124463

Haldar, S. (2019). Towards a conceptual understanding of sustainability – driven entrepreneurship. *Corporate Social Responsibility & Environmental Management, 26*(6), 1157–1170. Doi:10.1002/csr.1763

Hobson, K. (2013). "Weak" or "strong" sustainable consumption? Efficiency, degrowth, and the 10 year framework of programmes. *Environment and Planning C: Government and Policy, 31*(6), 1082–1098. Doi:10.1068/c12279

Hofenk, D., van Birgelen, M., Bloemer, J., & Semeijn, J. (2019). How and when retailers' sustainability efforts translate into positive consumer responses: The interplay between personal and social factors. *Journal of Business Ethics*. Doi:10.1007/s10551-017-3616-1

Hysa, E., Kruja, A., Rehman, N. U., & Laurenti, R. (2020). Circular economy innovation and environmental sustainability impact on economic growth: An integrated model for sustainable development. *Sustainability, 12*(12), 4831. Doi:10.3390/su12124831

Ingrao, C., Arcidiacono, C., Siracusa, V., Niero, M., & Traverso, M. (2021). Life cycle sustainability analysis of resource recovery from waste management systems in a Circular Economy perspective key findings from this special issue. *Resources, 10*(4), 32. Doi:10.3390/resources10040032

Ingrao, C., Faccilongo, N., Di Gioia, L., & Messineo, A. (2018). Food waste recovery into energy in a Circular Economy perspective: A comprehensive review of aspects related to plant operation and environmental assessment. *Journal of Cleaner Production, 184*, 869–892. Doi:10.1016/j.jclepro.2018.02.267

Jose, C., Jabbour, C., Seuring, S., Beatriz, A., Sousa, L. De, Jugend, D., . . . Colucci, W. (2020). Stakeholders, innovative business models for the Circular Economy and sustainable performance of firms in an emerging economy facing institutional voids. *Journal of Environmental Management, 264*, 110416. Doi:10.1016/j.jenvman.2020.110416

Kallis, G. (2011). In defence of degrowth. *Ecological Economics, 70*(5), 873–880. Doi:10.1016/j.ecolecon.2010.12.007

Kiser, B. (2016). Getting the circulation going. *Nature, 531*.

Korhonen, J., Honkasalo, A., & Seppälä, J. (2018). Circular economy: The concept and its limitations. *Ecological Economics*. Doi:10.1016/j.ecolecon.2017.06.041

Korhonen, J., Nuur, C., Feldmann, A., & Birkie, S. E. (2018). Circular economy as an essentially contested concept. *Journal of Cleaner Production, 175*, 544–552. Doi:10.1016/j. jclepro.2017.12.111

Kuzincow, J. (2018). Packaging spectrum: Mapa drogowa transformacji w kierunku gospodarki o obiegu zamkniętym. *Opakowanie*. Doi:10.15199/42.2018.2.2

Lazarevic, D., & Valve, H. (2017). Narrating expectations for the Circular Economy: Towards a common and contested European transition. *Energy Research & Social Science, 31*, 60–69. Doi:10.1016/j.erss.2017.05.006

Lin, C. J., Belis, T. T., & Kuo, T. C. (2019). Ergonomics-based factors or criteria for the evaluation of sustainable product manufacturing. *Sustainability*, *11*(18), 4955. Doi:10.3390/su11184955

Ljungberg, L. Y. (2007). Materials selection and design for development of sustainable products. *Materials & Design*, *28*(2), 466–479. Doi:10.1016/j.matdes.2005.09.006

Martino, J. D., Nanere, M. G., & Dsouza, C. (2019). The effect of pro-environmental attitudes and eco-labelling information on green purchasing decisions in Australia. *Journal of Nonprofit & Public Sector Marketing*, 1–25. Doi:10.1080/10495142.2019.1589621

Maxwell, D., & van der Vorst, R. (2003). Developing sustainable products and services. *Journal of Cleaner Production*, *11*(8), 883–895. Doi:10.1016/S0959-6526(02)00164-6

Millar, N., McLaughlin, E., & Börger, T. (2019). The Circular Economy: Swings and roundabouts? *Ecological Economics*, *158*, 11–19. Doi:10.1016/j.ecolecon.2018.12.012

Ministero dello Sviluppo Economico (MiSE) – Ministero dell'ambiente e della tutela del territorio e del mare (MATTM). (2017). *Towards a model of Circular Economy for Italy – overview and strategic framework*. Retrieved from https://circulareconomy.europa.eu/platform/en/strategies/towards-model-Circular-Economy-italy-overview-and-strategic-framework

Moraga, G., Huysveld, S., Mathieux, F., Blengini, G. A., Alaerts, L., Van Acker, K. . . . Dewulf, J. (2019). Circular economy indicators: What do they measure? *Resources, Conservation and Recycling*, *146*, 452–461. Doi:10.1016/j.resconrec.2019.03.045

Murray, A., Skene, K., & Haynes, K. (2017). The circular economy: An interdisciplinary exploration of the concept and application in a global context. *Journal of Business Ethics*. Doi:10.1007/s10551-015-2693-2

Nikolaou, I. E., Tsalis, T. A., & Evangelinos, K. I. (2018). A framework to measure corporate sustainability performance: A strong sustainability-based view of firm. *Sustainable Production and Consumption*, *18*, 1–18. Doi:10.1016/j.spc.2018.10.004

Nuryakin, N., & Maryati, T. (2020). Green product competitiveness and green product success. Why and how does mediating affect green innovation performance? *Entrepreneurship and Sustainability Issues*. Doi:10.9770/jesi.2020.7.4(33)

Ottman, J. A., Stafford, E. R., & Hartman, C. L. (2006). Avoiding green marketing myopia: Ways to improve consumer appeal for environmentally preferable products. *Environment*. Doi:10.3200/ENVT.48.5.22-36

Potting, J., Hekkert, M., Worrell, E., & Hanemaaijer, A. (2017). *Circular economy: Measuring innovation in the product chain*. Retrieved from https://www.pbl.nl/sites/default/files/downloads/pbl-2016-circular-economy-measuring-innovation-in-product-chains-2544.pdf

Qiu, L., Jie, X., Wang, Y., & Zhao, M. (2020). Green product innovation, green dynamic capability, and competitive advantage: Evidence from Chinese manufacturing enterprises. *Corporate Social Responsibility and Environmental Management*. Doi:10.1002/csr.1780

Rada Ministrów. (2019). *Mapa Drogowa: Transformacji w kierunku gospodarki o obiegu zamkniętym*. Retrieved from https://www.gov.pl/web/rozwoj-technologia/rada-ministrow-przyjela-projekt-mapy-drogowej-goz

Salvatori, G., Holstein, F., & Böhme, K. (2019). Circular economy strategies and roadmaps in Europe: Identifying synergies and the potential for cooperation and alliance building. *European Economic and Social Committee*. Doi:10.2864/554946

Sanyé-Mengual, E., Lozano, R. G., Farreny, R., Oliver-Solà, J., Gasol, C. M., & Rieradevall, J. (2014). *Introduction to the eco-design methodology and the role of product carbon footprint* (pp. 1–24). Singapore: Springer. Doi:10.1007/978-981-4560-41-2_1

Sauvé, S., Bernard, S., & Sloan, P. (2016). Environmental sciences, sustainable development and circular economy: Alternative concepts for trans-disciplinary research. *Environmental Development*, *17*, 48–56. Doi:10.1016/j.envdev.2015.09.002

Schulz, C., Hjaltadóttir, R. E., & Hild, P. (2019). Practising circles: Studying institutional change and circular economy practices. *Journal of Cleaner Production*, *237*, 117749. Doi:10.1016/j.jclepro.2019.117749

Sdrolia, E., & Zarotiadis, G. (2019). A comprehensive review for green product term: From definition to evaluation. *Journal of Economic Surveys*. Doi:10.1111/joes.12268

Shuaib, M., Seevers, D., Zhang, X., Badurdeen, F., Rouch, K. E., & Jawahir, I. S. (2014). Product sustainability index (ProdSI). *Journal of Industrial Ecology*, *18*(4), 491–507. Doi:10.1111/jiec.12179

Smol, M., Kulczycka, J., Czaplicka-Kotas, A., & Włóka, D. (2019). Zarządzanie i monitorowanie gospodarki odpadami komunalnymi w Polsce w kontekście realizacji gospodarki o obiegu zamkniętym (GOZ). *Zeszyty Naukowe Instytutu Gospodarki Surowcami Mineralnymi Polskiej Akademii Nauk*. Retrieve d from http://yadda.icm.edu.pl/baztech/element/bwmeta1.element.baztech-bf5854ef-cec1-4dbf-bc30-126bda36d51c

Tezer, A., & Bodur, H. O. (2021). The green consumption effect: How using green products improves consumption experience. *Journal of Consumer Research*. doi:10.1093/JCR/UCZ045

Thorpe, A. (2015). *Sustainable consumption and production: A handbook for policymakers* (Global ed.). Nairobi: United Nations Environment Programme.

Tunn, V. S. C., Bocken, N. M. P., Van Den Hende, E. A., & Schoormans, J. P. L. (2018). Business models for sustainable consumption in the circular economy: An expert study. *Journal of Cleaner Production*. doi:10.1016/j.jclepro.2018.11.290

UNEP. (2011). Decoupling natural resource use and environmental impacts from economic growth. In *International resource panel*. Nairobi, Kenya: Author.

Venkata Mohan, S., Modestra, J. A., Amulya, K., Butti, S. K., & Velvizhi, G. (2016). A circular bioeconomy with biobased products from CO2 sequestration. *Trends in Biotechnology*, *34*(6), 506–519. doi:10.1016/j.tibtech.2016.02.012

Webster, K. (2020). *The circular economy, a wealth of flows*. Cowes: Ellen MacArthur Foundation.

Zhou, C. C., Yin, G. F., & Hu, X. B. (2009). Multi-objective optimization of material selection for sustainable products: Artificial neural networks and genetic algorithm approach. *Materials & Design*, *30*(4), 1209–1215. doi:10.1016/j.matdes.2008.06.006

Zink, T., & Geyer, R. (2017). Circular economy rebound. *Journal of Industrial Ecology*, *21*(3), 593–602. doi:10.1111/jiec.12545

# 2 Challenges of eco-design of integrated products

*Agnieszka Cholewa-Wójcik and*
*Agnieszka Kawecka*

## Introduction

The diversity of products, including packaging, offered on the market is largely related to the different needs and requirements of individual links in the supply chain and is an extremely desirable feature from the perspective of the end consumer. However, this diversity, in particular in terms of material or construction, may pose a significant problem from the point of view of the possibility of reusing post-consumer products and their packaging. One extremely important aspect in this respect is thus the transformation in terms of introducing solutions to the market offer that support, in the broad sense, a pro-ecological approach to creating projects, while, at the same time, meeting a number of requirements regarding their quality, safety, usability, and ergonomics, including visual attractiveness.

The purpose of this chapter is to present the assumptions involved in the eco-design of products and to develop principles of eco-design of integrated products, indicating the benefits in terms of social, economic, and environmental issues.

This chapter presents the requirements concerning the eco-design of products, which are included in European directives and international standards. In addition, solutions in the field of the eco-design of products, including packaging, are also presented in this chapter. An analysis is also conducted of national programmes and mechanisms supporting activities in favour of the implementation of the principles of product design in companies, taking into account the minimisation of the negative impact on the environment in Poland and Italy. The aforementioned analysis was used to develop principles of eco-design of integrated products, making recommendations for companies in the area of the manufacturing of integrated sustainable products as an innovation strategy in favour of greater ecological efficiency. The chapter concludes with an indication of the benefits in terms of technological, social, economic, and environmental aspects.

## Review of the literature concerning eco-design and a critical examination of the process

In the literature, the issue of eco-design of products is analysed by authors representing different scientific areas and disciplines, which results in a wide range of

DOI: 10.4324/9781003179788-2

views on this issue. The diversity of views on eco-design is already to be encountered at the definition stage. According to Platcheck, Schaeffer, Kindlein, and Candido (2008), eco-design is a holistic view in that, starting from the moment we know the environmental problems and their causes, we begin to influence the conception, the materials' selection, the production, the use, the reuse, the recycling, and final disposition of those products. Bhamra (2004), Alonso (2006), and Plouffe, Lanoie, Berneman, and Vernier (2011) define eco-design as an approach that integrates environmental criteria into the design of products in order to reduce environmental impact taking all stages in their life cycle. In turn, Wimmer, Züst, and Kun-Mo (2004) and Borchardt, Wendt, Pereira, and Sellitto (2011) define eco-design as a set of practices or ways of taking environmental issues into account in product design and development. There is thus a lack of any one accepted definition of eco-design, which could be considered as binding. For this reason, in the literature on the subject, the idea of eco-design is analysed with reference to its different aspects.

The solutions proposed in the literature are dominated by analyses concerning the environmental as well as the social aspects of eco-design. The question of the impact of the eco-design of products, the purpose of which is to minimise their environmental burden across the whole life cycle has been addressed by many researchers and in relation to the products of different sectors (Maccioni, Borgianni, & Pigosso, 2020). For example an analysis of the environmental context related to products and the inclusion of such considerations in the design process already at the early stage of development of that product has been conducted in the automotive sector by, among others, Bracke, Yamada, Kinoshita, Inoue, and Yamada (2017); in the electrical and electronics sector, by Mendoza, Sharmina, Gallego-Schmid, Heyes, and Azapagic (2017); in the food sector, by Shamraiz, Wong, and Riaz (2019); in the cosmetics sector, by Civancik-Uslu, Puig, Voigt, Walter, and Fullana-Palmer (2019); and in the packaging sector, by Gavrilescu, Campean, and Gavrilescu (2018), Foschi, Zanni, and Bonoli (2020), and Sumrin et al. (2021).

In turn, the issue of the social aspect eco-design has been the subject of works conducted, among others, by authors such as MacDonald and She (2015), Cor and Zwolinski (2015), Komeijani, Ryen, and Babbitt (2016), Paparoidamis, Tran, Leonidou, and Zeri (2019), Ketelsen, Janssen, and Hamm (2020), and Zeng, Durif, and Robinot (2021). An analysis of the level of knowledge of needs and expectations in the area of eco-innovation in packaging has been conducted by Ketelsen et al. (2020) and Zeng, Durif, and Robinot (2021). Komeijani et al. (2016), in their studies, developed strategies to provide guidance in developing products that will encourage consumers to use them in an environmentally sustainable way. Sustainable Behaviour Design (SBD) frameworks link common design concepts (ergonomic, emotional, preventative, and interaction design) with core consumer needs to create features to make users aware of their behaviour and decisions (reflective thinking) or result in sustainable behaviours even when users are unaware (automatic thinking). However, MacDonald and She (2015) conducted an analysis of pro-environmental behaviours and, on that basis, formulated recommendations concerning the eco-design of products. In turn, relations between the creation of value both for the environment and for consumers have been described by Maccioni, Borgianni, Pigosso, and McAloone (2020). The results of these studies allowed the author to make more precise recommendations in the area of eco-design.

In turn, in their studies, Paparoidamis et al. (2019) identified existing barriers limiting purchases of environmentally friendly products while indicating potential means of overcoming such barriers. The results of the studies also provided useful information concerning mechanisms of consumer response to eco-innovative designs of products from different industries (Maccioni, Borgianni, & Pigosso, 2019).

The second area of research identified in the subject literature concerns the possibility of using methods and tools of eco-design to measure the environmental burden of products across their entire life cycle. Le Pochat, Yannou-Le Bris, and Froelich (2007) classified and characterised seven basic groups of eco-design support tools: LCA, Simplified LCA, Eco-matrix, Checklist, Eco-parametric Tool, Guidelines, and Manual. In addition, they also developed the EcoDesign Integration Method for SMEs (EDIMS) to facilitate the integration of eco-design in companies. Byggeth and Hochschorner (2016) performed a comparison of such methods as ABC-Analysis (Tischner, Schmincke, Rubik, & Prosler, 2000), the Environmentally Responsible Product Assessment Matrix (ERPA) (Graedel & Allenby, 1995), MECO (Wenzel, Hauschild, & Alting, 1997), MET-Matrix (Brezet & van Hemel, 1997), Comparing tools Philips Fast Five Awareness (Meinders, 1997), Funktionkosten [Function costs] (Schmidt-Bleek, 1998), Dominance Matrix and Paired Comparison, EcoDesign Checklist, Econcept Spiderweb (Tischner, Schmincke, Rubik, & Prosler, 2000), Environmental Objectives Deployment (EOD) (Karlsson, 1997), the LiDS-wheel, The Morphological Box (Brezet & Hemel, 1997), and Prescribing tools Strategy List (Le Pochat et al., 2007). In these studies, tools were analysed to assess their utility in assessing environmental impact (Varžinskas, Kazulytė, Grigolaitė, Daugėlaitė, & Markevičiūtė, 2020). By contrast, Yi et al. (2020) propose using a tool developed for energy performance assessment (incorporating a simulation tool for energy consumption predictions, as well as an assessment model for energy performance and general workflows) instead of LCA. In turn, Yokokawa, Masuda, Amasawa, Sugiyama, and Hirao (2020) developed tools for the design of products, including packaging, such as Life Cycle Asset Management (LCAM) and the Function Network Diagram (FND), which integrate functional and environmental aspects across the entire life cycle. An interesting proposal for a tool supporting the process of eco-design of products, including packaging for the food industry, was proposed Molina-Besch and Pålsson (2020). The authors point out that one of the clear benefits of the Environmental Evaluation Tool for Food Packaging (EEFP) is that there is no need to have expert knowledge to use it. Furthermore, in the results obtained, no significant differences were observed between the proposed solution and LCA. Moreover, many authors draw attention to software tools, which are grouped into two categories: autonomous and CAD system-integrated. These programmes are described as being an important part of eco-design support (Dostajni, 2018).

It is worth drawing attention to the fact that another important criterion differentiating the view of the process of eco-design of products is that of the point of view of individual entities on the market, such as, inter alia, producers, consumers, and recyclers. Their approach to design taking into account the functioning of product in the Circular Economy may be considered to be insufficiently coherent due to the differences in the interests of individual entities.

From a business perspective, eco-design is one of the key areas conditioning the transition to the Circular Economy. Companies above all draw attention to legal requirements and the economic viability of the implemented solutions, together with the possibility of generating profit in the event of conducting activity properly in a systemic manner. The favourable impact on company development resulting from the application of new technologies and innovative solutions is also not without significance. It is important that the eco-design, taking CE guidelines into account, essentially consists of creating a circular organisation within the framework of economic structures. It is also worth highlighting that the key factor, inextricably linked to the functioning of companies on the market, is that of competitiveness. After all, alongside responsible manufacturers that are developing in the direction of eco-design, there are companies on the market that are focussed on short-term benefits. The policy of such companies makes positive changes on the market impossible to enforce, which generates real competitive inequality (Ekoprojektowanie. Nowe spojrzenie na biznes, 2017).

From the point of view of the consumer, eco-design is an approach which should result in the provision of environmentally friendly products, which will be functional, with the availability of a systematic update function extending the product life cycle, together with the possibility of using low-cost (including no-cost) options, as a result of putting products with a longer life cycle into circulation, which can be revitalised and offered as used. Furthermore, consumers expect communicative product labelling, showing both the ecological characteristics of the product, as well as the ways of using it resulting therefrom, and the ways of handling post-consumer waste. Labels should thus clearly specify the materials used and indicate appropriate handling of waste, including suitability for recycling. The introduction of a system of labelling that is uniform and easy for the customer to understand thus requires industry cooperation. This is after all an activity which is especially important in including consumers in the process of closing the loop in the economy. Engagement in building consumer awareness is key to the implementation of the concept of eco-design (Magnier & Crié, 2015).

For recyclers, however, an important criterion in the evaluation of the approach to eco-design and its role in the Circular Economy is above all that of compliance with legal requirements to which market practice has to be adapted and based on which new technologies are created. One exceptionally important problem is the possibility of obtaining quality waste and the ease of processing raw materials, which are currently perceived to be difficult. Moreover, the impact of eco-design on technologies required for the recycling of eco-designed products is also non-negligible.

The eco-design aspect of products requires a broad outlook and taking the interests of all parties into account. This is also why, when designing products in accordance with the principles of eco-design, all stages of a product's life should be taken into account, including adapting the product to extend its useful life and determining its suitability for recycling. The economic dimension should also be taken into account in eco-design: the profitability of production and sales and also the depth of consumers' wallets. In addition to this, the following are important: real availability and convenience of adapted solutions and the possibility of easy recycling of individual elements. However, in communications, it is important to disseminate the concept of product modularity, understood as the ease with which individual modules (complex elements) can be separated and combined without compromising the integrity of the product. Another important factor is the familiarisation of consumers with new forms of product ownership and use, combined with new business models.

Based on an analysis of the subject literature in the area of eco-design, it can be said that research studies have most often focussed mainly on an analysis of its environmental or social aspects together with an indication of the benefits, as well as certain inconveniences related to eco-design for individual entities. In addition, the subject literature also includes studies conducted concerning the possibility of using methods and tools of eco-design to measure the environmental burden of products across their entire life cycle. Despite the fact that, in certain studies, the issue of eco-design is increasingly frequently perceived to be a necessary element in activities concerning products within the framework of the Circular Economy, it is however worth noting that, in all the studies analysed, the issue of eco-design was analysed separately for products and packaging without taking the entire logistics chain into account. Therefore, the approach to conceiving of the final product as an integrated product was not taken into account in the eco-design process. Hence, an analysis of the scope of the research studies carried out to date and presented in the literature pointed to a need to undertake research to show the role and significance of a holistic process of eco-design of an integrated product, understood as an inherent combination of a product and packaging, which together constitutes the market offer. After all, only the mutual interaction and interpenetration of the stage of the design of products and their packaging allow for the implementation of the CE for the largest number of products. Such an approach to design on the one hand guarantees an impact of the product that places as little burden as possible on society, the economy, and the environment. On the other hand, it will promote raw materials and products created from them, which have been created with the concept of closing material cycles in mind and which can remain in the cycle longer than with a classic approach to the life cycle of products.

## Legal and normative requirements in the field of the eco-design of products

Eco-design is becoming the subject of scientific considerations but is also finding applications in practice. The principles of eco-design are successively being implemented in binding legal and normative requirements in European countries.

Energy-related products account for a large proportion of consumption of natural resources and energy, and they also have another type of, often significant, impact on the environment. This is also why the European Parliament and the Council adopted directive 2009/125/EC of 21 October 2009 establishing a framework for the setting of eco-design requirements for energy-related products in order for them to be placed on the market and/or put into service. "Eco-design" means the integration of environmental aspects into product design with the aim of improving the environmental performance of the product throughout its whole life cycle. Directive 2009/125/EC is closely related to the document concerning the Commission Recommendation of 9 April 2013 on the use of common methods to measure and communicate the life-cycle environmental performance of products and organisations. The recommendation promotes the use of the environmental footprint methods in relevant policies and schemes related to the measurement or communication of the life-cycle environmental performance of products or organisations (Directive 2009/125/EU) (Commission Recommendation of April 9, 2013).

The legal acts in force in the European Union also include very detailed requirements for specific categories of products such as small, medium, and large power transformers (Commission Regulation (EU) No. 548/2014 of May 21, 2014).

All legal acts of the European Union concerning eco-design have their origin in the Communication from the Commission to the Council and the European Parliament – Integrated Product Policy – Building on Environmental Life-Cycle Thinking COM/2003/0302. The Integrated Product Policy (IPP) aims for businesses to look for ways of minimising the negative impact of products and services on the environment by considering all stages in the product's life cycle and taking measures where they are likely to be most effective. IPP attempts to stimulate each of the elements associated with each individual stage of the life cycle to improve environmental performance (Commission to the Council COM/2003/0302).

Beyond legal requirements, the issue of eco-design is also addressed in normative documents. The question of a general approach of the organisation in establishing, documenting, implementing, maintaining, and continually improving their management of eco-design as part of an Environmental Management System (EMS) is set out in European standard EN ISO 14006:2020–08. This document is intended to be used by organisations that have implemented an environmental management system in accordance with ISO 14001, but it can also help in integrating eco-design using other management systems (EN ISO 14006:2020–08).

Other standards currently in force for specific subject areas and applicable to specific products include, for example:

- EN 50645:2017;
- EN 61800–9–2:2017;
- EN 16524:2020.

Despite a major commitment on the part of European Union, government, and standardisation bodies on the issue of limiting the impact of economic activity on the environment, the problem of eco-design has not been discussed in sufficient details, and its principles have not been introduced in the form of legal and normative requirements.

## Analysis of European Union projects and programmes related to eco-design

The countries of the European Union are leading the way in terms of legislative measures to support the design and production of products, including packaging, in accordance with the principle of Sustainable Development.

Proposals for solutions taking into account the practical application of eco-design of products in European Union countries, based on the example of packaging according to the Round Table Guidelines for the Eco Design of Plastic Packaging, are being developed based on four key elements of a strategy (Eco Design of Plastic Packaging, 2019):

- Design for Optimised Resource Use;
- Design for Sustainable Sourcing;
- Design for Environmentally Sound Use;
- Design for Recycling.

Taking the aforementioned four strategy elements into account in innovative packaging designs allows the shape and weight of packaging to be optimised, along with the materials used, while also allowing additional unnecessary elements of packaging to be eliminated.

The design of innovative products, including packaging which takes legal requirements concerning eco-design into account, is to a large extent stimulated by European Union projects and programmes. To analyse EU projects and programmes related to eco-design from the point of view of the effects of implementation, two European countries were selected, providing examples from Central, Eastern, and Southern Europe. Italy was selected as one of the largest producers of materials used in the eco-design of products (bioplastics) and as a country that possesses a developed composting industry. Poland was however selected as a representative of countries of Central and Eastern Europe due the fact that Polish businesses, compared to those in other countries of the region, are the least active with regard to innovation in production and the economy. It should however be underlined that both Poland and Italy are countries with an average score in the

"Summary Innovation Index" which was 64.1 for Poland and 90.1 for Italy (European Innovation Scoreboard, 2020).

The analysis of EN projects and programmes related to eco-design covered the period from 2014 to 2020 and took different forms of support in the area of eco-innovation into account. A list of the programmes analysed is provided in Table 2.1.

*Table 2.1* List of EU projects and programmes related to the eco-design of products

| Project | Description |
| --- | --- |
| Holistic Innovative Solutions for an Efficient Recycling and Recovery of Valuable Raw Materials from Complex Construction and Demolition Waste (HISER) Budget: 7.7 M | Development and presentation of technological and non-technological solutions for a higher recovery of raw materials and to allow them to be reused. |
| Energy Efficiency Complaint Products 2014 (EEPLIANT) Budget: 2.5 million | Support for market surveillance authorities to improve their effectiveness |
| INdustrial and tertiary product Testing and Application of Standards (INTAS) Budget: 1.9 M | Support for market surveillance authorities to guarantee compliance of products with the directive on eco-design |
| Improvement and sustainability of box production's process for e-commerce (FAST BOX) Budget: 71 thousand | Development of a system for the production of packaging to meet the needs of e-commerce |
| Optimised moulded pulp for renewable packaging solutions (PULPACKTION) Budget: 11 million | Development of new packaging solutions (including cellulose trays and barrier films made of biopolymers and starch) |
| ANTI-Circumvention of Standards for better market Surveillance (ANTICSS) Budget: 2 M | Support for market surveillance authorities and business in the area of harmonisation of legislation (incl. directives on eco-design) |
| New Circular Economy Business Model for More Sustainable Urban Construction (CINDERELA) Budget: 7.6 M | Development of a new circular economy model based on secondary raw materials. |
| Circular Process for Eco-Designed Bulky Products and Internal Car Parts (ECO BULK) Budget: 11.8 million | Activities promoting the role of eco-design and ecological aspects of production (including recycling and reuse) |
| A circular economy approach for lifecycles of products and services Budget: 7.2 million | Development of three circular economy models |
| Towards circular economy in the plastic packaging value chain (CIRC-PACK) Budget: 9.2 million | Eco-design and production of new, biodegradable packaging materials |
| Improvement of the plastic packaging waste chain from a circular economy approach (PlastiCircle) Budget: 8.7 M | Development and implementation of methods of increasing the proportion of materials from recycling. |

(*Continued*)

*Table 2.1* (Continued)

| Project | Description |
| --- | --- |
| High-performance-polyhydroxyalkanoates-based packaging to minimise food waste (YPACK)<br>Budget: 7.3 million | Project concerns innovative packaging derived from by-products of the food industry |
| Best markets for the exploitation of innovative sustainable food packaging solutions (MYPACK)<br>Budget: 5.8 million | Promotion of the development of biodegradable materials, using secondary raw materials |
| Novel packaging films and textiles with tailored end of life and performance based on bio-based copolymers and coatings (BIOnTop)<br>Budget: 5.4 million | Design and production of shopping bags made from biodegradable bioplastics |
| The transition of multilayer/multipolymer packaging into more sustainable multilayer/single polymer products for the food and pharma sectors through the development of innovative functional adhesives (MANDALA)<br>Budget: 4.6 M | Project presents sustainable solutions for the packaging sector made from plastics, which covers 3 pillars: eco-design, dual functionality and final disposition. |
| Unlocking the potential of Sustainable BiodegradabLe Packaging (USABLE PACKAGING)<br>Budget: 6.5 M | Development of raw materials, materials and packaging from biomass |
| High performance sustainable bio-based packaging with tailored end of life and upcycled secondary use (PRESERVE)<br>Budget: 8 million | Development of optimised packaging made from bioplastics |

An analysis of the review of the scopes of EU programmes and projects implemented in Poland and Italy demonstrated (as shown in Table 2.1) that, despite the different potential for the implementation of innovations in individual countries, the issue of developing and implementing innovative solutions is one of the most important challenges for these countries.

The purpose of the analysed programmes and projects was above all to develop and implement solutions in the field of the eco-design of specific products or groups of products. The execution of these projects falls within the policy areas of European Union bodies, which aim to strengthen the position of industry.

Achievement of goals in the area of Sustainable Development and the Circular Economy related to the development eco-innovation of products, including packaging, is also integral to the leading principles for the development of countries in modern times. These principles include the efficient management of resources, economic development, as well as reducing the burden on the environment and guaranteeing safety.

A detailed exploration of the final outcomes of programmes showed that, despite major investments in the creation of production, there is a visible lack of any holistic view of the process of eco-design. In all of the projects presented in Table 2.1, eco-design is treated as a separate process in the design of products and packaging, which takes account of the environmental aspect but without taking the inherent relationship between these products into consideration. It is necessary to view the design process from the perspective of development of the concept of an integrated product in which the packaging is treated as an inherent element of the product, so that end products appear on the market, whose designs take different aspects into account, that is not only the environmental aspects but technical and social aspects too. Furthermore, interrelationships between factors with an influence on the components used in the creation of an integrated product (quality of raw materials, methods and technological conditions of production, techniques and systems of packing, type of packaging and the cryptoclimate within it, properties of the packaging material, conditions of storage/use) should also be taken into account in the design of an integrated product.

## Guidelines for the eco-design of an integrated product

Directive 2009/125/EC (concerning eco-design for energy-related products) characterises eco-design as a preventive approach, designed to optimise the environmental performance of products, while maintaining their functional qualities and providing genuine new opportunities for manufacturers, consumers, and society as a whole. One key aspect is thus to analyse the impact of the product and its packaging on the environment throughout the whole life cycle, extend the durability of products, and also avoid substances that are harmful (to the environment and health) in the production process and in the product itself. The eco-design of end products should thus be viewed as a method of preventing the generation of post-consumer waste, which means the use by the producer in the integrated product design phase of technological solutions which take into account an assessment of environmental impact and a perspective that extends throughout the whole life cycle of the integrated product. The application of the guidelines for the eco-design of an integrated product, in which the packaging is treated as an inherent element of the product and which together constitutes the market offer, must be true for the entire life cycle of the product and its packaging. It is therefore not possible to design a sustainable integrated product without implementing the principles of eco-design in the design process (Directive 2009/125/EC).

Guidelines for the eco-design of an integrated product, taking into account the stages in the product life cycle, are shown in Figure 2.1.

Analysing guidelines for the eco-design of an integrated product, the first question that should be considered is that of the sourcing of raw materials, which should come from reliable sources allowing their quality to be confirmed. Thus, as part of a sustainable policy for the procurement of raw materials, it is important that businesses obtain documents from their suppliers confirming their environmental performance.

| Sourcing of raw materials |
| --- |
| • Sourcing of renewable raw materials from sustainable sources,<br>• Limited sourcing of non-renewable raw materials from sustainable production processes,<br>• Sourcing of secondary raw materials. |

| Production of integrated product |
| --- |
| • Optimisation of the physical characteristics of products and their packaging while maintaining appropriate parameters and quality characteristics of integrated products,<br>• Use of materials with limited negative environmental impact,<br>• Use of energy-efficient, low-waste production technologies. |

| Distribution of end products |
| --- |
| • Optimisation of products and packaging from the point of view of facilitating logistics processes,<br>• Use of as short supply chains as possible by low-emissions means of transport.<br>• Use of materials with limited negative environmental impact,<br>• Use of energy-efficient, low-waste production technologies. |

| Use |
| --- |
| • Maximum prolongation of period of use of products taking account of their functional characteristics,<br>• Rationalisation of energy consumption while maintaining product efficiency. |

| Final disposition |
| --- |
| • Identification of type of post-consumer waste,<br>• Easy separation of different materials,<br>• Taking practical and profitable opportunities for waste management into consideration. |

*Figure 2.1* Guidelines for eco-design taking into account the stages in the product life cycle

At the stage of production of an integrated product, it is important to pay attention to the optimisation of the physical characteristics of products and their packaging while maintaining appropriate parameters and quality characteristics of the end products. Thus, as part of a sustainable production policy, it is important to use materials with limited negative environmental impact, including secondary materials and materials for which material/raw material recycling or organic recycling technologies are available, while using energy-efficient, low-waste production technologies.

In turn, in the distribution process, the overriding goal is to reduce $CO_2$ emissions into the atmosphere, as well as consumption of fossil fuels and energy. Hence, there is the need to use intermodal transport using low-emission modes of transport or to limit the transport of goods by preferring local products or giving priority to products where the distance between the sender and the recipient is the shortest.

The primary goal of use is however to seek to maximise the period of use of products taking account of their functional characteristics and minimise energy consumption while maintaining product efficiency. In practice, this means putting products into use, which will be combined with multiple-use packaging, moving away from practices of product obsolescence and guaranteeing economically justified product repairs.

The last stage in the product life cycle is final disposition, which should be carried out via a properly organised and transparent system of segregation, return, and collection of waste. It is also important to use efficient, low-cost processes of recovery and recycling, enabling the process of conversion of waste into a raw material with maximum use value for further processing.

Taking the guidelines for the eco-design of an integrated product described earlier into consideration, attention should be drawn to the necessity of including the 10R strategy (Refuse, Rethink, Reduce, Reuse, Repair, Refurbish, Remanufacture, Repurpose, Recycle, and Recover) in the product life cycle analysis. This strategy is taken from the Circular Economy and involves taking 10 forms of action, whose purpose is to (Potting, Hekkert, Worrell, & Hanemaaijer, 2017):

- Refuse, which means giving up excessive, unnecessary consumption (consumerism);
- Rethink by looking for innovative alternatives that are safer for the environment;
- Reduce – reduction of consumption of raw materials, materials, and energy;
- Reuse – ensuring the possibility of multiple use;
- Repair – designing with the possibility of repairing the product in mind;
- Refurbish – restoring an old or used product and bringing it up to date, so that it is able to perform its original function;
- Remanufacture – regeneration or renovation consists of renovation and reuse of part of the product in a new product with the same function;
- Repurpose – change in the intended use of the product to perform a different function;

- Recycle – production of a new material or product from waste; and
- Recover – use in whole or in part of selected substances, materials, or energy.

The guidelines for the eco-design of an integrated product described before, taking the stages in the product life cycle and the 10R strategy into consideration, should be applied in the developed, general model of the process of eco-design of an integrated product, presented in Figure 2.2.

The model developed makes it possible to provide a concise description of the process to be followed in the eco-design of an integrated product and indicates the need to treat packaging as an inherent element of the product and its indispensable direct packaging, emphasising the need for them still to be considered of equal importance. At the same time, an analysis of the stages in eco-design makes it possible to determine common and divergent stages in the performance of design activities. The developed model takes account of the necessity of taking social, technical, economic, and ecological aspects into account at every stage in the design process. Such a perspective ensures optimisation of decision and a comprehensive approach to following an iterative process in the eco-design of an integrated product.

## Conclusion

The implementation of the assumptions of the developed model of the process for the eco-design of an integrated product in the development of a concept and the introduction of new products onto the market may result in the following groups of benefits being achieved, as shown in Table 2.2.

The benefits resulting from the application of guidelines for the eco-design of an integrated product and the 10R strategy in the development of a concept and the introduction of new products onto the market were divided into four main groups of environmental, technological, economic, and social benefits. The group of environmental benefits includes reduced consumption of energy and materials, reduced greenhouse gas emissions, a reduction in the weight of generated waste, savings on natural resources (rational sourcing of renewable and non-renewable raw materials, recovery of secondary raw materials of appropriate quality, and easy processing of secondary raw materials).

A change in the method of design to conform to the eco-friendly format may bring a range of technological benefits, including a positive impact on the company's development resulting from the use of new technologies limiting the negative impact on the environment, optimisation of production processes (an increase in effectiveness and efficiency and an improvement in the quality of manufactured products), and the introduction of innovative products that satisfy the assumptions of circular business models.

A reduction in costs related to production, transport, distribution, and post-consumer waste management was however included in the group of economic benefits. Moreover, other economic benefits indicated include stimulation

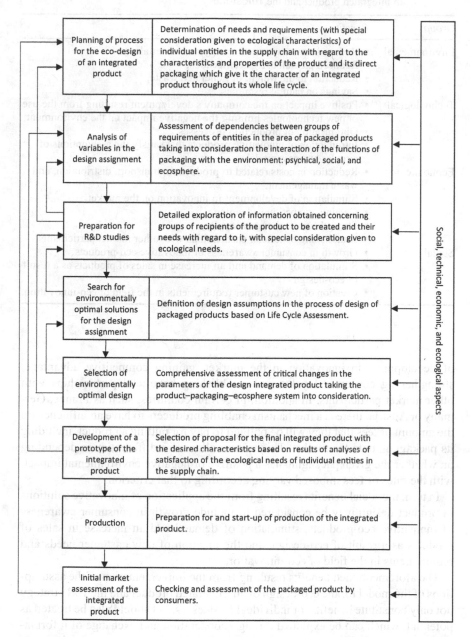

*Figure 2.2* Model of process for the eco-design of an integrated product
Source: Own elaboration based on (Cholewa-Wójcik 2018).

*Table 2.2* Groups of benefits for businesses implementing the model for the eco-design of an integrated product and the 10R strategy

| Group | Benefits |
|---|---|
| Environmental | • Reduced consumption of energy and materials,<br>• Reduced greenhouse gas emissions,<br>• Reduction in the weight of generated waste,<br>• Savings on natural resources. |
| Technological | • Positive impact on the company's development resulting from the use of new technologies limiting the negative impact on the environment,<br>• Optimisation of production processes,<br>• Introduction of innovative products that satisfy the assumptions of circular business models. |
| Economic | • Reduction in costs related to production, transport, distribution, and waste management,<br>• Stimulation of development in innovation on the market,<br>• Gaining competitive advantage,<br>• Strengthening consumer loyalty,<br>• Establishing beneficial partnerships with other market participants. |
| Social | • Growth in consumer awareness of innovative eco-products,<br>• Stimulation of demand and an increase in sales of products as a result of eco-design,<br>• Creation of new customer requirements in the field of eco-innovation. |

of development in innovation in the market, gaining competitive advantage, strengthening consumer loyalty, and establishing beneficial partnerships with other market participants. In addition, in certain countries, such as France, Germany or Austria, there is a mechanism enabling producers to have an influence on the amount of fees that they will be obliged to pay for putting a product, including its packaging, onto the market. This system allows for differentiation depending on whether the given packaging has a positive or negative environmental impact, with the rates of fees imposed varying according to that criterion.

In turn, the social benefits resulting from the application of innovative solutions in product design may be considered to include growth in consumer awareness of innovative eco-products, stimulation of demand and an increase in sales of products as a result of eco-design, and the creation of new customer needs and requirements in the field of eco-innovation.

The aforementioned benefits resulting from the implementation of the assumptions of the model for the eco-design of an integrated product and the 10R strategy not only constitute benefits for individual businesses, but should also be treated as potential, which can be exploited through cooperation and exchange of information between individual market participants. This is because cooperation between entities in the supply chain is essential when designing an integrated product in order for the social, technical, economic, and environmental impacts at every stage of its life cycle to be taken into consideration.

# References

Alonso, G. M. (2006). *La Norma de ecodiseño UNE1503001, CONAMA, Congreso Nacional del Medio Ambiente, en Los retos del desarrollo sostenible en España.* Madrid: CONAMA.

Bhamra, T. A. (2004). Ecodesign: the search for new strategies in product development. *Proceedings of the Institution of Mechanical Engineers, Part B, 218*, 557–569.

Borchardt, M., Wendt, M. H., Pereira, G. M., & Sellitto, M. A. (2011). Redesign of a component based on ecodesign practices: Environmental impact and cost reduction achievements. *Journal of Cleaner Production, 19*, 49–57.

Bracke, S., Yamada, S., Kinoshita, Y., Inoue, M., & Yamada, T. (2017). Decision making within the conceptual design phase of eco-friendly products. *Procedia Manufacturing, 8*, 463–470.

Brezet, H., & van Hemel, C. (1997). *Ecodesign: A promising approach to sustainable production and consumption.* France: United Nations Environment Programme, Industry and Environment, Cleaner Production.

Byggeth, S., & Hochschorner, E. (2016). Handling tradeoffs in ecodesign tools for sustainable product development and procurement. *Journal of Cleaner Production, 14*(15–16), 1420–1430. doi:10.1016/j.jclepro.2005.03.024

Cholewa-Wójcik, A. (2018). *Opakowanie i jego rola w projektowaniu produktu zintegrowanego w aspekcie potrzeb i wymagań konsumentów.* Kraków: Wydawnictwo PTTŻ.

Civancik-Uslu, D., Puig, R., Voigt, S., Walter, D., & Fullana-Palmer, P. (2019). Improving the production chain with LCA and eco-design: Application to cosmetic packaging. *Resources, Conservation and Recycling, 151*.

Communication from the Commission to the Council and the European Parliament – Integrated Product Policy – Building on Environmental Life-Cycle Thinking COM/2003/0302.

Commission Recommendation of April 9, 2013 on the use of common methods to measure and communicate the life cycle environmental performance of products and organisations, OJ L 124, 4.5.2013, pp. 1–210.

Commission Regulation (EU) No 548/2014 of May 21, 2014 on implementing Directive 2009/125/EC of the European Parliament and of the Council with regard to small, medium and large power transformers, OJ L 152, 22.5.2014, pp. 1–15.

Cor, E., & Zwolinski, P. (2015). A protocol to address user behavior in the eco-design of consumer products. *Journal of Mechanical Design, 137*(7).

Directive 2009/125/EC of the European Parliament and of the Council of October 21, 2009 establishing a framework for the setting of ecodesign requirements for energy-related product, OJ L 285, 31.10.2009, pp. 10–35.

Dostajni, E. (2018, September). Recycling-oriented eco-design methodology based on decentralised artificial intelligence. *Management and Production Engineering Review, 9*(3), 79–89.

Eco Design of Plastic Packaging Round Table and the Management Guidelines. (2019). *Bad Homburg, German association for plastics packagings and films.* Retrieved from https://ecodesign-packaging.org/wp-content/uploads/2019/10/ecodesign_core_guidelines_online.pdf

Ekoprojektowanie. Nowe spojrzenie na biznes, 2017. Circular Economy. Zamykamy obieg! Cykl warsztatów dla biznesu i jego otoczenia. Centrum UNEP/GRID-Warszawa.

EN ISO 14006:2020 Environmental management systems – guidelines for incorporating ecodesign.

EN 16524:2020 Mechanical products – methodology for reduction of environmental impacts in product design and development.

EN 50645:2017 Ecodesign requirements for small power transformers.

European Innovation Scoreboard. (2020). *European Commission, publications office of the European Union.* Luxembourg: Author.

Foschi, E., Zanni, S., & Bonoli, A. (2020). Combining eco-design and LCA as decision-making process to prevent plastics in packaging application. *Sustainability, 12*(22).

Gavrilescu, M., Campean, T., & Gavrilescu, D. (2018). Extending production waste life cycle and energy saving by eco-innovation and eco-design: The case of packaging manufacturing. In I. Visa & A. Duta (Eds.), *Nearly zero energy communities* (pp. 611–631). Cham: Springer.

Graedel, T. E., & Allenby, B. R. (1995). *Industrial ecology.* Englewood Cliffs, NJ: Prentice Hall.

IEC 61800-9-1:2017. *Adjustable speed electrical power drive systems – Part 9–1: Ecodesign for power drive systems, motor starters, power electronics and their driven applications – – General requirements for setting energy efficiency standards for power driven equipment using the extended product approach (EPA) and semi analytic model (SAM).* Retrieved from https://standards.iteh.ai/catalog/standards/clc/712bb10b-e932-4e4d-93d3-2f819f5c9990/en-61800-9-1-2017

Karlsson, M. (1997). *Green concurrent engineering: Assuring environmental performance in product development* (diss.), Lund University, Lund.

Ketelsen, M., Janssen, M., & Hamm, U. (2020). Consumers' response to environmentally-friendly food packaging – a systematic review. *Journal of Cleaner Production, 254.*

Komeijani, M., Ryen, E., & Babbitt, C. (2016). Bridging the gap between eco-design and the human thinking system. *Challenges, 7*(1), 5.

Le Pochat, S., Yannou-Le Bris, G., & Froelichm, D. (2007). Integrating ecodesign by conducting changes in SMEs. *Journal of Cleaner Production, 15*(7), 671–680.

Maccioni, L., Borgianni, Y., & Pigosso, D. (2019). Can the choice of eco-design principles affect products' success? *Design Science, 5*, E25. doi:10.1017/dsj.2019.24

Maccioni, L., Borgianni, Y., Pigosso, D., & McAloone, T. (2020). Are eco-design strategies implemented in products? A study on the agreement level of independent observers. *Proceedings of the Design Society: DESIGN Conference, 1*, 2039–2048.

MacDonald, E., & She, J. (2015). Seven cognitive concepts for successful eco-design. *Journal of Cleaner Production, 92*, 23–36.

Magnier, R., & Crié, D. (2015). Communicating packaging eco-friendliness: An exploration of consumers' perceptions of eco-designed packaging. *International Journal of Retail & Distribution Management, 43*(4–5), 350–366.

Meinders, H. (1997). *Point of no return: Philips EcoDesign guidelines, Philips Electronics N.V.* The Netherlands, Eindhoven: Corporate Environmental & Energy Office.

Mendoza, J., Sharmina, M., Gallego-Schmid, A., Heyes, G., & Azapagic, A. (2017). Integrating backcasting and eco-design for the circular economy: The BECE framework. *Journal of Industrial Ecology, 21*, 526–544.

Molina-Besch, K., & Pålsson, H. (2020). A simplified environmental evaluation tool for food packaging to support decision-making in packaging development. *Packaging Technology & Science, 33*, 141–157.

Paparoidamis, N., Tran, T., Leonidou, L., & Zeri, A. (2019). Being innovative while being green: An experimental inquiry into how consumers respond to eco-innovative product designs. *Journal of Product Innovation Management, 36*, 824–847.

Platcheck, E. R., Schaeffer, L., Kindlein, W., & Candido, L. H. A. (2008). Methodology of ecodesign for the development of more sustainable electro-electronic equipment. *Journal of Cleaner Production, 16*, 75–86.

Plouffe, S., Lanoie, P., Berneman, C., & Vernier, M. F. (2011). Economic benefits tied to eco-design. *Journal of Cleaner Production, 19*(6–7), 573–579.

Potting, J., Hekkert, M., Worrell, E., & Hanemaaijer, A. (2017). *Circular economy: Measuring innovation in the product chain*. The Hauge: PBL – Netherlands Environmental Assessment Agency.

Schmidt-Bleek, F. (1998). *Ecodesign, from the product to the service fulfillment machine (Ökodesign – Vom Produkt zur Dienstleistungserfüllungsmaschine)*. Vienna: Schriftenreihe Wirtschaftsf örderungsinstituts Österreich.

Shamraiz, A., Wong, K., & Riaz, A. (2019). Life cycle assessment for food production and manufacturing: Recent trends, global applications and future prospects. *Procedia Manufacturing, 34*, 49–57.

Sumrin, S., Gupta, S., Asaad, Y., Wang, Y., Bhattacharya, S., & Foroudi, P. (2021). Eco-innovation for environment and waste prevention. *Journal of Business Research, 122*, 627–639.

Tischner, U., Schmincke, E., Rubik, F., & Prosler, M. (2000). *How to do ecodesign? A guide for environmentally and economically sound design*. Berlin: German Federal Environmental Agency.

Varžinskas, V., Kazulytė, L., Grigolaitė, V., Daugėlaitė, V., & Markevičiūtė, Z. (2020). Eco-design methods and tools: An overview and applicability to packaging. *Environmental Research, Engineering and Management, 76*(4), 32–45.

Wenzel, H., Hauschild, M., & Alting, L. (1997). *Methodology, tools and case studies in product development, environmental assessment of products* (Vol. 1). London: Chapman Hall.

Wimmer, W., Züst, R., & Kun-Mo, L. (2004). *Ecodesign implementation: A systematic guidance on integrating environmental considerations into product development* (Vol. 6). Alliance for Global Sustainability Book Series. Berlin: Springer.

Yi, L., Glatt, M., Sridhar, P., de Payrebrune, K., Linke, B. S., Ravani, B., & Aurich, J. C. (2020). An eco-design for additive manufacturing framework based on energy performance assessment. *Additive Manufacturing, 33*, 101–120.

Yokokawa, N., Masuda, Y., Amasawa, E., Sugiyama, H., & Hirao, H. (2020). Systematic packaging design tools integrating functional and environmental consequences on product life cycle: Case studies on laundry detergent and milk. *Packaging Technology & Science, 33*, 445–459.

Zeng, T., Durif, F., & Robinot, E. (2021). Can eco-design packaging reduce consumer food waste? An experimental study. *Technological Forecasting and Social Change, 162*.

# 3 Verification of Circular Economy solutions and sustainability of products with Life Cycle Assessment

*Tomasz Nitkiewicz and Giulio Mario Cappelletti*

## Introduction

As mentioned in Chapter 1, the concept of the Circular Economy, at its core, refers to the recirculation of goods and materials through the application of different processes, such as the reuse of products, components, and materials through repair and refurbishment, remanufacturing, and recycling, respectively (EMF, 2013; Zink & Geyer, 2017).

Since the CE is being adopted as the current strategy for development, appropriate monitoring tools need to be able to capture different issues related to the implementation of CE solutions. Life Cycle Assessment (LCA) is very well suited to assess the sustainability impacts of sustainability and CE strategies. LCA is a science-based technique for assessing the impacts associated with entire product life cycles, which can provide technical support to CE decision-makers to assess trade-offs of impacts on a variety of environmental impact indicators and may also be applied to identify the most promising sustainability and CE strategies and options for improving the environmental performance of society's consumption and production patterns (Pena et al., 2020). The literature discusses three different levels on which CE strategies and solutions may be addressed: the micro-level (product), meso-level (eco-industrial park), and macro-level (city, nation or global), with significantly diversified approaches to quantify its means and measure its progress (Ghisellini, Cialani, & Ulgiati, 2016; Harris, Martin, & Diener, 2021; Saidani, Yannou, Leroy, Cluzel, & Kendall, 2019). This would certainly affect the reliability of use of LCA and the conducting of the LCA procedure, as well as the methodological choices and data sources needed for the assessment.

In order to provide support for CE decision-making, methods for assessing and quantifying the environmental benefits of CE strategies, such as LCA, are facing the challenge posed by the need to reflect the systemic context of an organisation (Schulz, Bey, Niero, & Hauschild, 2020). The objective of this chapter is to present the possible scope of use of LCA for the assessment of CE solutions. Since biopolymers are now regarded as the best possible replacements for fossil-based plastics, it is especially worth taking a look at possible manufacturing technologies that could contribute to the circularity of specific life cycles. The objective is achieved through the application of LCA to the experimental production of

DOI: 10.4324/9781003179788-3

biopolymers, namely mcl-PHA and P(3HB), from waste rapeseed oil. Biopolymer production through bacterial fermentation is used to close the loop, meet circularity requirements, and to explain important choices made as part of the assessment procedure.

## Potential of LCA use in Circular Economy solutions

LCA is currently the most commonly used framework for the verification and assessment of circular business models and sustainable products (Harris et al., 2021; Saidani et al., 2019; van Loon, Diener, & Harris, 2021; Walker, Coleman, Hodgson, Collins, & Brimacombe, 2018). The intention of LCA is to quantify the potential environmental impacts of a product system throughout its life cycle and, as such, is an appropriate tool for managing both sustainable products and CE solutions that are product-oriented. LCA obliges practitioners to derive appropriate conclusions and recommendations that reduce the impacts of a product and thus ensure greater sustainability and make micro-scale solutions circular (Dieterle, Schäfer, & Viere, 2018; ISO 14040:2006, 2006).

The key argument for the appropriateness of LCA is related to its holistic approach towards life-cycle orientation and its flexibility concerning both life-cycle phases and processes, as well as the approach taken to impact assessment. Despite its virtues, LCA is facing a big challenge as to how to include circularity assessment and provide an appropriate framework for data handling and allocation (Murphy, Detzel, Guo, & Krüger, 2011; Rufi-Salís, Petit-Boix, Villalba, Gabarrell, & Leipold, 2021). Since LCA is a decision-driven assessment, circularity requirements would mean this decision-making process would have to be placed in a new context.

The issue of the interrelationship between the Circular Economy and sustainability also has an influence on LCA and its scope. Additional variants of life-cycle-based assessment should be considered while CE and sustainable solutions need to be verified in a complex way. Life cycle costing (LCC) and Social Life Cycle Assessment (S-LCA) could provide additional insights, allowing for a better understanding of the circumstances and consequences of the implementation of CE and sustainability solutions. The economic perspective, as introduced by the LCC methodology, is usually considered within the framework of CE-related studies (Dieterle & Viere, 2021; Joachimiak-Lechman, 2014). The social perspective seems to be more peripheral as far as the LCA approach is concerned but could be beneficial for avoiding problem shifting between different stakeholders and to contribute to a more holistic sustainability assessment in a potential CE (Rusch & Baumgartner, 2020). To address complex issues, Niero and Hauschild (2017) recommend using the framework of Life Cycle Sustainability Assessment (LCSA) to evaluate Circular Economy solutions and to provide the most comprehensive and yet still operational framework for the assessment. Such an approach uses the major functionalities of LCA, LCC, and S-LCA and enables the prevention of burden shifting between stakeholders in the value chain.

LCA is in most of the cases confirmed to play a vital role in measuring the environmental benefits of CE solutions (Niero & Rivera, 2018). Although several attempts exist in the literature to ensure consensus and the standardised application of LCA for the different CE solutions, there are still some doubts as to methodological choices and the appropriateness of LCA for more complex scenarios, with repurposing, multiple end-of-life streams, or uncertainty over a number of potential life cycles (Ardente & Mathieux, 2014; Cooper & Gutowski, 2017; Richa, Babbitt, & Gaustad, 2017; Schulz et al., 2020). The complexity of CE solutions and its multiplication through sectoral applications causes the real problem with standardising the LCA approach to the assessment. Nonetheless, LCA methodologies are considered to be well placed to quantify the environmental implications and provide good coverage to feed circularity indicators (Harris et al., 2021; Walker et al., 2018).

## Methodological approaches to LCA

Within LCA, two general approaches to the assessment can be used (Chen & Fukushima, 2012; Earles & Halog, 2011; Ekvall, 2020; Ekvall et al., 2016):

- an attributional approach, where the assessment includes precisely every area of the environmental impacts that could be observed for a certain location and moment in time (static approach) and
- a consequential approach, where the focus is on those aspects of the environmental impacts that are changing due to the change in the life-cycle approach, including both direct and indirect outcomes and possibly, enlarging the system boundaries in order to include all the impacted areas.

Peters (2016) argues that a long-term consequential LCA offers the most appropriate, realistic and accurate view because it looks at the environmental effects of end-of-life scenarios related to the expansion of system boundaries to calculate the environmental effect of reuse and recycling yield rates and compare them against a system without the reverse streams. The condition for making the comparison with linear business models reliable is to include new services and their related impacts in the LCA (Kjaer, Pigosso, McAloone, & Birkved, 2018).

Concerning the process of allocation in the assessment, Wernet et al. (2016) propose the following models:

- A "cut-off" approach, where by-products' management and recycling processes, with their possible environmental impacts, are excluded from the assessment,
- Allocation at the point of substitution, where all the end-of-life processes are included in the assessment and contribute to its overall impact score,
- A small-scale and long-term consequential approach, where the system boundaries of the approach given before are extended in order to also cover indirect impacts.

LCA enables the use of a holistic perspective, as well as one focussed on specific phases. Depending on the scope and goals of the assessment, the system boundaries can be defined as (Chen & Fukushima, 2012; Nitkiewicz, 2017) "cradle-to-grave", "cradle-to-cradle", "cradle-to-gate", "gate-to-grave", "gate-to-cradle", or "gate-to-gate". In order to provide full coverage of the possible impacts of Circular Economy solutions and business models, the approaches that cover end-of-life processes and are denoted as "to cradle" should be considered. Of course, the allocation of specific end-of-life flows, which are directed towards other life cycles, would be a critical issue here. If we combine "cradle-to-cradle" boundaries with a small-scale consequential approach, we obtain the best possible coverage for circular solutions.

## *Different opportunities for the inclusion of circular flows in LCA*

In order to classify the possibilities of LCA use for the assessment of Circular Economy strategies, we should examine up-to-date experiences of its applications that fall into such categories as: waste management, recycling, reuse, and its different forms such as remanufacturing, refurbishing, or repurposing. It is important to note that specific circular set-ups could consist of processes that fall into all three of the aforementioned categories simultaneously.

Waste management approaches to assessing the environmental impacts of CE solutions are usually part of wider studies including different streams of wastes. Such studies are quite numerous, but the coverage for complex CE solutions is rather limited since most of the cases covered include open loops and non-manageable streams that are ultimately land filled (Hiloidhari et al., 2017; Ripa, Fiorentino, Vacca, & Ulgiati, 2017). More recent studies apply a waste management approach in LCA to cover actually circular, real-life, or potential scenarios but usually focus on the micro-scale (Christensen et al., 2020; Niero & Rivera, 2018; Rufí-Salís et al., 2021; Santagata, Ripa, Genovese, & Ulgiati, 2021; Schwarz et al., 2021).

A recycling approach to CE-oriented LCA studies is much more relevant. There are some product categories, like waste electronic equipment (Sangprasert & Pharino, 2013), home appliances (Nitkiewicz & Starostka-Patyk, 2017; Xiao, Zhang, & Yuan, 2016), batteries (Czerniak, Gacek, Rychwalski, & Nitkiewicz, 2019; Richa et al., 2017; Schulz et al., 2020), and materials, like paper (Corcelli, Fiorentino, Vehmas, & Ulgiati, 2018; Horne, Grant, & Verghese, 2009), aluminium (Adamczyk, Nitkiewicz, Rychwalski, & Wojnarowska, 2015; Sevigné-Itoiz, Gasol, Rieradevall, & Gabarrell, 2014), and plastics (Frischknecht, 2010; Schwarz et al., 2021) or food and agricultural products (Colley, Birkved, Olsen, & Hauschild, 2020; Galli et al., 2015; Ingrao et al., 2021; Renzulli et al., 2015; Santagata et al., 2021) that are well covered with this approach.

Reuse and its possible extensions such as remanufacturing, refurbishing, and repurposing are also widely covered in the literature and usually are oriented towards closed-loop CE solutions. The approaches used here are very diversified and cover issues ranging from single-reuse scenarios (Hatcher, Ijomah, &

*Table 3.1* Methodological issues when handling different CE solutions in LCA

| *Areas of LCA use with CE solutions at micro-scale* | *LCA methodological aspects* | | |
|---|---|---|---|
| | *Approach to the assessment* | *Approach to allocation* | *Approach to setting of systems boundaries* |
| Waste management | Consequential or attributional | Cut-off approach or allocation at the point of substitution or consequential model | Cradle-to-grave, cradle-to-cradle, or gate-to-grave |
| Recycling | Consequential | Cut-off approach or allocation at the point of substitution | Cradle-to-cradle or gate-to-cradle |
| Reuse | Attributional or consequential | Allocation at the point of substitution | Cradle-to-cradle or gate-to-cradle |

Windmill, 2013; Milios, Beqiri, Whalen, & Jelonek, 2019; Schau, Traverso, Lehmannann, & Finkbeiner, 2011), through to a comparative analysis of these scenarios (Colley et al., 2020; Cottafava et al., 2021; Schulz et al., 2020), and where reuse refers to other options within CE end-of-life scenarios (King, Burgess, Ijomah, & McMahon, 2006; Nitkiewicz & Starostka-Patyk, 2017; Starostka-Patyk, 2015). The dilemmas that are referred to are often related to the allocation of products and reused products to LCA (Peters, 2015; Plevin, Delucchi, & Creutzig, 2014; Schulz et al., 2020).

The major differences between the approaches could be explained through different methodological choices that are made within the LCA procedure. Table 3.1 presents the choices that are made in the LCA of CE solutions when using different approaches. The differences between them are based on decisions made by researchers but could be also classified to some extent. Each choice would have significant consequences for the assessment and its results. Using a consequential approach would imply the use of market-specific allocation of raw materials versus the use of recycled materials in the production process. Allocation decisions in the assessment of CE solutions could transfer the environmental benefits or burdens inside or outside the life cycle considered. At the end, setting boundaries on a life cycle will define the scope, information needs, allocation mechanisms, and dilemmas or even the possible level of detail of the assessment (Nitkiewicz, 2019).

## LCA of Circular Economy solutions in biopolymers

In order to illustrate the possible use of LCA and its methodological choices for micro-scale CE solutions, the case of biopolymers will be used. Biopolymers are currently perceived to be the best alternative to conventional fossil-based plastics (Ciardelli, Bertoldo, Bronco, & Passaglia, 2019) and are expected to slowly replace them (Gironi & Piemonte, 2011; Kookos, Koutinas, & Vlysidis, 2019). Biopolymers can be derived from a wide range of biomass types that includes agricultural products such as corn or soybeans as well as algae or food waste (Nitkiewicz et al., 2020).

Among the various types of biopolymers, the group of PHAs will be investigated. Unlike conventional polymers or poly(lactic) acid and starch polymers, PHAs are characterised by useful properties that do not need to be improved. These features are comparable with those of poly(ethylene) and poly(propylene), but while offering many features in terms of environmentally friendly processing and biodegradability. Furthermore, all possible PHAs have a potential to be utilised in a wide range of applications due to the ease of achieving desired properties and quality (Koller, 2019). Microbial fermentation is one of the most widespread ways of producing PHAs (Heimersson, Morgan-Sagastume, Peters, Werker, & Svanström, 2014). Inputs for the production of PHAs are refined materials like monosaccharides, cellulose, and starch which can be extracted from a variety of crops and waste flows like biomass residues, post-process industrial waste streams, and wastewaters (Wojnowska-Baryła, Kulikowska, & Bernat, 2020). PHA production technologies are being developed not only in order to make them more affordable but also in order to accelerate their possible role in introducing CE solutions to the polymer manufacturing sector (Ingrao et al., 2018). The wide use of biopolymers does not guarantee a lack of environmental impact, and different factors such as production processes, material feed, technical performance, as well as disposal must be carefully considered throughout the life of the product (Ingrao & Siracusa, 2017; Ingrao et al., 2015).

For the purpose of searching for possible CE solutions with regard to environmental impact, the experimental process of producing of two types of PHAs, namely, polyhydroxybutyrate (P(3HB)) and an amorphous medium-chain-length PHA (mcl-PHA), was investigated with the use of LCA. Since the objective is to illustrate the choices made in the LCA procedure and to check whether open loops are closed with PHA production, the scenario of production of mcl-PHA/P(3HB) from used cooking oil will be considered. Details of this innovative method of production of mcl-PHA/P(3HB), based on bacterial fermentation with *Pseudomonas putida* (mcl-PHA) and *Zobellella denitrificans* (P(3HB)) strains, have been presented elsewhere (Nitkiewicz et al., 2020). Here, we will focus on a production process that is fed with used, post-consumer rapeseed oil. The process is based on experimental data, but its testing prior to its scaling-up is related to the possibility of introducing CE solutions. The main life cycle refers to the production of rapeseed oil while PHA production is one of the options for closing down the open loops with used oil flows. The life cycle considered is also a good example of the difficulties that CE designers or LCA practitioners could face.

The general overview of the life cycle considered is presented in Figure 3.1. It is important to mention that the considered life cycle could be divided into two different parts due to the integrity and complexity of the phases included. The first part is a "cradle-to-grave" life cycle of primary product that is rapeseed oil. Denoting it as "cradle-to-grave" life cycle shows that it excludes end-of-life processes. Such an approach could be justified in linear economy context, where there is no material flow of used oil at the end of its life cycle. In fact, this is often

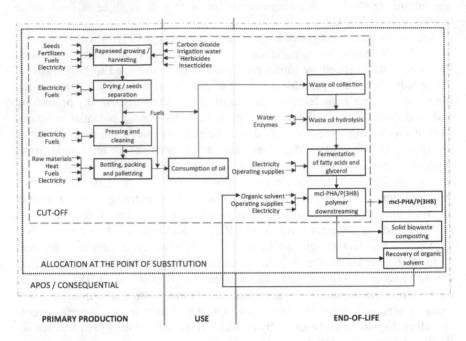

*Figure 3.1* System boundaries for mcl-PHA/P(3HB) production from used rapeseed oil
Source: Own elaboration.

the case, and this is related to the lack of support for appropriate reverse flows and low consumer awareness. The second part of the life cycle is related to the end-of-life processing that could be vital for closing the loop with used oil by using it as by-product. The objective would be to recover its value and provide economically and environmentally sound solutions to meet CE requirements. Combining these two parts results in a cradle-to-cradle life cycle.

At first, it should be considered whether the assessment refers to the micro- or meso-level of implementation of CE solutions. At the micro-level, the CE refers to a specific product and its life cycle, which is also the case here. In contrast, meso-level assessments focus on symbiotic systems of an industrial or industrial/urban nature. Meso-level assessment focusses on by-product and utility exchanges between different stakeholders, which is also the case here. In the end, the scenario considered was defined as a micro-level approach. This is due to the omission of possible reverse flows in the first part of rapeseed oil production. For example the by-product rapeseed meal is frequently used as animal fodder and accounted for by system expansion, replacing soymeal (Arvidsson, Persson, Fröling, & Svanström, 2011). Additionally, rapeseed oil is commonly used as a source for the production of biofuels, and its output could be allocated to its different potential products (i.e. rapeseed oil vs. biodiesel) (Ozturk, 2014; Shah, Arslan, Cirucci, O'Brien, & Moss, 2016). Certainly, in such a context, the case of a complex life cycle should be considered from the meso-level.

## Goal and scope definition of LCA

Since the objective of the book and this chapter is to provide business-ready solutions in the context of sustainable products, we will focus on the micro-level and, as part of our approach, will mainly consider the impacts of specific end-of-life phases. The functional unit for the study is defined as a life cycle of 1 kg of biopolymer produced from used rapeseed oil using an experimental manufacturing process. We take biopolymer to refer to both possible outcomes of the bacterial fermentation of used oil: mcl-PHA and P(3HB). These are quite different biopolymers with regard to their properties and to the final products that can be obtained using them, but the efforts involved in transforming used oil to produce them are similar, and they can thus be considered to be substitutes from the perspective of the end-of-life phase of rapeseed oil. The study therefore investigates two functional units that are the same with regard to the life-cycle phases included but which differ with regard to the final outcome (1 kg of mcl-PHA and 1 kg of P(3HB) biopolymers).

The life-cycle phases included in the study and the system boundaries are presented in Figure 3.1.

The study does not refer to possibilities concerning the transformation of the biopolymer obtained into a final product. This is to keep the study focussed on the methodological choices related to the use of LCA and not on different options relating to the use of PHAs for delivering final products.

## Life-cycle inventory analysis

The analysis of material and energy flows within considered life cycles is based on primary experimental production data as presented by Nitkiewicz et al. (2020). Primary data refers to the end-of-life phase, namely the biopolymer production process only, while the production and provision of raw materials for production are modelled with secondary data from the literature and the ecoinvent 3.1 database. Table 3.2 presents the inventory data for both life cycles included in the study.

## Life-cycle impact assessment

In order to provide full coverage for possible life-cycle impacts, the ReCiPe method was used for the assessment. The impacts were calculated with SimaPro software and the endpoint variant of the ReCiPe (H) V1.08 indicator. It is important to mention that the Life Cycle Assessment was approached with three different strategies: (1) a cut-off approach, (2) allocation at the point of substitution approach (APOS), and (3) consequential assumptions applied to the allocation at the point of substitution approach (APOS/consequential). The differences between the approaches are listed here:

1   The cut-off approach uses all the life-cycle data but does not account for environmental benefits from providing mcl-PHA/P(3HB) biopolymers; the database used for life-cycle modelling is attributional-oriented;

*Table 3.2* Inventory data for investigated life cycles

| Specification | Units | Quantities in life cycles | |
|---|---|---|---|
| | | *mcl-PHA* | *P(3HB)* |
| *Hydrolysis* | | | |
| Used rapeseed oil** | kg | 2.38 | 59.09 |
| Fatty acids/glycerine | kg | 2.11 | 6.28 |
| *Fermentation* | | | |
| Fatty acids/glycerine* | kg | 1.05 | 1.57 |
| Ammonium | ml | 378.95 | 605.82 |
| Fermentation efficiency | % | 0.60 | 0.60 |
| *Salt medium* | | | |
| $Na_2HPO_4 \cdot 12H_2O$ | g/l | 151.58 | 436.19 |
| $KH_2PO_4$ | g/l | 25.26 | 72.70 |
| $NH_4Cl$ full/lim | g/l | 16.84 | 48.47 |
| $MgSO_4 \cdot 7H_2O$ | g/l | 3.37 | 9.69 |
| Solution (total amount) | l | 33.07 | 819.70 |
| $CaCl_2$ | g/l | 0.000337 | 0.000969 |
| Fe(III)$NH_4$ citrate | g/l | 0.000020 | 0.000058 |
| $ZnSO_4 \cdot 7H_2O$ | g/l | 0.000067 | 0.000194 |
| $FeSO_4 \cdot 7H_2O$ | g/l | 0.000168 | 0.000485 |
| $CuCl_2 \cdot 2H_2O$ | g/l | 0.000017 | 0.000048 |
| $MnCl_2 \cdot 4H_2O$ | g/l | 0.000017 | 0.000048 |
| $Na_2B4O_7 \cdot 10H_2O$ | g/l | 0.000017 | 0.000048 |
| $NiCl_2 \cdot 6H_2O$ | g/l | 0.000003 | 0.000010 |
| $Na_2MoO_4 \cdot 2H_2O$ | g/l | 0.000005 | 0.000015 |
| Air flow | l/min | 5 | 5 |
| $CO_2$ emissions | l | 14895 | 14895 |
| *Separation* | | | |
| Solvent | kg | 1.26 | 1.26 |
| Alcohol | kg | 12.63 | 12.63 |
| Fatty acids/glycerine | kg | 1.05 | 1.05 |
| Extraction efficiency | % | 0.95 | 0.95 |
| Product | kg | 1 | 1 |

* Rate of glycerine fermentation to P(3HB) is taken from Ibrahim and Steinbüchel (2009)
** Rates of mcl-PHA production from used oil from Tufail et al. (2017)

Source: Based on Nitkiewicz et al. (2020).

2    The APOS approach uses all the life-cycle data including the benefits from producing mcl-PHA/P(3HB) biopolymers; the database used for life-cycle modelling is attributional-oriented;

3    The APOS/consequential approach uses all the life-cycle data including the benefits from producing mcl-PHA/P(3HB) biopolymers; the database used for life-cycle modelling is consequential-oriented.

## *LCIA results and discussion*

Figure 3.2 presents the results for all functional units assessed within considered calculation scenarios. Due to the low yield of P(3HB) in comparison to

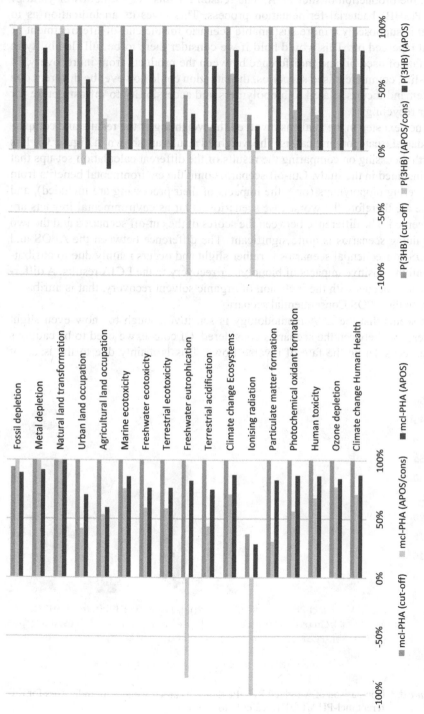

*Figure 3.2* Comparison of characterisation results for ReCiPe impact category endpoint indicators

mcl-PHA production, PHB production cycles have significantly higher impacts than the production of mcl-PHA. The reason for this lies in the lower yield of the P(3HB) bacterial fermentation process. This gives us an indication as to what is supposedly a more sustainable scenario for the end-of-life treatment of used rapeseed oil. This would hold if we consider there to be full flexibility in the fate of used oil and indifference between the products from its recovery. In real-life, commercialised scenarios, the situation could however be different due to demand factors related to biopolymers and accessibility to infrastructure for their recycling.

The next step in the analysis involved the weighting of the results and comparing damage category indicators. The weighted results are shown in Figure 3.3. It is worth focussing on comparing the results of the different calculation set-ups that are included in the study. Cut-off scenarios omit the environmental benefits from recovering biopolymers (only the impacts of their processing are included), and these are therefore the worst-case scenarios as far as environmental impacts are concerned. The difference between the scores of the cut-off scenarios and the two remaining scenarios is quite significant. The difference between the APOS and APOS/consequential scenarios is rather slight and occurs mainly due to attributing all the positive impacts of biopolymer recovery to the LCIA results. A difference also occurs with the inclusion of organic solvent recovery, that is attributed only to the APOS/Consequential scenario.

It seems that the LCA methodology is sensitive enough to show even slight differences between the scenarios considered. Of course, we need to be cautious when considering the fate of reverse flows. This is mainly due to the issue of

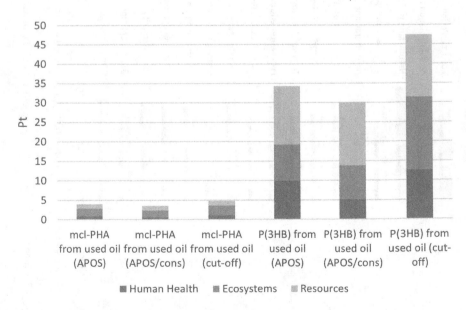

*Figure 3.3* Comparison of weighted ReCiPe damage category endpoint indicators for different mcl-PHA/P(3HB) calculation set-ups

whether to attribute them to the objective life cycle or to split them between the different cycles that are benefactors of reverse flows.

## Conclusion

The results of the study showed high sensitivity to the different assumptions of LCA calculation parameters. The example used here was not so complex as to cover all the nuances of the methodological choices involved in LCA, but it has certainly shed light on its consequences in terms of the results. With regard to the choice between an attributional or consequential approach towards LCA for Circular Economy-oriented studies, our chapter concurs with the majority of evidence in the literature that it should be dependent on the possible coverage of reverse flows (Cullen, 2017; van Loon et al., 2021; Walker et al., 2018).

However, we cannot fully subscribe to the argument that a consequential approach is more suitable for the assessment of CE solutions due to its high dependence on market information on closed loops. In that sense, we agree with authors who indicate that an attributional approach, or even its combination with a consequential one, could be also useful within a specifically defined framework (Hottle, Bilec, & Landis, 2017; Renzulli et al., 2015). In a consequential LCA, the system boundaries often need to be expanded, which requires environmental data on more processes and also economic data on the markets affected by the production and use of the product investigated (Ekvall, 2020). In our case, the expansion is made through a consequential variant of ecoinvent database and by expansion of the modelled system. Therefore, our results are partly based on rough marginal data from ecoinvent and should be regarded as being unfeasible for a specific location, with reference to geographic and market characteristics. However, using a consequential approach does allow us to limit the raw data provisions to those processes that are affected by a proposed change.

Another point that should be mentioned by way of conclusion is the capability of the LCA methodology with reference to the assessment of CE solutions and sustainable products. The critical issue here is the complexity of the object of assessment and its level of complexity (micro-, meso-, or macro-level). The example presented here proves that LCA could itself be a good framework for CE solution assessment. But, for more demanding cases, using LCA may not be sufficient to present both circularity and sustainability at the same time. The recommendation is to build upon an existing CE and sustainability evaluation framework with LCA as its central point (Harris et al., 2021; Rufí-Salís et al., 2021; van Loon et al., 2021).

## References

Adamczyk, W., Nitkiewicz, T., Rychwalski, M., & Wojnarowska, M. (2015). "Gate-to-grave" life cycle assessment of different scenarios for handling used PV cells. In K. Michocka & M. Tichoniuk (Eds.), *Current trends in commodity science: Development and assessment of non-food products* (pp. 263–282). Poznań: Poznań University of Economics and Business.

Ardente, F., & Mathieux, F. (2014). Environmental assessment of the durability of energy-using products: Method and application. *Journal of Cleaner Production, 74*, 62–73. doi:10.1016/j.jclepro.2014.03.049

Arvidsson, R., Persson, S., Fröling, M., & Svanström, M. (2011). Life cycle assessment of hydrotreated vegetable oil from rape, oil palm and Jatropha. *Journal of Cleaner Production, 19*(2–3), 129–137. doi:10.1016/j.jclepro.2010.02.008

Chen, I. C., & Fukushima, Y. (2012). A graphical representation for consequential life cycle assessment of future technologies. Part 1: Methodological framework. *International Journal of Life Cycle Assessment, 17*(1), 119–125. doi:10.1007/s11367-011-0356-9

Christensen, T. H., Damgaard, A., Levis, J., Zhao, Y., Björklund, A., Arena, U., . . . Bisinella, V. (2020). Application of LCA modelling in integrated waste management. *Waste Management, 118*, 313–322. doi:10.1016/j.wasman.2020.08.034

Ciardelli, F., Bertoldo, M., Bronco, S., & Passaglia, E. (2019). Environmental Impact. *Polymers from Fossil and Renewable Resources*, 161–187. doi:10.1007/978-3-319-94434-0_7

Colley, T. A., Birkved, M., Olsen, S. I., & Hauschild, M. Z. (2020). Using a gate-to-gate LCA to apply circular economy principles to a food processing SME. *Journal of Cleaner Production, 251*, 119566. doi:10.1016/j.jclepro.2019.119566

Cooper, D. R., & Gutowski, T. G. (2017). The environmental impacts of reuse: A review. *Journal of Industrial Ecology, 21*, 38–56. doi:10.1111/jiec.12388

Corcelli, F., Fiorentino, G., Vehmas, J., & Ulgiati, S. (2018). Energy efficiency and environmental assessment of papermaking from chemical pulp – a Finland case study. *Journal of Cleaner Production, 198*, 96–111. doi:10.1016/j.jclepro.2018.07.018

Cottafava, D., Costamagna, M., Baricco, M., Corazza, L., Miceli, D., & Riccardo, L. E. (2021). Assessment of the environmental break-even point for deposit return systems through an LCA analysis of single-use and reusable cups. *Sustainable Production and Consumption, 27*, 228–241. doi:10.1016/j.spc.2020.11.002

Cullen, J. M. (2017). Circular economy: Theoretical benchmark or perpetual motion machine? *Journal of Industrial Ecology, 21*(3), 483–486. doi:10.1111/jiec.12599

Czerniak, J., Gacek, A., Rychwalski, M., & Nitkiewicz, T. (2019). Optymalizacja środowiskowa zasilania urządzeń bateryjnych w gospodarstwach domowych jako realizacja idei zrównoważonej konsumpcji. In R. Salerno-Kochan (Ed.), *Nauki o zarządzaniu i jakości wobec wyzwań zrównoważonego rozwoju (Management and quality studies facing challenges of sustainable development)* (pp. 63–74). Radom: Instytut Technologii Eksploatacji – Państwowy Instytut Badawczy.

Dieterle, M., Schäfer, P., & Viere, T. (2018, May). Life cycle gaps: Interpreting LCA results with a circular economy mindset. *Procedia CIRP, 69*, 764–768. doi:10.1016/j.procir.2017.11.058

Dieterle, M., & Viere, T. (2021). Bridging product life cycle gaps in LCA & LCC towards a circular economy. *Procedia CIRP, 98*, 354–357. doi:10.1016/j.procir.2021.01.116

Earles, J. M., & Halog, A. (2011). Consequential life cycle assessment: A review. *International Journal of Life Cycle Assessment, 16*, 445–453. doi:10.1007/s11367-011-0275-9

Ekvall, T. (2020). Attributional and consequential life cycle assessment. *Sustainability Assessment at the 21st Century*. doi:10.5772/intechopen.89202

Ekvall, T., Azapagic, A., Finnveden, G., Rydberg, T., Weidema, B. P., & Zamagni, A. (2016). Attributional and consequential LCA in the ILCD handbook. *International Journal of Life Cycle Assessment, 21*(3), 293–296. doi:10.1007/s11367-015-1026-0

EMF. (2013). *Towards the circular economy: Economic and business rationale for an accelerated transition*. Ellen MacArthur Foundation. Retrieved May 15, 2021, from https://emf.thirdlight.com/link/x8ay372a3r11-k6775n/@/preview/1?o.

Frischknecht, R. (2010). LCI modelling approaches applied on recycling of materials in view of environmental sustainability, risk perception and eco-efficiency. *International Journal of Life Cycle Assessment, 15*(7), 666–671. doi:10.1007/s11367-010-0201-6

Galli, F., Bartolini, F., Brunori, G., Colombo, L., Gava, O., Grando, S., & Marescotti, A. (2015). Sustainability assessment of food supply chains: An application to local and global bread in Italy. *Agricultural and Food Economics, 3*(1). doi:10.1186/s40100-015-0039-0

Ghisellini, P., Cialani, C., & Ulgiati, S. (2016). A review on circular economy: The expected transition to a balanced interplay of environmental and economic systems. *Journal of Cleaner Production, 114*, 11–32. doi:10.1016/j.jclepro.2015.09.007

Gironi, F., & Piemonte, V. (2011). Bioplastics and petroleum-based plastics: Strengths and weaknesses. *Energy Sources, Part A: Recovery, Utilization and Environmental Effects, 33*(21), 1949–1959. doi:10.1080/15567030903436830

Harris, S., Martin, M., & Diener, D. (2021). Circularity for circularity's sake? Scoping review of assessment methods for environmental performance in the circular economy. *Sustainable Production And Consumption, 26*, 172–186. doi:10.1016/j.spc.2020.09.018

Hatcher, G. D., Ijomah, W. L., & Windmill, J. F. C. (2013). Integrating design for remanufacture into the design process: The operational factors. *Journal of Cleaner Production, 39*, 200–208. doi:10.1016/j.jclepro.2012.08.015

Heimersson, S., Morgan-Sagastume, F., Peters, G. M., Werker, A., & Svanström, M. (2014). Methodological issues in life cycle assessment of mixed-culture polyhydroxy-alkanoate production utilising waste as feedstock. *New Biotechnology, 31*(4), 383–393. doi:10.1016/j.nbt.2013.09.003

Hiloidhari, M., Baruah, D. C., Singh, A., Kataki, S., Medhi, K., Kumari, S., . . . Shekhar Thakur, I. (2017). Emerging role of geographical information system (GIS), life cycle assessment (LCA) and spatial LCA (GIS-LCA) in sustainable bioenergy Planning. *Bioresource Technology*. doi:10.1016/j.biortech.2017.03.079

Horne, R., Grant, T., & Verghese, K. (2009). *Life cycle assessment: Principles, practice and prospects* (Vol. 1). Collingwood: CSIRO Publishing. doi:10.1017/CBO9781107415324.004

Hottle, T. A., Bilec, M. M., & Landis, A. E. (2017). Biopolymer production and end of life comparisons using life cycle assessment. *Resources, Conservation and Recycling*. doi:10.1016/j.resconrec.2017.03.002

Ibrahim, M. H. A., & Steinbüchel, A. (2009). Poly(3-hydroxybutyrate) production from glycerol by Zobellella denitrificans MW1 via high-cell-density fed-batch fermentation and simplified solvent extraction. *Applied and Environmental Microbiology, 75*(19), 6222–6231. doi:10.1128/AEM.01162-09

Ingrao, C., Bacenetti, J., Bezama, A., Blok, V., Goglio, P., Koukios, E. G., . . . Huisingh, D. (2018, December). The potential roles of bio-economy in the transition to equitable, sustainable, post fossil-carbon societies: Findings from this virtual special issue. *Journal of Cleaner Production, 204*, 471–488. doi:10.1016/j.jclepro.2018.09.068

Ingrao, C., Matarazzo, A., Gorjian, S., Adamczyk, J., Failla, S., Primerano, P., & Huisingh, D. (2021). Wheat-straw derived bioethanol production: A review of life cycle assessments. *Science of the Total Environment, 781*, 146751. doi:10.1016/j.scitotenv.2021.146751

Ingrao, C., & Siracusa, V. (2017). Quality- and sustainability-related issues associated with biopolymers for food packaging applications: A comprehensive review. *Biodegradable and Biocompatible Polymer Composites: Processing, Properties and Applications*, 401–418. doi:10.1016/B978-0-08-100970-3.00014-6

Ingrao, C., Tricase, C., Cholewa-Wójcik, A., Kawecka, A., Rana, R., & Siracusa, V. (2015). Polylactic acid trays for fresh-food packaging: A carbon footprint assessment. *Science of the Total Environment*, *537*, 385–398. doi:10.1016/j.scitotenv.2015.08.023

ISO 14040:2006. (2006). *Environmental management – life cycle assessment – principles and framework*. International Organization fo r Standardization. Retrieved from https://www.iso.org/standard/37456.html

Joachimiak-Lechman, K. (2014). Środowiskowa ocena cyklu życia (LCA) i rachunek kosztów cyklu życia (LCC): Aspekty porównawcze [Environmental life cycle assessment and life cycle cost: Comparative aspects]. *Ekonomia i Środowisko*, *1*(48).

King, A. M., Burgess, S. C., Ijomah, W., & McMahon, C. A. (2006). Reducing waste: Repair, recondition, remanufacture or recycle? *Sustainable Development*, *14*, 257–267. doi:10.1002/sd.271

Kjaer, L. L., Pigosso, D. C. A., McAloone, T. C., & Birkved, M. (2018). Guidelines for evaluating the environmental performance of product/service-systems through life cycle assessment. *Journal of Cleaner Production*, *190*, 666–678. doi:10.1016/j.jclepro.2018.04.108

Koller, M. (2019). Polyhydroxyalkanoate biosynthesis at the edge of water activity-haloarchaea as biopolyester factories. *Bioengineering*, *6*(2), 34. doi:10.3390/bioeng ineering6020034

Kookos, I. K., Koutinas, A., & Vlysidis, A. (2019). Life cycle assessment of bioprocessing schemes for poly(3-hydroxybutyrate) production using soybean oil and sucrose as carbon sources. *Resources, Conservation and Recycling*, *141*, 317–328. doi:10.1016/j.resconrec.2018.10.025

Milios, L., Beqiri, B., Whalen, K. A., & Jelonek, S. H. (2019). Sailing towards a circular economy: Conditions for increased reuse and remanufacturing in the Scandinavian maritime sector. *Journal of Cleaner Production*, *225*, 227–235. doi:10.1016/j.jclepro.2019.03.330

Murphy, R., Detzel, A., Guo, M., & Krüger, M. (2011). Comment on sustainability metrics: Life cycle assessment and green design in polymers. *Environmental Science and Technology*, *45*(11), 5055–5056. doi:10.1021/es103890v

Niero, M., & Hauschild, M. Z. (2017). Closing the loop for packaging: Finding a framework to operationalize circular economy strategies. *Procedia CIRP*, *61*, 685–690. doi:10.1016/j.procir.2016.11.209

Niero, M., & Rivera, X. C. S. (2018, May). The role of life cycle sustainability assessment in the implementation of circular economy principles in organizations. *Procedia CIRP*, *69*, 793–798. doi:10.1016/j.procir.2017.11.022

Nitkiewicz, T. (2017). *Wykorzystanie środowiskowej oceny cyklu życia w analizie procesów i przepływów logistycznych*. Częstochowa: Wyd. Wydziału Zarządzania Politechniki Częstochowskiej.

Nitkiewicz, T. (2019). Possible scopes of life cycle assessment (LCA) use in the management of returned products. In K. S. Soliman (Ed.), *Vision 2025: Education excellence and management of innovations through sustainable economic competitive advantage, proceedings of the 34th international business information management association conference (IBIMA)* (pp. 10785–10792). Madrid: IBIMA. ISBN: 978-0-9998551-3-3

Nitkiewicz, T., & Starostka-Patyk, M. (2017). Contribution of returned products handling scenarios to life cycle impacts – research case of washing machine. *Environmental Engineering and Management Journal*, 1–29.

Nitkiewicz, T., Wojnarowska, M., Sołtysik, M., Kaczmarski, A., Witko, T., Ingrao, C., & Guzik, M. (2020). How sustainable are biopolymers? Findings from a life cycle assessment of polyhydroxyalkanoate production from rapeseed-oil derivatives. *Science of the Total Environment, 749,* 141279. doi:10.1016/j.scitotenv.2020.141279

Ozturk, H. H. (2014). Energy analysis for biodiesel production from rapeseed oil. *Energy Exploration and Exploitation, 32*(6), 1005–1031. doi:10.1260/0144-5987.32.6.1005

Pena, C., Civit, B., Gallego Schmid, A., Druckman, A., Caldeira-Pires, A., Weidema, B., . . . Motta, W. (2020). *Using life cycle assessment to achieve a circular economy* (Vol. 10). Cham: Springer.

Peters, K. (2015). Methodological issues in life cycle assessment for remanufactured products: A critical review of existing studies and an illustrative case study. *Journal of Cleaner Production, 126,* 21–37. doi:10.1016/j.jclepro.2016.03.050

Plevin, R. J., Delucchi, M. A., & Creutzig, F. (2014). Using attributional life cycle assessment to estimate climate-change mitigation benefits misleads policy makers. *Journal of Industrial Ecology, 18,* 73–83. doi:10.1111/jiec.12074

Renzulli, P. A., Bacenetti, J., Benedetto, G., Fusi, A., Ioppolo, G., Niero, M., . . . Supino, S. (2015). *Life cycle assessment In the agri-food sector.* Cham: Springer. doi:10.1007/978-3-319-11940-3

Richa, K., Babbitt, C. W., & Gaustad, G. (2017). Eco-efficiency analysis of a lithium-ion battery waste hierarchy inspired by circular economy. *Journal of Industrial Ecology, 21,* 715–730. doi:10.1111/jiec.12607

Ripa, M., Fiorentino, G., Vacca, V., & Ulgiati, S. (2017). The relevance of site-specific data in life cycle assessment (LCA): The case of the municipal solid waste management in the metropolitan city of Naples (Italy). *Journal of Cleaner Production, 142,* 445–460. doi:10.1016/j.jclepro.2016.09.149

Rufi-Salís, M., Petit-Boix, A., Villalba, G., Gabarrell, X., & Leipold, S. (2021). Combining LCA and circularity assessments in complex production systems: The case of urban agriculture. *Resources, Conservation and Recycling, 166,* 105359. doi:10.1016/j.resconrec.2020.105359

Rusch, M., & Baumgartner, R. J. (2020). *Social life cycle assessment in a circular economy – a mixed-method analysis of 97 SLCA publications and its CE connections,* 7th International Social Life Cycle Assessment Conference, Sweden (Online), 73–76.

Saidani, M., Yannou, B., Leroy, Y., Cluzel, F., & Kendall, A. (2019, January 10). A taxonomy of circular economy indicators. *Journal of Cleaner Production, 207,* 542–559. doi:10.1016/j.jclepro.2018.10.014

Sangprasert, W., & Pharino, C. (2013, January 8–9). *Environmental impact evaluation of mobile phone via life cycle assessment.* 3rd International Conference on Chemical Biological and Environment Sciences (ICCEBS'2013), Kuala Lampur.

Santagata, R., Ripa, M., Genovese, A., & Ulgiati, S. (2021, March 1). Food waste recovery pathways: Challenges and opportunities for an emerging bio-based circular economy. A systematic review and an assessment. *Journal of Cleaner Production, 286,* 125490. doi:10.1016/j.jclepro.2020.125490

Schau, E. M., Traverso, M., Lehmannann, A., & Finkbeiner, M. (2011). Life cycle costing in sustainability assessment-A case study of remanufactured alternators. *Sustainability, 3*(11), 2268–2288. doi:10.3390/su3112268

Schulz, M., Bey, N., Niero, M., & Hauschild, M. (2020). Circular economy considerations in choices of LCA methodology: How to handle EV battery repurposing? *Procedia CIRP, 90,* 182–186. doi:10.1016/j.procir.2020.01.134

Schwarz, A. E., Ligthart, T. N., Godoi Bizarro, D., De Wild, P., Vreugdenhil, B., & van Harmelen, T. (2021). Plastic recycling in a circular economy; determining environmental performance through an LCA matrix model approach. *Waste Management, 121,* 331–342. doi:10.1016/j.wasman.2020.12.020

Sevigné-Itoiz, E., Gasol, C. M., Rieradevall, J., & Gabarrell, X. (2014). Environmental consequences of recycling aluminum old scrap in a global market. *Resources, Conservation and Recycling, 89*, 94–103. doi:10.1016/j.resconrec.2014.05.002

Shah, J., Arslan, E., Cirucci, J., O'Brien, J., & Moss, D. (2016). Comparison of oleo- vs petro-sourcing of fatty alcohols via cradle-to-gate life cycle assessment. *Journal of Surfactants And Detergents, 19*(6), 1333–1351. doi:10.1007/s11743-016-1867-y

Starostka-Patyk, M. (2015). New products design decision making support by SimaPro software on the base of defective products management. *Procedia Computer Science, 65*(Iccmit), 1066–1074. doi:10.1016/j.procs.2015.09.051

Tufail, S., Munir, S., & Jamil, N. (2017). Variation analysis of bacterial polyhydroxyalkanoates production using saturated and unsaturated hydrocarbons. *Brazilian Journal of Microbiology, 48*(4), 629–636. doi:10.1016/j.bjm.2017.02.008

van Loon, P., Diener, D., & Harris, S. (2021). Circular products and business models and environmental impact reductions: Current knowledge and knowledge gaps. *Journal of Cleaner Production, 288*, 125627. doi:10.1016/j.jclepro.2020.125627

Walker, S., Coleman, N., Hodgson, P., Collins, N., & Brimacombe, L. (2018). Evaluating the environmental dimension of material efficiency strategies relating to the circular economy. *Sustainability, 10*(3), 666. doi:10.3390/su10030666

Wernet, G., Bauer, C., Steubing, B., Reinhard, J., Moreno-Ruiz, E., & Weidema, B. (2016). The ecoinvent database version 3 (part I): Overview and methodology. *International Journal of Life Cycle Assessment, 21*(9), 1218–1230. doi:10.1007/s11367-016-1087-8

Wojnowska-Baryła, I., Kulikowska, D., & Bernat, K. (2020). Effect of bio-based products on waste management. *Sustainability, 12*(5), 2088. doi:10.3390/su12052088

Xiao, R., Zhang, Y., & Yuan, Z. (2016). Environmental impacts of reclamation and recycling processes of refrigerators using life cycle assessment (LCA) methods. *Journal of Cleaner Production, 131*, 52–59. doi:10.1016/j.jclepro.2016.05.085

Zink, T., & Geyer, R. (2017). Circular economy rebound. *Journal of Industrial Ecology, 21*, 593–602. doi:10.1111/jiec.12545

# 4 Significance and adjustment of environmental certification schemes in the Circular Economy

*Piotr Kafel and Paweł Nowicki*

## Introduction

The most popular standardised management systems are those built based on the ISO 9001 standard, especially its latest edition, created based on the High Level Structure (HLS) clauses, which allow the system to be adapted the format of many industries. For the needs of the Circular Economy, two standards specifically devoted to that issue have been created to date: the British standard – BS 8001: 2017 and the French standard X PX 30–901:2018. Apart from the aforementioned standards, there are also standards whose requirements refer to a large extent to the CE, as presented in the ISO 14009:2020 and ISO 20400:2017 standards. The goal of this chapter is to characterise, as well as analyse, the newest management standards dedicated for use in enterprises operating in accordance with a Circular Economy approach. One of the barriers to the implementation and certification of standardised management systems are the high costs, especially in the case of small- and medium-sized enterprises. One of the key benefits is that of obtaining a certificate, and thus the possibility of being able to provide reliable proof of implementation, which may be an important marketing benefit.

## Meta-standards (management system standards – MSS) and their structure

There are many methods and techniques which have made it possible to manage organisations more effectively over the years. This group of methods may be considered to include standards containing requirements for different management systems developed by private organisations, such as the International Organization for Standardization (ISO) for example. In the group of generic standards, intended for implementation in any organisation, the most popular standards may be considered to be: ISO 9001, ISO 14001, and ISO 45001.

Due to the specificity of different industries and sectors of the economy, it was necessary to develop requirements corresponding to these needs. Examples of sector-specific standardised management systems are to be found in ISO 22000, ISO 13485, and ISO/IEC 17025. They are intended as management systems for the food industry, medical devices, and testing and calibration laboratories.

DOI: 10.4324/9781003179788-4

The wide variety of management systems does however mean that it is possible to identify four basic groups of management systems, presented in Table 4.1.

Management system standards (type A) were designed to formalise, systematise, and legitimise a very diverse set of managerial activities (Heras-Saizarbitoria, Boiral, & Ibarloza, 2020). A measure of the popularity of that approach is the number of systems implemented and certified, which is estimated to amount to nearly 2 million certified sites (ISO, 2019).

The majority of management system standards have a similar structure and use the same definitions. These similarities result from the principles introduced by the ISO in 2012 concerning the HLS setting out the requirements for any management system. The clause numbers of individual chapters and sub-chapters of management system standards are presented in Table 4.2 (Kafel, 2017).

The similarities between different management systems, sometimes known as meta-standards, are characterised by a similar approach resulting from the use of the Plan–Do–Check–Act (PDCA) model. The requirements of the standards themselves do not provide a basis for the assessment of performance but rather propose guidelines as to how the goal can be accomplished by adopting appropriate policies, plans, organisational practices, and control mechanisms (Heras-Saizarbitoria et al., 2020).

In addition to the most popular standardised management systems, it is possible to point to other standards which may facilitate the implementation of the CE at the organisational level. Corporate social responsibility (CSR) guidelines share some common ground with activities aiming to bring about the transition from a linear economy to a CE model. Such connections are in particular to be found in the area of environmental requirements, which are integral to the concept of the CE. In the ISO 26000 standard, it is the third issue addressed under the core subject of the environment, namely climate change mitigation and adaptation.

*Table 4.1* Breakdown of management system standards

|  | Type A MSS<br>*MSS providing requirements* | Type B MSS<br>*MSS providing guidelines* |
|---|---|---|
| **Generic MSS**<br>MSS designed to be widely applicable across economic sectors | Option 1<br>(certification possible) | Option 2 |
| **Sector-specific MSS**<br>MSS that provides additional requirements or guidance for the application of a generic MSS to a specific economic or business sector | Option 3<br>(certification possible) | Option 4 |

Source: Elaboration based on ISO/IEC (2020) and Kafel, Nowicki, and Wojnarowska (2021).

*Table 4.2* Structure of management system standards proposed by ISO (type A/generic)

| *Clause numbers for management system standards* | | |
|---|---|---|
| Introduction | 6 | Planning |
| 1   Scope | 7 | Support |
| 2   Normative references | 8 | Operation |
| 3   Terms and definitions | 9 | Performance evaluation |
| 4   Context of the organisation | 10 | Improvement |
| 5   Leadership | 10.2 | Continual improvement |

Source: Own elaboration based on ISO/IEC (2020).

The transition from a linear economy to a CE requires activities on various levels, that is at macro-, meso-, and micro-levels, as has been noted by, among others, Prieto-Sandoval, Jaca, and Ormazabal (2018) and Kirchherr, Reike, and Hekkert (2017). Management system standards aid with these practices mainly at the micro-level or, in other words, at that of individual organisations. There are also examples of standardised requirements which subscribe to Sustainable Development goals in the form of management systems aimed, for example, at cities or agglomerations, such as ISO 37101, though their popularity is limited. Analysing requirements related to the concept of the CE and published standards relating to management systems, it is currently possible to identify some solutions which may provide support in the implementation of the CE. Individual groups of such standards are presented in Figure 4.1.

The individual systems related to the environment shown in Figure 3.1 have a long tradition of implementation within organisations. The first international environmental management system was developed by the ISO/TC 207 Technical Committee as far back as in 1994. After many changes, these requirements are currently contained in the ISO 14001 standard and are used as a basis for certification (Kafel, 2017).

Environmental management covers the efforts of an organisation to monitor its activities and the effects of those activities on the environment in order to minimise negative impacts on the environment and increase positive impacts (ISO, 2017a). The main aim of the system is to protect the environment and respond to changing environmental conditions in balance with socio-economic needs. Such an approach should contribute to the implementation of the concept of the CE by, among other things, preventing or mitigating adverse environmental impacts and assisting organisations in the fulfilment of compliance obligations (ISO, 2015).

It should be emphasised, that in a way similar to other meta-standards, fulfilment of the requirements of the ISO 14001 standard does not amount to confirmation that the organisation has a low negative impact on the environment. Proper implementation of the system is rather a guarantee of a genuine willingness to act to make one's activity more sustainable. However, the actual pace at which changes are made, or the specific solutions used to achieve that goal, can be chosen by the organisation on a voluntary basis.

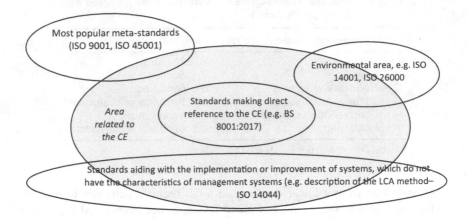

*Figure 4.1* Standardisation of CE activities in the context of management systems

The presented approach related to the environmental activity of an organisation, and in particular the pace and effectiveness of the results achieved, has led to the creation of other solutions, both in the area of standardisation and outside of it. One example of a voluntary system, whose requirements are written into legal regulations at EU, is the eco-management and audit scheme EMAS (EU, 2009). Due to the additional requirements in relation to the environmental management system (EMS) compliant with ISO 14001 and the regional character of European Union legislation, the popularity and recognition of the EMAS system are lower than that of the system compliant with ISO 14001. There are also significant differences in the popularity of this system between countries of the EU. An example of this imbalance is to be found in the number of EMAS certificates issued (as of August 2020) for individual sites in Poland and Italy, amounting to 4,933 and 574, respectively (EC, 2021; Canestrino, Ćwiklicki, Kafel, Wojnarowska, & Magliocca, 2020).

Another system belonging to the group of environmental systems shown in Figure 3.1 is the energy management system. Its requirements are presented in the ISO 50001 standard. The aim of the system is to design processes in organisation and management, which make it possible to improve energy performance, including energy efficiency, energy use, and energy consumption (ISO, 2018). This system develops upon and goes into greater detail regarding activities in one of the areas of the EMS in accordance with the requirements of the ISO 14001 standard.

The last of the areas shown in Figure 4.1 is that of standards aiding with the implementation or improvement of management systems. Often, such standards are referred to as standards belonging to the ISO 14001 family of standards.

Examples of standards which at the same time provide support in selected areas of the CE in organisations through the use of various methods are:

- ISO 14006. Environmental management systems. Guidelines for incorporating ecodesign;
- ISO 14020. Environmental labels and declarations. General principles;
- ISO 14044. Environmental management. Life Cycle Assessment. Requirements and guidelines;
- ISO 14063. Environmental management. Environmental communication. Guidelines and examples;
- ISO 14009. Environmental management systems. Guidelines for incorporating material circulation in design and development;
- ISO 14046. Environmental management. Water footprint. Principles, requirements and guidelines;
- ISO 14067. Greenhouse gases. Carbon footprint of products. Requirements and guidelines for quantification.

## Normalised standards dedicated to the Circular Economy

### BS 8001:2017 Framework for implementing the principles of the Circular Economy in organisations – guide

In 2017, the BSI (British Standards Institution) published the world's first standard concerning the Circular Economy. This standard is intended to help organisations and individuals consider and implement more circular and sustainable practices within their businesses, whether through improved ways of working, providing more circular products and services, or redesigning their entire business model and value proposition. The standard is divided into two areas: The first of these describes what the CE is and why it is worth making the move to a sustainable mode of operation, while the second describes the way in which the requirements of this standard should be implemented (Nowicki, 2020).

This standard focuses on the CE, which promotes optimal use of resources, reuse, repair, refurbishing, remanufacture, and the recycling of materials and products, as well as the preservation and regeneration of natural capital by returning biological nutrients into the biosphere. Process and product or service design and innovation (e.g. for repair, reuse, recyclability) can be complemented by business model design and innovation using approaches such as performance-based models to manage how products and materials circulate within the system (BSI, 2017).

Implementing the principles of the CE offers organisations an opportunity to re-think how they do business, potentially enabling them to be more circular, sustainable, and competitive. For example business opportunities could arise from understanding resource use and adopting new ways of working, both internally and across the value chain.

The standard consists of three main elements: Guiding principles (Table 4.3), a flexible framework (Table 4.4), and supporting guidance, which are presented later (BSI, 2017).

The standard outlines the principles of the Circular Economy, which underpin the flexible framework, and provides a strategic frame of reference to know how closely decision-making and activities align with the guidance provided.

As part of its structure, the standard provides organisations with a flexible framework to determine the extent to which they intend to implement the principles of the CE and transition to a more circular and sustainable mode of operation. This part of the requirements provides a guide divided into eight stages of implementation.

The standard also provides supporting guidance mechanisms and business models that can support the transition to a more circular and sustainable mode of operation as well as key factors which might be relevant to consider when using the framework. The selected business model and its value proposition must be underpinned by the principles of the CE and the flexible framework (Pomponi & Moncaster, 2019). A business model comprises an organisation's chosen system of interconnected and interdependent decisions and activities, which determines how it creates, delivers, and captures value over the short, medium, and long term. However, business model innovation for the CE goes beyond advances in processes and/or products or services. Implementing any one business model does not necessarily equate to a shift to a more circular and sustainable mode of operation. From the perspective of this standard, this is best delivered through a systems approach.

*Table 4.3* Principles of the Circular Economy

| Guiding principles | Characteristics |
| --- | --- |
| Principle 1 Systems thinking | Organisations take a holistic approach to understand how individual decisions and activities interact within the wider systems they are part of. |
| Principle 2 Innovation | Organisations continually innovate to create value by enabling the sustainable management of resources through the design of processes, products/services, and business models. |
| Principle 3 Stewardship | Organisations manage the direct and indirect impacts of their decisions and activities within the wider systems they are part of. |
| Principle 4 Collaboration | Organisations collaborate internally and externally through formal and/or informal arrangements to create mutual value |
| Principle 5 Value optimisation | Organisations keep all products, components, and materials at their highest value and utility at all times. |
| Principle 6 Transparency | Organisations are open about decisions and activities that affect their ability to transition to a more circular and sustainable mode of operation and are willing to communicate these in a clear, accurate, timely, honest, and complete manner. |

Source: Own elaboration based on BSI (2017).

*Table 4.4* Stages of implementation covering the framework of the Circular Economy system

| Stages | Characteristics |
| --- | --- |
| Stage 1. Framing | Organisations should determine the relevance of the circular economy to their business and identify where to begin. |
| Stage 2. Scoping | In considering a vision, strategic plan, and direction for circular economy activity, organisations should look carefully at what is possible and/or required within the context of the CE. |
| Stage 3. Idea generation | Organisations should develop a list of ideas/options to tackle problems and/or opportunities identified in Stage 2 and prioritise these accordingly within the context of their CE vision, strategic plan, and objectives. |
| Stage 4. Feasibility | Organisations should assess the practicality of progressing their prioritised ideas/options identified in Stage 3. |
| Stage 5. Business case | Organisations should develop a business case to secure the necessary resources to pilot new ideas/options and then implement, scale up, and roll out. |
| Stage 6. Piloting and prototyping | Organisations should experiment with ideas/options on a small scale to determine practical viability. |
| Stage 7. Delivery and implementation | Organisations should scale/roll out the adoption and integration of proven approaches to transition to a more circular and sustainable mode of operation. |
| Stage 8. Monitor, review, and report | Organisations should track performance to ensure ongoing success and continual and transformational improvement. |

Source: Own elaboration based on BSI (2017).

## XP X 30–901:2018 Circular Economy. Circular Economy project management system. Requirements and guidelines

The next normalised standard concerning topics related to the CE is the French standard published by AFNOR in 2018. This standard specifies the requirements relative to a project management system led by an organisation to improve its environmental, economic, and societal performance and to contribute to the development of a CE. Like the British standard described earlier, it is applicable to organisations of all sizes, operating in all sectors of business. In this document, the term project refers to any aim for change fostered by an organisation with the aim of making its activity evolve in part or in whole towards a model that is more efficient in its utilisation of resources, thereby limiting the environmental impact of its activities. The project can touch on the activities, products, or services that the organisation identifies (AFNOR, 2018; Nowicki, Kafel, Balon, & Wojnarowska, 2020).

The standard provides requirements and practical recommendations for the initiation, planning, implementation, measurement, and management of projects by adopting an open and holistic approach. It in particular defines the various steps organisations shall follow to ensure that their project contributes to a transition towards a CE. The standard was built in a way similar to management of meta-standards and in accordance with the guidelines of the HLS model. In addition to system issues arising from the HLS structure, the most important element of the standard are the requirements for a project management system in a CE. The standard includes requirements to simultaneously take into account the contribution of projects to the three dimensions of Sustainable Development, taking into account direct and indirect effects. In addition to the requirements and recommendations for applying the approach to the management system, this standard also provides organisations with a broader methodological framework based on cross-analysis taking into account the seven areas of project management activities in the CE (Table 4.5) (AFNOR, 2018).

The aforementioned areas of activities should be analysed with regard to three dimensions of environmental, economic, and social, which result from the assumption of the standard and the proposed cross-analysis.

### ISO 14009:2020 Environmental management systems. Guidelines for incorporating material circulation in design and development

The necessary transition from a linear to a Circular Economy to achieve Sustainable Development has been spearheaded by the EU and is one of the elements of the new European Green Deal (EU, 2019). One of the methods to consider for supporting the transition to a Circular Economy is implementing a design that facilitates the material circulation of products and their constituent parts. Considering that products are largely composed of raw materials, the material circulation of products plays an important role in the sustainable use of resources. The widely held perception is that strategy/planning for the material circulation of products and their constituent parts should precede their design and development. In 2020, the ISO 14009:2020 standard was therefore published containing guidelines for incorporating material circulation in design and development. This standard contains guidelines and guidance on analysing existing products prior to redesign, identifying measures for improvement, and reflecting those measures into the redesign of the products and components with a focus on material circulation (ISO, 2020).

This standard fits well with the idea of using normalised standards, which can help and facilitate the implementation of systematised activities in the area of the CE. ISO 14009 contains, among other things, guidelines concerning the application of ISO 14001 in order to improve material circulation. Aspects of material circulation for the product in the CE are considered from the point of view of two design strategies – use of materials and ease of disassembly considered over the individual stages of the product life cycle. This international standard provides guidelines to an organisation for managing design and development of products in a systematic manner using the framework and requirements of the environmental management system. This standard contains

*Table 4.5* Areas of project management activities in the CE

| Area | Description |
| --- | --- |
| Sustainable procurement of supplies | Sustainable procurement involves taking into account the environmental and societal impacts of the production cycle of the resources, whether renewable or not, necessary for a goods or service production process. This area of action can concern the processes of extraction and utilisation of natural resources, the acquisition of the components necessary for a production process or the replacement of non-renewable raw materials by renewable materials, or secondary raw materials or recycled raw materials. |
| Eco-design | Eco-design is implemented by the systematic integration of environmental aspects from the design stage and development of products (goods and services, systems), with the aim of reducing adverse environmental impacts throughout their cycle while achieving equivalent or higher performance (adapted from ISO 14006). |
| Industrial symbiosis | Industrial symbiosis is materialised by the pooling and/or the interrelation of different production phases or of several processes for the manufacture of goods or services, with the aim of enabling the shared management of certain functions, goods, stocks, flows of materials, and energy in order to optimise them. Symbiosis can encompass substitution synergies set up between several economic entities or synergies pooling flows of materials, energy, water, infrastructures, goods, or services in order to optimise the use of resources. |
| Functional service economy | The functional or service economy is oriented towards fostering usage rather than ownership. The current trend in this area is to sell services rather than the products themselves, with the aim of ensuring usage performance. |
| Responsible, sustainable consumption | Sustainable consumption takes into account the economic, social, and environmental impacts of purchasing and using a product or a service. This applies to the changing of consumption practices by, for example, examining the possibility of reuse of a product for a different purpose, repair, reuse for the same purpose, management of production waste, etc. It is based on the effectiveness of informing buyers, who are committed to and focussed on the environmental and social aspects of the product life cycle. |
| Product lifetime extension | Product lifetime extension results in the provision of a product or service under conditions that enable its duration of use to be extended compared to an equivalent product or service while guaranteeing that the initial performance or clearly specified performance characteristics shall be maintained. This covers the measures taken to guarantee the availability of spare parts and product modularity to facilitate performance, updates, maintenance, reuse, repairability, and conformity. |
| Efficient management of end-of-life products and materials | Efficient management of end-of-life products and materials is materialised by the transformation, including recycling, of post-consumption residues into substances, materials, or products to fulfil their initial function or for other purposes. This domain touches on all the techniques for transforming waste – including organic waste – with the aim of reintroducing all or part of it into a production cycle. |

Source: Own elaboration based on AFNOR (2018).

guidelines concerning a strategy for material circulation. The specific areas of this strategy are (ISO, 2020):

- Material use: Simplify and minimise the number of types of materials used for existing products and/or components.
- Ease of disassembly: Simplify assembly and manufacturing of products and components to promote easy dismantling and separation of components, which enhance materials' reuse, remanufacturing, and recyclability of existing products.

Currently, limitation and uneven distribution of resources are a key concern for organisations, businesses, regions, or countries that rely significantly on specific resources. The added value of this standard for businesses which decide to implement it is minimisation of the risk of supply shortages, which smoothens the transition to a Circular Economy (ISO, 2020).

The concept of the CE used in this standard in the context of material circulation is characterised as follows (ISO, 2020):

- An economy which aims to keep products, materials, and resources at their highest value for as long as possible and to minimise the generation of waste.
- An economy that is restorative and regenerative by design and which aims to keep products, components, and materials at their highest utility and value at all times.
- An economy that closes the loop between different life cycles through the application of designs that allow for the enhancement of recycling and reuse for the more efficient use of raw materials, goods, and waste and improved energy efficiency.

Organisations implementing this standard should take into consideration the benefits derived from the application of a material circulation strategy. The material circulation strategy adopted should be the result both of actual material circulation policy and prior analyses (analysis of existing situation and environmental review). This strategy also enables a set of economic, technical, and human measures to be established. In order to implement material circulation with regard to the life cycle of manufactured products, the material circulation strategy should take the following two tasks into consideration (ISO, 2020):

1    The first task concerns the strategic aspects of material circulation.
2    The second task is to manage internal processes after establishing a material circulation strategy.

These tasks lead to the implementation of a programme of activities guided by a methodology encompassing goals, obligations, time frames, and resources. Risks and opportunities related to material circulation concerning products and their constituent parts should be taken into consideration in the implemented processes.

When planning the environmental management system, the organisation should take up issues related to the material circulation of products and their specific elements into consideration. Examples of risks and opportunities related to the material circulation of products and components are provided in Table 4.6.

## ISO 20400:2017 Sustainable procurement – guidance

One of the elements of the modern approach to the CE is sustainable procurement, which represents an opportunity to provide more value to the organisation by improving productivity; assessing value and performance; enabling communication between purchasers, suppliers, and all stakeholders; and by encouraging innovation. To enable organisations to achieve the aforementioned goals and to assist them in doing so, a standard containing guidelines while also serving as a guide to sustainable procurement was drawn up in 2017. This standard is intended to help improve supplier relations and also to make the efforts of businesses in the area of sustainable supply chain development more effective and secure them an advantage over the competition (Yeleyko & Zamojski, 2017).

The requirements are divided into four main thematic areas. The first of these provides general information about sustainable procurement. This includes a description of the principles and core subjects of sustainable procurement. The second area contains guidelines on how sustainability considerations are

*Table 4.6* Examples of risks and opportunities related to the material circulation of products and components

| Area | Issue | Risks | Opportunities |
| --- | --- | --- | --- |
| Environmental aspects of the product | Use of recycled materials | Requires more detailed management of product quality Customer complaints | Reputation of the organisation |
| | Lack of flame retardant | Risk of fire | Product recycling rate |
| Obligation of compliance | Regulations concerning the recyclability of products | Limited freedom in the design of products | Ease of entry onto market |
| Principles | Improvement in the quality of recycled materials | Restriction of the function of products or components | Securing of recycled materials of good quality Creation of a new market |
| Other issues | Ease of disassembly of elements suitable for reuse/ remanufacturing | Decrease in new product sales | Creation of new business opportunities |

Source: ISO (2020).

integrated at a strategic level within the procurement practices of an organisa-tion to ensure that the intention, direction, and key sustainability priorities of the organisation are achieved. The third area describes the organisational conditions and management techniques needed to successfully implement and continually improve sustainable procurement. The fourth area addresses the procurement process and is intended for individuals who are responsible for procurement within their organisation. It is also of interest to those in associated functions, as it describes how sustainability considerations are integrated into existing pro-curement processes (ISO, 2017b).

The requirements include a complete set of sustainable procurement principles that have a significant influence on how the organisation functions *inter alia* in a pro-environmental manner and assists organisations in implementing a Circular Economy strategy and model (Table 4.7) (ISO, 2017b).

This standard also places a great deal of emphasis on analysing the organisa-tion's needs for specific products or services. It recommends that the organisation should consider what alternative options might exist to deliver the same outcome

*Table 4.7* Principles of sustainable procurement in the context of the Circular Economy

| Principle | Description |
| --- | --- |
| Accountability | Taking responsibility for the organisation's impact on society, the economy, and the environment. |
| Transparency | Being transparent in decisions and activities that impact the environment, society, and the economy. |
| Ethical behaviour | Behaving ethically and promoting ethical behaviour throughout the organisation's supply chains. |
| Respect for the rule of law and international norms of behaviour | Striving to be aware of any violations throughout the organisation's supply chains. |
| Innovative solutions | Seeking solutions to address sustainability objectives and encourage innovative procurement practices to promote more sustainable outcomes throughout the entire supply chain. |
| Focus on needs | Reviewing demand, buying only what is needed, and seeking more sustainable alternatives. |
| Integration | Ensuring that sustainability is integrated into all existing procurement practices to maximise investment in sustainable solutions. |
| Analysis of costs | Analysis of possible costs incurred over the life cycle and an analysis of the costs and benefits for society, the environment, and the economy resulting from the organisation's procurement activities. |
| Continual improvement | Working towards continually improving its sustainability practices and outcomes and encouraging organisations in one's supply chains to do the same. |

Source: Own elaboration based on ISO (2017b).

in a better way, which is an excellent illustration of the concept of the CE. Such activities may be (ISO, 2017b):

- Eliminating the demand by reviewing the need;
- Reducing the frequency of use/consumption;
- Identifying alternative methods of fulfilling demand, such as outsourcing services or leasing rather than owning;
- Aggregating and/or consolidating the demand;
- Sharing use between divisions or organisations;
- Encouraging recycling, repairing, reusing, or repurposing of older goods;
- Determining whether outsourcing is required and how to extend the scope of responsibility for environmental and labour practices throughout supply chains;
- Using recycled/renewable materials.

## Certification of management systems

Currently, organisations, to be successful, have to be able to offer their employees the feeling of having prospects and of having a mission. The implementation and certification of environmental management systems or requirements concerning the CE are one of the possible solutions for businesses on the road to Sustainable Development .

Certification is an activity as a result of which the fulfilment of specific requirements – for example those contained in standards – is confirmed by an independent certifying body. Unfortunately, consumers do not always understand the difference between different types of certification and the meaning behind the certification. As a result, this may lead to a lack of confidence. As Zhang, Joglekar, and Verma (2014) point out, customers often do not understand the meaning of an eco-certification or what it guarantees.

The requirements of management meta-standards contain guidelines concerning management within an organisation and not the products themselves. This is also why such certification is not a direct confirmation that products from the certified organisation have less of an environmental impact than those of the same class manufactured by the competition. For this purpose, the product markings described in the next chapter, which concerns environmental labelling, should instead be used.

There are many studies which point to the advantages of the implementation and certification of management systems, such as (Al-Kahloot, Al-Yaqout, & Khan, 2019; Boiral & Henri, 2012; D'Aveni, Dagnino, & Smith, 2010; de Jong, Paulraj, & Blome, 2014; Nowicki, Ćwiklicki, Kafel, & Wojnarowska, 2021; Santos, Rebelo, Lopes, Alves, & Silva, 2016):

- Prevention and reduction of environmental risks;
- Improved company image;

- Compliance with legal requirements;
- Rational use of natural resources;
- Promotion of recycling and other pro-environmental activities;
- Improved employee awareness on environmental issues;
- Savings on direct and indirect costs resulting from increased efficiency; reduction in the costs of waste removal; and avoidance of costs, such as insurance premiums, costs of utilisation, administrative penalties;
- Better market access and lowering of barriers to trade;
- More formalised, systematic, rigorous, and effective practices;
- Greater employee engagement;
- Improved relations with stakeholders.

The majority of the advantages given before result not from certification itself, but from the correct implementation of the management system and the effects of its operation. Advantages related to the certification process itself will result from:

- Interaction between employees of the organisation and external auditors, who will assess the functioning of management processes. These will mainly be external benefits contributing to the improvement of the system.
- The use of certificates issued by certifying bodies, which will mainly translate into achieving external benefits.

Certifications processes also have some related disadvantages, which need to be taken into account before taking any decision on certification. Certification of management systems related to the area of Sustainable Development can and often is used by owners of the organisation as a way of improving its image. Unfortunately, in extreme cases, such practices lead to a distortion of the basic assumptions underlying the systems described in management meta-standards and lead to a phenomenon referred to by researchers as 'greenwashing'. In that case, the process of implementation of the system and its certification is conducted mainly in order for the organisation to communicate about its engagement on environmental issues, but in reality its activities are minimal and are focussed solely on obtaining a positive result in certification (Dahl, 2010; Di Noia & Nicoletti, 2016; EC, 2015). More and more people and organisations are becoming aware of such conduct, and the benefits of certification are diminishing accordingly.

Other important disadvantages of the certification of a system include its costs, which may be a major barrier especially for small- and medium-sized organisations. Another important problem may also be the willingness of management to focus on obtaining a formal confirmation of the functioning of the system, instead of on the improvement of processes (Ejdys, 2010). In extreme cases, activities may be carried out with the sole purpose of satisfying external auditors, with the sense of such action being questioned by employees and managers. In such a situation, it is difficult to achieve engagement and any real effects of improvement.

# Summary

This chapter has presented the most popular management systems providing support for the practical implementation of the CE, which can be used in the process of implementation, improvement, and certification. Certification of the management systems presented in this chapter only confirms the fulfilment of minimal system requirements. Moreover, these activities are voluntary. As a result, improvement of the system and the real environmental benefits which could be achieved in this area will depend on the engagement and willingness on the part of those managing the organisation to conduct such activities. The time for which the system has functioned within the organisation is also of major significance. The Deming cycle, which is integral to the requirements of the management systems presented, means that the effects will only be visible after at least several cycles of improvement related to the implementation of the programme of audits and management reviews. Considering the advantages and disadvantages of the certification of management systems described before, it should be remembered that certification itself only confirms, on the basis of the selected audit sample, that the management system has been effectively implemented and is functioning within the organisation. From the point of view of the expected effects – that is of added value, what is most important is the manner in which the management system is implemented and its continual improvement, and not its certification. Certification often helps with the functioning and improvement of the management system but should not be an end-in-itself.

# References

AFNOR. (2018). *XP X 30–901 circular economy – circular economy project managem ent system – requirements and guidelines*. Retrieved from https://standards.globalspec.com/std/13096170/XP%20X30-901

Al-Kahloot, E., Al-Yaqout, A., & Khan, P. B. (2019). The impact of ISO 14001 standards certification on firms' performance in the state of Kuwait. *Journal of Engineering Research (Kuwait)*, *7*, 286–303.

Boiral, O., & Henri, J. F. (2012). Modelling the impact of ISO 14001 on environmental performance: A comparative approach. *Journal of Environmental Management*, *99*, 84–97. doi:10.1016/j.jenvman.2012.01.007

BSI. (2017). *BS 8001:2017 Framework for implementing the principles of the circular economy in organizations – guide*. London: BSI Standards Limited.

Canestrino, R., Ćwiklicki, M., Kafel, P., Wojnarowska, M., & Magliocca, P.(2020). The digitalization in EMAS-registered organizations: Evidences from Italy and Poland. *The TQM Journal*, *32*(4), 673–695. doi:10.1108/TQM-12-2019-0301

Dahl, R. (2010). Green washing: Do you know what you're buying? *Environmental Health Perspectives*. doi:10.1289/ehp.118-a246

D'Aveni, R. A., Dagnino, G. B., & Smith, K. G. (2010). The age of temporary advantage, *Strategic Management Journal*, *31*, 1371–1385. doi:10.1002/smj.897

de Jong, P., Paulraj, A., & Blome, C. (2014). The financial impact of ISO 14001 certification: Top-line, bottom-Line, or both? *Journal of Business Ethics*, *119*, 131–149. doi:10.1007/s10551-012-1604-z

Di Noia, A. E., & Nicoletti, G. M. (2016). ISO 14001 certification: Benefits, costs and expectations for organization. *Studia Oeconomica Posnaniensia, 4*, 94–109. doi:10.18559/SOEP.2016.10.7

EC. (2015). *Communication of 2.12.2015 on an EU action plan for the circular economy.* Brussels: Author. doi :10.1017/CBO9781107415324.004

EC. (2021). Retrieved from https://ec.europa.eu/environment/emas/emas_registrations/statistics_graphs_en.htm

Ejdys, J. (2010). Za i przeciw normalizacji systemów zarządzania. *Zarządzanie Zasobami Ludzkimi, 3–4*, 67–80.

EU. (2009). *Regulation (ec) no 1221/2009 of the European Parliament and of the Council of November 25, 2009 on the voluntary participation by organisations in a community eco-management and audit scheme (EMAS), repealing Regulation (EC) No 761/2 001 and Commission Deci. Official Journal.* Retrieved from https://eur-lex.europa.eu/legal-content/en/ALL/?uri=CELEX%3A32009R1221

EU. (2019). *Communication from the commission to the European Parliament, the European Council, the Council, the European Economic and Social Committee and the Committe e of the Regions. The European Green Deal.* Retrieved from https://eur-lex.europa.eu/legal-content/EN/TXT/?uri=CELEX%3A52019DC0640

Heras-Saizarbitoria, I., Boiral, O., & Ibarloza, A. (2020). ISO 45001 and controversial transnational private regulation for occupational health and safety. *International Labour Review, 159*, 397–421. doi:10.1111/ilr.12163

ISO. (2010). *ISO 26000:2010 Guidance on social responsibility.* Geneva, Switzerland: Author.

ISO. (2015). *ISO 14001:2015 Environmental management systems: Requirements with guidance for use.* Geneva, Switzerland: Author.

ISO. (2017a). *ISO 14001:2015 Environmental management systems: A practical guide for SMEs.* Geneva, Switzerland: Author.

ISO. (2017b). *ISO 20400:2017 sustainable procurement: Guidance.* Retrieved from https://www.iso.org/standard/63026.html#:~:text=ISO%2020400%3A2017%20provides%20guidance,by%2C%20procurement%20decisions%20and%20processes

ISO. (2018). *ISO 50001:2018 Energy management sy stems: Requirements with guidance for use.* Retrieved from https://www.iso.org/standard/69426.html

ISO. (2019). *The ISO survey of management system standard certifications.* Retrieved from https://www.iso.org/the-iso-survey.html

ISO. (2020). *ISO 14009:2020 environmental management systems: Guidelines for incorporating mater ial circulation in design and development.* Retrieved from https://www.iso.org/standard/43244.html

ISO/IEC. (2020). *ISO/IEC. Directives, part 1. Consolidated ISO supplement – procedures specific to ISO, annex SL, proposals for management system standards, appendix 2 – high level structure, identical core text, com mon terms and core definitions* (11th ed.). Retrieved from https://www.iec.ch/members_experts/refdocs/iec/Consolidated_JTC1_Supplement_2020_publication.pdf

Kafel, P. (2017). *Integracja systemów zarządzania: Trendy, zastosowania, kierunki doskonalenia.* Kraków: Wydawnictwo UEK Kraków.

Kafel, P., Nowicki, P., & Wojnarowska, M. (2021). Assumptions of a circular economy management standard for the food industry: Choosing the best structure. In *Key challenges and opportunities for quality, sustainability and innovation in the fourth industrial revolution* (pp. 489–503). Singapore: World Scientific. doi:10.1142/9789811230356_0023

Kirchherr, J., Reike, D., & Hekkert, M. (2017). Conceptualizing the circular economy: An analysis of 114 definitions. *Resources, Conservation and Recycling.* doi:10.1016/j. resconrec.2017.09.005

Nowicki, P. (2020). Gospodarka o obiegu zamkniętym a wykorzystanie znormalizowanych systemów zarządzania. In Z. Wojciechowski & P. Zaskórski (Eds.), *Czwarta Rewolucja Przemysłowa. Mity, Paradygmaty i Zastosowania. Tom 2 – Wybrane Obszary Zastosowań Idei Przemysłu 4.0* (pp. 297–310). Warszawa: Wojskowa Akademia Techniczna.

Nowicki, P., Ćwiklicki, M., Kafel, P., & Wojnarowska, M. (2021). Credibility of certified environmental management systems: Results from focus group interviews. *Environmental Impact Assessment Review, 88,* 106556. doi:10.1016/j.eiar.2021.106556

Nowicki, P., Kafel, P., Balon, U., & Wojnarowska, M. (2020). Circular economy's standardized management systems: Choosing the best practice. Evidence from Poland. *International Journal for Quality Research.* doi:10.24874/IJQR14.04-08

Pomponi, F., & Moncaster, A. (2019). BS 8001 and the built environment: A review and critique. *Proceedings of the Institution of Civil Engineers – Engineering Sustainability, 161.*

Prieto-Sandoval, V., Jaca, C., & Ormazabal, M. (2018). Towards a consensus on the circular economy. *Journal of Cleaner Production, 179,* 605–615. doi:10.1016/j.jclepro. 2017.12.224

Santos, G., Rebelo, M., Lopes, N., Alves, M. R., & Silva, R. (2016). Implementing and certifying ISO 14001 in Portugal: Motives, difficulties and benefits after ISO 9001 certification. *Total Quality Management & Business Excellence, 27,* 1211–1223. doi:10.10 80/14783363.2015.1065176

Yeleyko, V., & Zamojski, J. (2017). ISO 20400 – pierwsza międzynarodowa norma o zrównoważonych zamówieniach. *Studia i Materiały: Miscellanea Oeconomicae,* 105–115.

Zhang, J. J., Joglekar, N., & Verma, R. (2014). Signaling eco-certification. *Journal of Service Management, 25,* 494–511. doi:10.1108/JOSM-01-2014-0035

# 5   Impact of environmental labelling upon popularisation of the Circular Economy

*Bartłomiej Kabaja*

## Introduction

International standards of certification (e.g. ISO standards) have already been implemented in areas of practice and science concerning quality management, environmental management, or occupational safety. However, for standards and systems in the field of the Circular Economy, the work is only beginning. As presented in the earlier chapter on the *Significance and Adjustment of Environmental Certification Schemes in the Circular Economy*, the first attempts at developing standards have already been made at local and global levels.

The chosen direction of changes to economic models now seems to be inevitable. The progressive decline in the value and productivity of the environment is having a negative impact on quality of life and seems to be the cause of many social conflicts (Asteria, Suyanti, Utari, & Wisnu, 2014).

This purpose of this chapter is to discuss the impact of product labelling systems on popularisation of the Circular Economy model. The study discusses the process of corporate environmental communication, the most popular environmental labelling systems and, on the basis of the subject literature, also proposes guidelines for future systems to promote the Circular Economy model.

## Role of environmental information provided in market communication between business and stakeholders

Issues of social responsibility, climate change, or pro-environmental thinking are contributing to development of environmental communication, which has become one of the fastest-growing areas within scientific and technical communication (Platonova, 2016). According to the ISO 14063:2020 international standard on Environmental management – Environmental communication, environmental communication is a process that an organisation conducts to provide and obtain information and to engage in dialogue with internal and external interested parties (3.5) to encourage a shared understanding on environmental issues, aspects, and performance. Though the standard does not make a direct reference to the Circular Economy model, it is a valuable source of recommendations and guidance

DOI: 10.4324/9781003179788-5

concerning communication, which can very much be of use in the popularisation of the Circular Economy.

The ISO 14000 series of standards defines environmental communication policy as overall intentions and directions of an organisation related to its environmental communication (3.1) as formally expressed by top management. This policy can be a separate policy or part of other more general policies within the organisation (ISO 14063:2020). A graphic model of the process of corporate environmental communication is presented in Figure 5.1.

Many authors emphasise that corporate environmental communication is one of the most important activities for a business to which it should devote a great deal of engagement (Altinay & Williams, 2019; Singleton, 2019; Preziosi, Tourais, Acampora, Videira, & Merli, 2019; Unde, Arianto, Bahfiarti, Pulubuhu, & Arsyad, 2020; Gulliver, Fielding, & Louis, 2021). Its presence is a key constituent of the value and maturity of the organisation.

The organisation (see Figure 5.1) is the transmitter of the environmental message and should develop a communication policy which determines its objectives, target groups, form and means of transmission, and budget. It is the organisation that analyses the factors which may have an impact on the process of environmental communication, selects information, and transforms it into verbal or graphic signs.

In accordance with the ISO 14063 standard, the environmental communication process should be parallel to and integrated into the organisation's general system

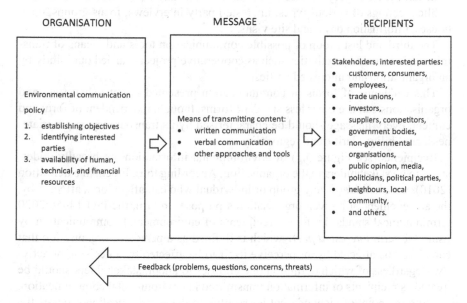

*Figure 5.1* Diagram of environmental communication for an organisation

of communication. Such an approach guarantees greater efficiency. Establishing the objectives of environmental communication is very useful, as it allows the effectiveness of this process to be monitored by verifying the achievement of those objectives.

The second very important element of the environmental communication process is the message (see Figure 5.1), which the organisation transmits to recipients. In principle, the message should be communicated in a form which makes use of various means of information provision: labels, brochures, advertising in traditional and social media, creating coherent content which is compliant with the adopted environmental communication policy.

Environmental information communicated to recipients should take the social, cultural, educational, economic, and political interests of stakeholders into consideration (Gregory-Smith, Manika, & Demirel, 2017). The language used should not contain jargon or excessively specialist or technical information, but should be appropriately adapted to the level of the recipients.

According to the ISO 14063 standard, we can distinguish between the following types of means of environmental communication: written, verbal, and other communication tools.

The following should be included in the category of written communication: websites; environmental or sustainability reports; printed material such as reports, brochures, and newsletters; and product or service information labels or declarations.

In turn, the category of verbal communication may be considered to include public meetings of various types, interested party interviews, focus groups, open houses, information days, and site visits.

The third and last group of possible communication tools and means of transmission is formed by activities such as cooperative projects carried out jointly by an organisation and interested parties.

This catalogue of forms of communication presented is not exhaustive, and organisations can use other less standard forms, though, independent of form and content, every message should be characterised by its transparency, appropriateness, credibility, clarity, and regionality.

Recipients (see Figure 5.1) of environmental information are broadly understood to be stakeholders in the organisation. According to E. Freeman's definition (2010), a *stakeholder* is any group or individual who can affect, or is affected by, the achievement of a given organisation's purpose. In turn, the ISO 14050:2020 terminological standard refers to recipients of environmental communication by using the term *interested party* which is defined as a person or organisation that can affect, be affected by, or perceive itself to be affected by a decision or activity. Regardless of which definition one chooses to adopt, stakeholders should be treated as recipients of information transmitted in environmental communication.

From the point of view of what is essential to the communication process, the number of recipients and the importance to the organisation, customers, and consumers should be considered to be the most important stakeholders (Brauweiler, 2013). They have a clear and significant influence on the organisation's activity.

Customers and consumers are stakeholders, who achieve their pro-ecological demands by means of mutual encouragement through their everyday purchasing decisions.

The true essence of the communication process is feedback (see Figure 5.1). It is the response of recipients to the message received (Wiktor, 2013). Feedback is an expression of the recipient's response, formed as a result of understanding and assimilation of the communicated information. The existence of feedback is evidence of the transactional and interactive character of environmental communication (Kabaja, 2018). The response time of recipients depends on the form of communication (e.g. direct or indirect).

Information received as feedback is an exceptionally valuable part of environmental communication. It can be analysed to verify how the message was received and if the recipients understood it properly. Furthermore, information gathered from stakeholders allows the organisation to improve its performance and increase the effectiveness of its environmental activities (Tam, Shen, Yau, & Tam, 2007).

### *Explanation of standards and recommendations concerning the placement of environmental information on product packaging*

The use of environmental marking (labelling) confirming the implementation of the assumptions of the Circular Economy is exceptionally important in order to build awareness of the concept and popularise it in society and on the business-to-business (B2B) market. To date, this specific type of environmental communication, which consists of applying labels to products, has often been used a means of promoting the implementation of different systems.

A systematic approach to the issue of environmental labels and declarations is set out by the International Organization for Standardization (ISO). This non-governmental international organisation proposed harmonised criteria, definitions, and guidelines for environmental labels.

Standards adopted and validated by the ISO/TC 207 Technical Committee of the International Organization for Standardization cover:

- Type I – Environmental labelling – these are ecolabelling programmes, where there are clearly defined criteria of assessment for products or services.
- Type II – Self-declared environmental claims – these concern products and services, for which there are no generally accepted criteria or systems of labelling.
- Type III – Environmental declarations – for specific aspects of products with the application of a life-cycle-based approach.

Type I of the ecolabelling programme was published in the ISO 14024:2018 standard – Environmental labels and declarations – Type I environmental labelling – Principles and procedures. This document contains the principles and procedures for the selection of categories for the products to be assessed and their environmental and functional criteria in particular. It also establishes the certification procedures for awarding the label, the principles for keeping documentation and for periodic verification of the fulfilment of criteria by the entity issuing the ecolabel. Environmental criteria used in the assessment must take the environmental impact of the product across its whole Life Cycle Assessment (LCA) into account, including life cycle stages such as resource extraction, production, distribution, use, and disposal. It is worth emphasising that programmes resulting from type I environmental labelling are voluntary and based on requirements of a third party or, in other words, a person or body that is recognised as being independent of the parties involved, as concerns the issue in question.

Type II environmental labels and declarations were addressed in the ISO 14021:2016 standard – Environmental labels and declarations – Self-declared environmental claims. This standard is helpful in situations where there are thematic areas related to ecolabelling in which there are no organisations in operation which can verify the validity of the use of such labels. Then producers or service providers who want to make use of the competitive advantage of using such labels and declarations on the market may themselves present information about the pro-environmental properties of their products or services. It is however necessary to have evidence proving the reliability of the information declared. The ISO 14021:2016 standard defines a range of terms, providing the necessary definitions for the correct use of environmental declarations. It also contains detailed information about methods of assessment and verification of the quantities and measures that are the subject of environmental labelling and declarations. These include terms such as compostable, degradable, designed for disassembly, or recyclable. Unfortunately, the ISO 14021:2016 standard currently does not provide any guidelines which could be of particular assistance in the application of labelling supporting the Circular Economy model.

Type III of the environmental labelling programme has a scope which encompasses environmental labels and declarations related to the product life cycle and was included in the ISO 14025:2006 standard. The set of principles set out in the aforementioned standard assumes the voluntary application of its provisions and verification of the fulfilment of requirements by an independent programme operator, which is the body that conducts the adopted declaration programme. The objectives of a Type III environmental programme are to provide LCA-based information and additional information on the environmental aspects of products, thus enabling purchasers and users of goods to make informed purchasing decisions based on being able to compare different products.

In summary, the ISO 14000 family of standards, including the standards presented before concerning labelling, put forward a coherent and adequate proposal for entities wishing to pursue the path of Sustainable Development. The recommendations which they contain are of a general nature, but, thanks to that, they can be easily implemented to meet the needs of different sectors of the economy (Patón-Romero, Baldassarre, Rodríguez, & Piattini, 2019). However, the content of the standards does not address issues related to the labelling of products or services based on the Circular Economy model.

## Selected systems of labelling in traditional trade and e-commerce

Though the ISO 14000 series of standards does not resolve issues relating to the principles of labelling of products and services based on the Circular Economy model, it establishes a certain framework upon which it may take shape in future. It also establishes criteria and guidelines for current and planned systems of labelling for the Circular Economy.

One of the few initiatives in labelling confirming the implementation of the Circular Economy, but one which is nevertheless very interesting, is the Cradle to Cradle certificate. Its goal is to recognise products and motivate companies to improve their methods of production, use, and reuse while at the same time having a beneficial impact on society and the environment.

To obtain the certificate, products are assessed for environmental performance and social responsibility across five sustainability categories. These categories are material health, material reutilisation, renewable energy, water stewardship, and social fairness (Cradle to Cradle Certified Assessment Categories, 2021).

Cradle to Cradle certification takes account of many aspects defined in the concept of Sustainable Development (Cradle to Cradle Certified and UN Sustainable Development Goals, 2021). According to data from the Cradle to Cradle Institute, this programme enables the implementation of 11 out of 17 of the Sustainable Development Goals defined by the United Nations (United Nations, 2021).

The shortcomings of the system of the Cradle to Cradle system are related to limitations on the products to which it may be applied. Products from sectors such as food, drink, and pharmaceuticals are not eligible to take part in the certification programme. Other weaknesses of the Cradle to Cradle Products Program are its generic nature, with certification criteria not focusing on specific products; the lack of LCA analysis; and the incompletely transparent analysis of interested parties (Minkov, Bach, & Finkbeiner, 2018).

Another label which aims to promote the Circular Economy is the OrganiTrust label. It is a global certification programme initiated by the Organics Council. This label is not very popular or widely recognised. The program is however guided by the idea of comprehensive application of the regulations concerning the Circular Economy and ecological products and services. The process of joining the programme is initiated at the request of the interested party. An audit is then conducted to assess the compliance of the organisation's procedures with OrganiTrust standards. The audit takes samples of products to confirm that they are compliant with declarations. After going through the verification process, the OrganiTrust certification sign is awarded and is valid for a period of 1 year.

The Cradle to Cradle and OrganiTrust labels presented promote the Circular Economy model, but they are not very popular or widely recognised. The Cradle to Cradle Institute's databases feature 622 products covered by that certificate (Cradle to Cradle Certified Products Registry, 2021). The programme discussed are in the development phase and currently do not have the potential to have a significant impact on customers.

Some of the much more widely recognised labelling systems include pro-grammes conducted under the supervision of the European Commission. One of these is the EU Ecolabel, commonly known as the EU Flower. This is an initiative of European Union bodies, which established criteria for individual groups of products and services on the basis of Regulation (EC) No. 66/2010 of the European Parliament and of the Council of 25 November 2009 on the EU Ecolabel and the decision of the European Commission. These include such sub-groups as cosmetics, cleaning products, textiles and clothing, furniture, paper products, and tourist accommodation. The programme plans to develop, and products and services to be eligible for certification in the future may include financial products and food or office buildings for which guidelines are currently being developed.

The EU Ecolabel is referred to by the European Commission as "a label of environmental excellence". Products and services to which it has been awarded have to meet high environmental standards throughout their life cycle: from raw material extraction, to production, distribution, and disposal. The EU Ecolabel promotes the Circular Economy by encouraging companies to develop products that are durable and easy to repair.

The European organic logo is another important and recognised label. It is used for food products that are produced using natural substances and processes of cultivation. Such farming has less of an impact on the environment. It encourages the responsible use of energy and natural resources, the maintenance of biodiversity, preservation of regional ecological balances, enhancement of soil fertility, and maintenance of water quality. By fulfilling these principles, the agricultural production of organic foods comes close to implementing the assumptions of Circular Economy creating a biological loop of chemical elements and compounds, from which new food products can be created after their decomposition. Thanks to the regulation of this area by European regulations (Regulation No. 2018/848) (Commission Regulation No. 889/2008), the possibility of fair competition on the market for trustworthy organic products has been guaranteed.

Another label supporting the idea of the Circular Economy, though not directly, is the Forest Stewardship Council (FSC) logo. It is an international not-for-profit organisation, which establishes and implements standards for sustainable forest management, taking social, ecological, and economic aspects into account. FSC certification is confirmed by three types of environmental labels. These include the FSC 100%, FSC Recycled, and FSC Mix labels (www.fsc.org). The certificates mentioned and their presence on products give a guarantee of rationally conducted management of forests, from which the raw materials for many products are obtained.

Together with the observed changes in consumer purchasing habits, the sales of goods via the Internet is playing an increasingly important role (Barska & Wojciechowska-Solis, 2020; Šaković et al., 2000). The COVID-19 epidemic which began in 2020 has further accelerated the preparations of companies for this type of sales and had an impact on the development of logistics services and increased their capacity (E-commerce, 2020). The effects in the form of the development of e-commerce can also be seen in Italy and Poland.

Sales of products via the Internet differs from traditional (physical) sales above all with regard to being able to present and come into contact with the product offered. In e-commerce, consumers cannot "take the product in hand" and familiarise themselves with it. In internet sales, there is a decrease in the importance of factors such as the role of the promotional value of packaging. The possibility of the consumers familiarising themselves with the offer is limited to way in which the owner of the website chooses to present it. The opportunities created by e-commerce can be used or missed by sellers.

Large companies in the food retail sector have adopted an active attitude in this respect, based on the opportunities and possibilities offered by presenting a product on websites. Examples of these include retail chains such as Conad, Crai, Lidl, Carrefour, or Auchan, which are reinforcing the environmental message in communication with consumers through the use of their own ecolabels.

The Conad (Consorzio Nazionale Dettaglianti) chain has introduced several of its own labels to distinguish and promote products with outstanding ecological characteristics. It has done this by introducing four product lines inspired by an awareness of Sustainable Development. Together, these products form the Verso Natura line: BIO, EQUO, VEG, and ECO, and their purpose is to promote products with a reduced environmental impact. Each of these groups has its own individual label, which is displayed on the packaging and is used to distinguish the product on Conad's website.

Very similar practices are also being employed by the Crai chain to expand its environmental communication with the addition of further labels. Voluntary labels promoting sustainable modes of consumption include Crai Bio, Filiera Garantita, or Senza antibiotici. The first of those mentioned – Crai Bio – provides information that the given product belongs to the group of organic products. Each of these products is produced in way which preserves natural resources and biodiversity. The guaranteed supply chain (in Italian: *filiera garantita*) is a mark which guarantees the Italian origin of fruit, vegetables and meat. The antibiotic-free (in Italian: *senza antibiotici*) label guarantees that no antibiotics are used in poultry, informing customers as to the high quality of poultry meat and eggs. At the same time, the retail chain declares free-range farming, the use of natural feed, and the Italian origin of those products.

Another very interesting ecolabel used by this chain is the CRAI – We and you for the environment (in Italian: *noi e Voi per l'ambiente*) label. The label is displayed on products and provides information concerning the separate collection of packaging waste. This voluntary label indicates in a way that is as clear and comprehensible as possible how to separate empty packaging and dispose of it in accordance with rules for separate waste collection, thus contributing to the Circular Economy. This is a very innovative and rarely encountered example of environmental communication.

Similar actions on the Polish market include practices adopted by Carrefour, which also sells its products online. This retailer uses the "Jakość z Natury" (English translation: quality from nature) label. Products marked with this symbol are produced by trusted producers in an environmentally friendly way. In the case of plants, this means plants sold in accordance with their natural ripening cycle and, in the case of animal products, with respect for their well-being. Products marked with this label are sourced from carefully selected and long-standing partners of the Carrefour chain.

In making an assessment of the labelling systems presented, it should be noted that, on both the Italian and the Polish market, labels related to the Circular Economy such as Cradle to Cradle or the Circular Living Label are not the most popular ones. For this reason, it would require, for example, promotional initiatives to be conducted in order to raise their level of recognition. The most well-known and recognised labels on those markets include the EU Ecolabel and the FSC certificates.

According to data from the European Commission, there are more products labelled with the EU Ecolabel in Italy – 9,703 units. In Poland, the figure is only 2,989 products (European Commission, 2021). The situation is similar in the case of the FSC certificates, which should also be considered to be popular. These certificates (2,952 in number) were awarded to entities registered in Poland, while 3,695 were awarded to companies from Italy.

## Effectiveness and recognition of environmental labelling systems and the shaping of consumer behaviour

Issues of the impact of environmental labelling on the popularisation of sustainable behaviours such as the Circular Economy, for example, have been the subject of numerous scientific studies. A review of the literature on this subject from the past few years (2019–2021) is presented later in the chapter.

The positive attitude of consumers towards the concept of the Circular Economy has been determined by R. Boyer, Hunka, Linder, Whalen, and Habibi (2021). His team analysed the possibility of consumers paying a higher price for products (mobile phones and robot vacuum cleaners) based on their compliance with principles of circularity. It showed that labels providing information about production of the product based on the Circular Economy have a positive influence on consumers. And they are thus willing to pay more for that product (Boyer et al., 2021).

The effectiveness of environmental label policy has been studied by H. Xiao and Wang (2020). Their work conducted in China consisting of a financial analysis of companies on that market led to the conclusion that firms using China Environmental Labeling experience better growth than companies which did not obtain that labelling due to the use of practices that were not environmentally friendly. Heavily polluting firms have more difficulty in obtaining loans and other sources of financing which makes it more difficult for them to develop. In this case, ecolabelling as an element of environmental policy had a major impact on limiting the expansion of polluting firms (Xiao & Wang, 2020).

The most important consumer studies conducted from 2019 to 2021 are listed in Table 5.1, along with their most important conclusions.

*Table 5.1* Impact of environmental labelling systems in shaping consumer behaviour – a review of the literature

| Authors | Sample size and country of study | Most important conclusions |
| --- | --- | --- |
| C. Rossi & F. Rivetti (2020) | N = 315 Italy | In order for consumers to make use of environmental labels, they must consider them to be credible. Consumers are willing to pay more for a product with a type III label or displaying self-declared environmental claims made by the producer. |
| J. Wojciechowska-Solis and A. Barska (2021) | N =1,067 Poland | The ecological values of a food product rank highly in the hierarchy of purchasing factors. Consumers make purchases based on information provided on labels such as certification marks, no additives, content of nutrients, health benefits. |
| R. Boyer, A. Hunka, M. Linder, and K. Whalen (2021) | N = 864, United Kingdom | Consumers are willing to pay more for products with labels confirming compliance with the circular economy model. |
| S. Cooper, L. Butcher, S. Scagnelli, J. Lo, M. Ryan, A. Devine, and T. O'Sullivan (2020) | N = 1,024, Australia | Surveyed grocery shoppers are willing to pay more for a product with the Health Star Rating label. System limited by the lack of guarantee of its accuracy. Potentially the program could have a positive impact on the health of consumers. |
| D. Menozzi, T. Nguyen, G. Sogari, D. Taskov, S. Lucas, J. S. Castro-Rial, and C. Mora (2020) | N = 2,509 France, Germany, Italy, Spain, the United Kingdom | The presence of sustainability labels on fish was valued in all countries in which the study was conducted. Among all the nationalities taking part in the survey, Italians were prepared to pay the most for fish with a sustainability label. The willingness to pay more for nutrition and health claims was greatest in the case of Italian and Spanish consumers. |
| M. Shabbir, M. Sulaiman, N. Al-Kumaim, A. Mahmood, and M. Abbas (2020) | N = 359 United Arab Emirates | Consumers have a positive attitude to green marketing, and it has an influence on their purchasing decisions. Environmentally friendly business strategies are a good and promising direction for the promotion of products. |
| T.-Ch. Liang, R. Situmorang, M.-Ch. Liao, and S.-Ch. Chang (2020) | N = 411 Taiwan | Consumers with knowledge about carbon labelling are willing to choose products with that label. At the same time, the level of knowledge about such labels is low and is a barrier to the recognition of organic products and ultimately to purchases of them. |

As can be seen from the studies discussed, environmental labels are an effective and recognised tool for communication between businesses and consumers. Moreover, they produce the desired responses with customers encouraging them to purchase or even pay more to purchase a labelled product.

Along with consumer satisfaction, environmental labelling also brings benefits for companies. Work conducted by Slamet, Nakayasu, and Bai (2016) on an increase in organic vegetable consumption led to the conclusion that encouraging companies to use communication tools in the form of organic labelling and certification will be one of the best ways of ensuring consumer interest (Slamet et al., 2016).

The literature review conducted clearly demonstrates that the impact of using environmental labels has a major influence on consumers. As the largest group of decision-makers in the economy, it is they who decide whether or not products or services will be successful when making their purchasing choices. Thanks to this, environmental labelling may become one of the main factors in enabling the Circular Economy model to be achieved in the economy. This is why it is so important to understand consumers and the determinants of their behaviour with regard to their preferences, reasons for purchasing, means of usage and disposal of used products, and consumer involvement in the repair and recycling of devices in order to implement a Circular Economy strategy.

## Determination of directions of improvement of Circular Economy communication – practical recommendations

Modern society has been developing an increasingly positive attitude to greening in every sphere of human activity, is interested in climate change, and wants to take care of the environment and the survival of the planet. The problem of pollution of the environment is being raised in nearly every political, social, and cultural forum. These factors all contribute to there being a focus on environmental problems in public opinion and awareness.

While it is true that many studies conducted emphasise the role of consumers in the transformation of the economy to the Circular Economy, much still remains to be done to build their involvement and awareness (Cordova-Pizarro, Aguilar-Barajas, Rodriguez, & Romero, 2020). Research by G. Orzan, Cruceru, Balaceanu, and Chivu (2018) identifies factors in the lack of consumer activity in this field. According to the authors of the study, the factors which explain this state include a lack of appropriate information and awareness. The main implication of this study is thus the need to educate consumers about the long-term benefits of using, in this case, ecological packaging. This message may be delivered via communication campaigns that will sensitise consumers and focus on specific pro-environmental activities (Orzan et al., 2018). In turn, Tseng et al. (2020) have shown that persuasive communication is the most effective factor in convincing consumers to transition to sustainable consumption. Other key determining factors include educating consumers and augmenting their knowledge about sustainable consumption (Tseng, Sujanto, Iranmanesh, Tan & Chiu, 2020).

Also of interest are the conclusions reached in the study by Altinay and Williams (2019). Based on the research conducted, it postulates that imagery and graphics emphasising environmental and economic losses have an influence on consumers. This form of message proved to be more engaging for recipients (Altinay & Williams, 2019). These indications may be used in future to design visual messages in environmental communication. In addition to the visual form of information, its scope is also important.

J. Cantillo et al. (2021) postulate that, apart from environmental issues, consumer messages should also include information about respect for social and ethical issues as well as animal welfare (Cantillo, Martín, & Román, 2021).

The discussed environmental communication process of the organisation as well as issues related to labelling systems and their influence on consumer decisions make it possible to put forward some practical recommendations to enable the popularisation and improvement of labelling programmes promoting the Circular Economy model. These include:

- The creation of a new Circular Economy certification programme for products and services at a supranational level;
- The establishment of clear and transparent certification criteria and the impartiality of that process;
- Guaranteeing that Circular Economy certification is supervised by government institutions – public authorities and governments should be involved in the control of the created system, thus increasing its importance and the level of stakeholder trust;
- Increasing the involvement of the International Organization for Standardization in the creation of international standards supporting the Circular Economy labelling system – possibly through the modification of existing standards to take the specifics of the Circular Economy model into account;
- The graphic design of labels for the Circular Economy system should be intuitive, have a visual impact on the recipient, and use simple language without involving excessively specialist or technical information;
- The graphic label for the Circular Economy should be relatively standardised for all groups of products undergoing certification;
- To differentiate between levels of circularity of products and services, a scale of colours and letters should be used for easy comparison of products;
- The Circular Economy label will be most visible on product packaging if it is positioned in the main field of view along with such information as price, product name, nutrition table so as to include the criterion of circularity in purchase decision factors,
- In proximity to the label, a link or QR code should be provided to allow consumers to use their mobile phone to be directed to more comprehensive information about the programme,
- In the case of e-commerce, the label should be presented together with the product;

- The high interest of stakeholders in social and ethical issues means these issues should be taken into consideration in the criteria of the Circular Economy certification programme;
- To guarantee the success of the programme, the level of recognition of the certificate should be raised by educational initiatives and information campaigns in the media;
- The programme should be promoted using different means of messaging, including both traditional and more innovative means;
- Promotional and educational campaigns should involve opinion leaders and authorities in the field of environmental protection in providing support for the Circular Economy labelling programme.

## Summary

A change in economic model and consumption habits is an inevitable necessity for every entrepreneur and consumer in all areas of activity and everyday life. The topic presented in this chapter allows readers to familiarise themselves with environmental labels and declarations as a tool in the corporate communication process. The presented review of the literature confirmed that they are effective instruments in influencing the purchasing decisions of consumers and in shaping their behaviours. However, to date, they have not led to the formation of a recognised system for the labelling of products meeting the principles of the Circular Economy.

The environmental labels, presented in this chapter, which function on different markets, are a manifestation of activities undertaken by individual independent organisations and commercial enterprises. They are a visible manifestation of actions taken as a result of differing motivations but with the common goal of promoting sustainable products in the Circular Economy. The biggest challenge in the coming years will thus be to develop a standard for the labelling of products and services, which should, to the extent that it is possible to do so, be harmonised for different products and recognised on a global scale.

From a practical point of view, this chapter presents the labelling systems already present on different markets, with a particular emphasis on the Italian and Polish markets. The analyses conducted clearly confirm the necessity and effectiveness of using labels in companies' market communication. In principle, it can be assumed that the popularity of ecolabelling concerning product and services is so large that it is hard to imagine that entities putting new products onto the market will neglect that form of communication with those around them. At the same time, the environmental labels designed should meet the communication criteria mentioned before and the practical recommendations presented in this chapter.

The environmental communication system will provide the confirmation of the organisation's value as well as its trustworthiness and responsibility on environmental issues. Efforts put into the transformation of the organisation towards a Circular Economy will be rewarded by a growing number of conscious customers, business partners, and investors, as well as the accompanying market benefits.

# References

Altinay, Z., & Williams, N. (2019, May). Visuals as a method of coastal environmental communication. *Ocean and Coastal Management, 178*, 104809. doi:10.1016/j.ocecoaman.2019.05.011

Asteria, D., Suyanti, E., Utari, D., & Wisnu, D. (2014). Model of environmental communication with gender perspective in resolving environmental conflict in urban area (study on the role of women's activist in sustainable environmental conflict management). *Procedia Environmental Sciences, 20*, 553–562. doi:10.1016/j.proenv.2014.03.068

Barska, A., & Wojciechowska-Solis, J. (2020). E-consumers and local food products: A perspective for developing online shopping for local goods in Poland. *Sustainability, 12*(12). doi:10.3390/su12124958

Boyer, R. H. W., Hunka, A. D., Linder, M., Whalen, K. A., & Habibi, S. (2021). Product labels for the circular economy: Are customers willing to pay for circular? *Sustainable Production and Consumption, 27*, 61–71. doi:10.1016/j.spc.2020.10.010

Brauweiler, J. (2013). Znaczenie interesariuszy strategicznych. In A. Kryński, M. Kramer, & A. F. Caekelbergh (Eds.), *Zintegrowane zarządzanie środowiskiem: Systemowe zależności między polityką, prawem, zarządzaniem i techniką*. Warszawa: Oficyna a Wolters Kluwer Business.

Cantillo, J., Martín, J. C., & Román, C. (2021). A hybrid fuzzy topsis method to analyze the coverage of a hypothetical EU ecolabel for fishery and aquaculture products (FAPs). *Applied Sciences, 11* (1), 1–21. doi:10.3390/app11010112

Carrefour. Retrieved February 15, 2021, from www.carrefour.pl

Commission Regulation (EC) No 889/2008 of September 5, 2008 laying down detailed rules for the implementation of Council Regulation (EC) No 834/2007 on organic production and labelling of organic products with regard to organic production, labelling and control, OJL 250, September 18, 2008.

Conad. Retrieved February 15, 2021, from www.conad.it

Cooper, S. L., Butcher, L. M., Scagnelli, S. D., Lo, J., Ryan, M. M., Devine, A., & O'Sullivan, T. A. (2020). Australian consumers are willing to pay for the health star rating front-of-pack nutrition label. *Nutrients, 12*(12), 1–16. doi:10.3390/nu12123876

Cordova-Pizarro, D., Aguilar-Barajas, I., Rodriguez, C. A., & Romero, D. (2020). Circular economy in Mexico's electronic and cell phone Industry: Recent evidence of consumer behavior. *Applied Sciences, 10*(21), 1–21. doi:10.3390/app10217744

Cradle to Cradle Certified Assessment Categories. Retrieved February 15, 2021, from www.c2ccertified.org/get-certified/product-certification

Cradle to Cradle Certified Products Registry. Retrieved February 15, 2021, from www.c2ccertified.org/%20products/registry

Cradle to Cradle Certified & UN Sustainable Development Goals. Retrieved February 15, 2021, from www.c2ccertified.org/get-cer tified/un-sustainable-development-goals

Crai. Retrieved February 15, 2021 , from www.crai-supermercati.it

C2ccertified. Retrieved February 15, 2021, from www.c2ccertified.org

E-commerce. Trade and the COVID-19 Pandemic. Information note. World Trade Organization. (2020, May 4). Retrieved February 15, 2021, from www.wto.org/english/tratop_e /covid19_e/ecommerce_report_e.pdf

European Commission. (2021). EU ecolabel key figures. Retrieved February 15, 2021, from https://ec.europa.eu/environment/ecolabel/facts-and-figures.html

Freeman, E. R. (2010). *Strategic management: A stakeholder approach*. Cambridge: Cambridge University Press.

FSC. Retrieved February 15, 2021, from www.fsc.org

Gregory-Smith, D., Manika, D., & Demirel, P. (2017). Green intentions under the blue flag: Exploring differences in EU consumers' willingness to pay more for environmentally-friendly products. *Business Ethics*, *26*(3), 205–222. doi:10.1111/beer.12151

Gulliver, R., Fielding, K. S., & Louis, W. R. (2021). Assessing the mobilization potential of environmental advocacy communication. *Journal of Environmental Psychology*, *74*, 101563. doi:10.1016/j.jenvp.2021.101563

ISO 14021:2016 Environmental labels and declarations. Self-declared environmental claims.

ISO 14024:2018 Environmental labels and declarations. Type I environmental labelling. Principles and procedures.

ISO 14025:2006 Environmental labels and declarations. Type III environmental declarations. Principles and procedures.

ISO 14050:2020 Environmental management. Vocabulary.

ISO 14063:2020 Environmental management. Environmental communication.

Kabaja, B. (2018). *Kryteria oceny znakowania opakowań jednostkowych suplementów diety*. Kraków: Prace Doktorskie, Wydawnictwo Uniwersytetu Ekonomicznego w Krakowie.

Liang, T. C., Situmorang, R. O. P., Liao, M. C., & Chang, S. C. (2020). The relationship of perceived consumer effectiveness, subjective knowledge, and purchase intention on carbon label products – a case study of carbon-labeled packaged tea products in Taiwan. *Sustainability*, *12*(19), 7892. doi:10.3390/su12197892

Menozzi, D., Nguyen, T. T., Sogari, G., Taskov, D., Lucas, S., Castro-Rial, J. L. S., & Mora, C. (2020). Consumers' preferences and willingness to pay for fish products with health and environmental labels: Evidence from five European countries. *Nutrients*, *12*(9), 2650. doi:10.3390/nu12092650

Minkov, N., Bach, V., & Finkbeiner, M. (2018). Characterization of the cradle to cradle certified™ products program in the context of eco-labels and environmental declarations. *Sustainabilit y*, *10*(3). doi:10.3390/su10030738

Organitrust. Retrieved February 15, 2021, from https://organitrust.org/

Orzan, G., Cruceru, A. F., Balaceanu, C. T., & Chivu, R. G. (2018). Consumers' behavior concerning sustainable packaging: An exploratory study on Romanian consumers. *Sustainability*, *10*(6). doi:10.3390/su10061787

Patón-Romero, J. D., Baldassarre, M. T., Rodríguez, M., & Piattini, M. (2019). Application of ISO 14000 to information technology governance and management. *Computer Standards and Interfaces*, *65*, 180–202. doi:10.1016/j.csi.2019.03.007

Platonova, M. (2016). Applying emotive rhetorical strategy to environmental communication in English and Latvian. *Procedia – Social and Behavioral Sciences*, *236*, 107–113. doi:10.1016/j.sbspro.2016.12.044

Preziosi, M., Tourais, P., Acampora, A., Videira, N., & Merli, R. (2019). The role of environmental practices and communication on guest loyalty: Examining EU-ecolabel in Portuguese hotels. *Journal of Cleaner Production*, *237*. doi:10.1016/j.jclepro.2019.117659

Regulation (EC) No. 66/2010 of the European Parliament and of the Council of November 25, 2009 on the EU Ecolabel, OJ L 27, January 30, 2010.

Regulation (EU) 2018/848 of the European Parliament and of the Council of May 30, 2018 on organic production and labelling of organic products.

Rossi, C., & Rivetti, F. (2020). Assessing young consumers' responses to sustainable labels: Insights from a factorial experiment in Italy. *Sustainability*, *12*(23), 1–23. doi:10.3390/su122310115

Šaković Jovanović, J., Vujadinović, R., Mitreva, E., Fragassa, C., & Vujović, A. (2000). The relationship between e-commerce and firm performance: The mediating role of internet sales channels. *Sustainability*, *12*(17), 6993. doi:10.3390/su12176993

Shabbir, M. S., Sulaiman, M. A. B. A., Al-Kumaim, N. H., Mahmood, A., & Abbas, M. (2020). Green marketing approaches and their impact on consumer behavior towards the environment – a study from the UAE. *Sustainability*, *12*(21), 1–13. doi:10.3390/su12218977

Singleton, B. E. (2019). The evolution of the super-whale: Complexity and simplicity in environmental communication. *Marine Policy*, *99*, 170–172. doi:10.1016/j.marpol.2018.10.018

Slamet, A., Nakayasu, A., & Bai, H. (2016). The determinants of Organic vegetable purchasing in Jabodetabek region, Indonesia. *Foods*, *5*(4), 85. doi:10.3390/foods5040085

Tam, V. W. Y., Shen, L. Y., Yau, R. M. Y., & Tam, C. M. (2007). On using a communication-mapping model for environmental management (CMEM) to improve environmental performance in project development processes. *Building and Environment*, *42*(8), 3093–3107. doi:10.1016/j.buildenv.2006.10.035

Tseng, M. L., Sujanto, R. Y., Iranmanesh, M., Tan, K., & Chiu, A. S. (2020, April). Sustainable packaged food and beverage consumption transition in Indonesia: Persuasive communication to affect consumer behavior. *Resources, Conservation and Recycling*, *161*, 104933. doi:10.1016/j.resconrec.2020.104933

Unde, A., Arianto, Bahfiarti, T., Pulubuhu, D. A. T., & Arsyad, M. (2020). Strategy on family communication and the extent of environmental health awareness in coastal area. *Enfermería Clínica*, *30*, 64–68. doi: 10.1016/j.enfcli.2019.09.004

United Nations. Retrieved February 15, 2021, from https://sdgs.un.org/goals

Wiktor, J. W. (2013). *Komunikacja marketingowa*. Warszawa: Wydawnictwo PWN.

Wojciechowska-Solis, J., & Barska, A. (2021). Exploring the preferences of consumers' organic products in aspects of sustainable consumption: The case of the Polish consumer. *Agriculture*, *11*(2), 138. doi:10.3390/agriculture11020138

Xiao, H., & Wang, K. M. (2020). Does environmental labeling exacerbate heavily polluting firms' financial constraints? Evidence from China. *China Journal of Accounting Research*, *13*(2), 147–174. doi:10.1016/j.cjar.2020.05.001

# 6 Interrelationship between sustainable manufacturing and Circular Economy in the building sector

*Agnieszka Nowaczek and Joanna Kulczycka*

## Introduction

A detailed analysis of the scientific literature of strategic documents developed by the European Commission (EC) and individual countries and regions regarding the Circular Economy (CE) shows that many methods and indicators have already been proposed for monitoring activities related to the CE. They mainly focus on the use of resources (raw materials), seeking to maximise their value and increase the efficiency of management of resources and the durability of products, and on methods of preventing and minimising waste in accordance with the waste hierarchy [EC (European Commission), 2014]. Such activities may be carried out not only by means of eco-innovative technological solutions, but also by means of organisational solutions, which take account of maximisation of value in the value chain or, in other words, from designer to consumer. The related monitoring of the transformation towards the CE concerns not only its technological aspects, but also its economic, social, and environmental aspects. It is important to make an assessment that takes the whole value chain into account, that is from the design phase, production, consumption, repair, and regeneration through to waste management and the recovery of secondary raw materials that can be put back into circulation in the economy. Moreover, it is not only economic transformation which is taken into account in CE indicators, but also the impact on socio-economic development, hence there is the necessity of taking into account aspects such as the impact on the level of eco-innovativeness, the development of the services sector, employment, and changes in consumer behaviours. Such a broad approach requires the analysis and availability of a great deal of statistical data, which has often not been assessed to date. This applies in particular to new business models proposed in the CE, for instance in the area of industrial symbiosis, the reuse of products, eco-design, and virtualisation (Kulczycka, Bączyk, & Nowaczek, 2020). Even though the EC has proposed a set of indicators and Eurostat monitors the CE in Member States, proposals and claims as to how they may be improved and concerning the possibility of transposition to regions or sectors continue to be made [EC (European Commission)] (2018). This applies for example to the taking into consideration of water consumption in CE indicators, or the management of areas, which were not taken into account in the original version and goals

DOI: 10.4324/9781003179788-6

of the CE. Moreover, the CE clearly recommends that individual Member States develop their own documents and frameworks for monitoring of the CE [EESC (European Economic and Social Committee), 2018]. Many countries, regions, cities, international organisations (UNEP, OECD), and non-governmental or sector-specific organisations have already developed their own sets of indicators. There are usually several or several dozen of these, which often tie activities in the CE together with Sustainable Development Goals (SDGs), reporting in the area of corporate social responsibility (GRI indicators) or CE and eco-innovation indicators developed and monitored at the EU forum. Many of these underline the need to take consumption and environmental impact into consideration across the entire value chain and to use indicators of carbon or environmental footprint, which employ the method of Life Cycle Assessment (LCA) and Material Flow Analysis (MFA). This is because such integrated indicators allow progress and changes to be quickly quantified and assessed and in specific cases also allow for comparative assessment (Walker, Coleman, Hodgson, Collins, & Brimacombe, 2018; Akerman, 2016). Their strength lies in being able to take account of the impact across the whole value chain, which promotes eco-efficient and zero-waste solutions, including those which make use of secondary raw materials and limit transport or the use of hazardous substances. The CE at company level requires a structural transformation in business model and appropriate adaptation of various management functions, such as production, sales, and logistics offering direct and indirect benefits (Wang, 2014). The most important of these are as follows (Liu, 2014):

- Reduction in the level of consumption of materials – companies implementing the CE seek to reduce total material inputs and improve reuse of waste and recycling in order to extend and prolong the lifetime of resources, as well as to reduce the scale of initial investment in resources and, as a result, to reduce the level of investment in raw materials.
- Increase in the efficiency of resource use – the CE uses green technology and recovery logistics, allowing for a holistic perception of the relationship between consumption of materials and the products obtained and for improving resource efficiency.
- Reduction in the level of emissions related to the production process thanks to the use of green technologies, more effective use of available resources, and reduction in total material input. This allows for increased availability of high-quality raw materials, which may have an impact in terms of a decrease in demand for imported raw materials (Bukowski & Sznyk, 2019). In theory, the full set of indicators for CE monitoring can be applied to all sectors of the economy. However, due to the specifics of each sector of industry, the authors of this chapter believe that the evolutionary nature of the undertaking of the transition to a circular model in the economy requires a more well thought-out approach. It is important that the planning and assessment of activities should take an assessment across the whole supply chain into consideration in order to identify so-called "hot spots" and thereby to guide activities towards their minimisation.

The implementation of the CE has an impact on the company's activity, since it leads to a necessity to employ new business models. The CE at company level requires a structural transformation in business model and appropriate adaptation of management functions. Based on the research conducted, this chapter proposes a set of possible indicators for monitoring CE activities for the building sector and conducts a detailed analysis of activities carried out by companies in the analysed sector of industry seeking to make the transformation to the CE.

## Analysis of the building sector in Poland and worldwide in the context of environmental protection requirements

The EU28 has introduced climate legislation that aims to achieve at least a 40% reduction in GHG emissions by 2030 compared to 1990. In the context of the EEA Agreement, they will, as of 2021, implement the Effort Sharing Regulation and the LULUCF Regulation and have been taking part in the EU ETS since 2008. Moreover, in November 2018, the European Commission presented its Strategic Vision "A Clean Planet for all" for the European economy to become climate-neutral by 2050. In EU policy, the minimisation of the impact of building on the environment has been proposed in many documents, such as:

*   Directive 2010/31/EU of 19 May 2010 on the energy performance of buildings and its subsequent amendments;
*   Directive 2012/27/EU of 25 October 2012 on energy efficiency;
*   Communication COM/2014/445 on resource efficiency opportunities in the building sector.

As a result, from 1 January 2019, all new buildings occupied and owned by public authorities must be "nearly zero-energy buildings" (Directive 2010/31/EU). Some innovative solutions could receive financial support for investment, for example, via the European Investment Bank program Elena – which provides support for investments in energy efficiency and sustainable transport or for research, where, for example, SPHERE is a 4-year, Horizon 2020 project that aims to provide a BIM-based Digital Twin Platform to optimise the building lifecycle, reduce costs, and improve energy efficiency in residential buildings. Many new ideas have been created for the exchange of best practices and solutions worldwide, such as building SMART International (http://buildingsmart.org) or a proposal for worldwide cooperation put forward by the World Green Building Council (www.worldgbc. org). However, Europe's buildings emit 36% of $CO_2$ emissions in Europe, and energy consumption by this sector has been increasing in recent decades. Some fluctuations in total emissions due to energy consumption in buildings may be observed from year to year due to weather-related changes in heat demand. However, an assessment of $CO_2$ emissions from fossil fuels by sector shows that Poland has seen a 7% increase in the construction sector in recent years.

The building industry is also responsible for 40% of $CO_2$ emissions and had a heavy eco-footprint at a global aggregate level. It represents one of the biggest problems, or alternatively, can be seen as one of the biggest opportunities for change, innovation, and assuming our social responsibilities. That process of transformation comes with a need for a change in the roles of the traditional design professions. As the world has become more complex and time is valued in money, more and more building professionals have become specialists in their field – but not in the industry as a whole. The range and variety of individual (computing) tools that are being used in the building industry (and across its related disciplines) in the phases of construction themselves are an expression of that tendency. However, knowledge about materials and the correct use of materials remains incomplete and usually results in the usage of already well-known and accepted materials that perhaps are not so durable, sustainable, or resource-efficient after all.

As has been underlined by the International Resource Panel (IRP, 2020):

> "with buildings, the primary focus of climate change policy has been on energy efficiency, with modest efforts at material efficiency reductions. . . . As a result, there is less attention to measurement on a life-cycle basis, but tools have emerged and efforts are expanding. Material and GHG measurement must go beyond individual building components. . . . The EU has developed guidelines and rules for how an LCA should be used for GPP" [green public procurement], as well as environmental product declarations (EPDs) or for certification systems for buildings, such as BREEAM, the German Sustainable Building Certificate (DGNB); and Haute Qualité Environnementale (HQE) certification in France).

The main products of the building sector are buildings and structures, characterised by high durability as well as the possibility of modernisation and reuse. As such, they are not really prime targets for the implementation of the CE concept. Analysing the sector, it should be noted that the new business model of the CE is something that has been in use there for many years. Building renovations, trade in real estate between users, and sharing are all parts of this process and are activities which have been going on for many years. However, to apply the new economic model of the CE, the approach is to focus only on the earlier stages in the life cycle of buildings (usually there is no key closing of the loop), and often this is a process which has not been optimised. Moreover, looking at more recent economic history, it is possible to observe a tendency for the sector to turn away from the idea of the CE, because:

• Non-renewable materials are being used on an ever-greater scale;
• Materials are being reused less and less; and
• The durability of buildings is becoming shorter.

These changes have led to a situation in which the building sector is now responsible for the greatest damage to the natural environment of any sector of the economy, both in the EU and worldwide. In the case of Poland, the situation is slightly different. The building sector generates approximately 10.4% of all waste, a share of the total amount of waste generated which is over three times less compared to statistics for Europe as a whole (Deloitte, 2020). Despite this, the impact of the sector on the natural environment is non-negligible. Of course, while taking those indicators into consideration, the benefits which originate from the building sector should also be borne in mind. In 2019, the Polish building market grew by 9% compared to the previous year. Thanks to investments made by small- and medium-sized enterprises in the building sector, the value of output in the building sector reached 224.3 billion Polish złoty (compared to 206.2 Bn PLN the previous year). Such a large increase in output is above all the result of high demand for building services, mainly in the infrastructure sector, which was caused, among other things, by subsidies from EU funds and government programmes for the construction of roads and railways. Figure 6.1 presents the size of the Polish construction market broken down by segment. As at the end of 2019, the largest share of the Polish construction market is accounted for by non-residential buildings (32.6%), transport infrastructure (29.1%), and residential buildings (16.9%). The remainder is accounted for by pipelines, telecommunications and power lines, buildings in industrial areas, and other civil engineering works.

The main motor driving the construction market in Poland is provided by investments made by the largest investors in road and rail infrastructure. Many infrastructure projects are currently in the construction phase, which is contributing to further growth in construction production. The peak in investment in road and rail projects is posing a challenge for contracting companies due to the costs of materials and payroll which have been on the rise for several years. Due to a decrease in demand for building services in the private sector as a result

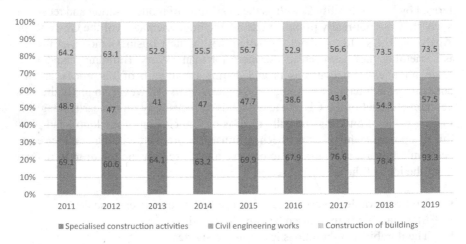

*Figure 6.1*  Size of the Polish construction market broken down by segment over the period from 2011 to 2019 (Bn PLN)

Source: Statistics Poland (GUS) - Business tendency in manufacturing, construction, trade and services 2000-2020.

of uncertainty related to the COVID-19 pandemic, there has been a slowing in growth in prices in recent months (Deloitte, 2020).

In recent years, it has become clear that the building sector has to respond actively not only to the environmental, but also to the social and economic aspects of Sustainable Development. Following the global increase in environmental activity, established sustainability certification systems such as BREEAM (Building Research Establishment Environmental Assessment Method), LEED (Leadership in Energy and Environmental Design), and Passivhaus (Passive building certificate) have started to move from being forms of best practice to being market requirements. According to the latest data from the British Building Research Establishment (BRE), there are over 546,600 projects with the BREEAM certificate in 77 countries worldwide. Both BREEAM and other Sustainable Development certification systems consist of several categories, which are a point of reference for all aspects of sustainable construction. At the level of the product, both quantitative and qualitative information is used to determine if the building fulfils specific sustainability criteria. Producers are thus being gradually obliged to provide verified, substantive information on the environmental friendliness of products (their environmental performance/properties) in order to gain a competitive advantage. At the same time, recipients of products are being encouraged to choose products which are the least hazardous to the environment and to human health and safety and which are most responsible with regard to the energy efficiency of buildings.

## Research methodology

Nearly every global organisation currently dealing with issues of the CE, and previously also with activities to promote Sustainable Development, has proposed its own solutions for the monitoring of the CE. These include the OECD, the World Bank, EUROSTAT, the Ellen MacArthur Foundation, and EURES. Indicators developed by these institutions focus on analyses of climate change, depletion of the ozone layer, waste management, consumption of natural resources (air, land, water, biodiversity), and environmental management (OECD, 2002).

When selecting indicators for monitoring the transformation towards the CE in the area of sustainable production which may be used in the building sector, indicators existing in Polish public statistics were analysed first. In particular, existing indicators analysed by Statistics Poland (*Główny Urząd Statystyczny – GUS*) in the area of Sustainable Development, waste management, and eco-innovation, as well as indicators proposed in strategies at EU and national level were assessed. In total, over 100 indicators grouped together in different thematic areas closely related to the CE were identified. The individual steps in the research process are presented in Figure 6.2.

Assessment of the importance of those indicators and the rate and direction of changes in them were the subject of a social consultation. In order to verify the proposed set of CE monitoring indicators, a survey questionnaire was developed with a set of 24 provisionally selected CE-monitoring indicators in the area of the

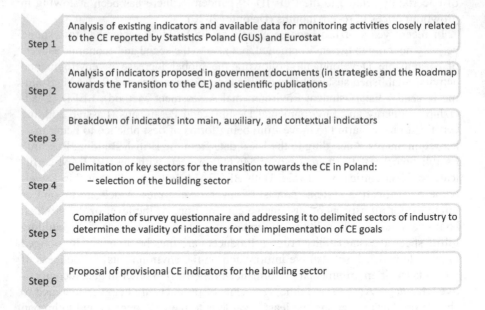

*Figure 6.2* Steps in the research process to identify CE monitoring indicators in the building sector

three fundamental aspects of sustainability: environmental, economic, and social (Nowaczek, Kulczycka, & Bączyk, 2020). The set of indicators was prepared based on consultation and verification with experts in the field of the CE, consisting of representatives of science, business, and non-governmental organisations. For the selection of the research sample, the form of simple random selection was used, where the required number of sample elements is drawn at random directly from the population. Sixteen companies from the building sector in Poland took part in the survey. The majority of these were large (seven) and medium-sized (six) companies responsible for the building of structures and construction works. Three small companies took part in the survey. To obtain a comprehensive picture and overviews that are representative of the entire industry, it is necessary to distinguish between companies with regard to their size and the specifics of their activities.

With regard to such specifics, companies were represented from the following areas:

• Production;
• Service activities; and
• Commercial activities.

*Table 6.1* CE indicators included in the survey

| Indicator number | Indicator name | Unit of measure |
|---|---|---|
| 1. | Amount of consumption of secondary raw materials/ amount of revenue | [Mg/PLN] |
| 2. | Share of renewable energy in total energy consumption | [%] |
| 3. | Method of handling waste in accordance with the waste hierarchy: Amount of waste generated | [t] |
| 4. | Quantity of hazardous waste generated | [t] |
| 5. | Amount of by-products generated | [Mg] |
| 6. | Number of people trained in the CE | [unit] |
| 7. | Quantity of waste reused | [t] |
| 8. | Quantity of waste recycled | [t] |
| 9. | Quantity of waste recovered by other processes | [t] |
| 10. | LCA environmental footprint | [Pt/MG] |
| 11. | Amount of consumption of primary raw materials/amount of revenue | [Mg/PLN] |
| 12. | Amount of water consumption/amount of revenue | [l/PLN] |
| 13. | Quantity of waste sent for disposal | [t] |
| 14. | Carbon footprint | [$CO_2$e/ Mg] |
| 15. | Amount of $CO_2$ emissions/amount of production | [Mg] |
| 16. | Share of costs in total costs of materials and energy | [%] |
| 17. | Share of fees for economic use of the environment/total costs | [%] |
| 18. | Amount invested in CE projects | [PLN] |
| 19. | Amount of consumption of critical raw materials/amount of revenue | [Mg/PLN] |
| 20. | Number of CE patents obtained | [unit] |
| 21. | Number of investment orders applied to the CE | [%] |
| 22. | Number of certificates held, e.g. EMAS, environmental declarations | [unit] |
| 23. | Number of industrial symbioses to use/manage waste | [unit] |
| 24. | Does the company have a developed CE strategy? | [YES/NO] |

The addressees of the surveys were people from at least middle managerial level and thematically related to the implementation of the CE within the company at the strategic or development level. The five-point Likert scale was used in development of the survey, where 1 means unimportant, 2 means rather unimportant, 3 means don't know, 4 means rather important, and 5 means important. In addition, respondents were asked for a short-term prognosis for the indicators over a time period of the next 5 years. Table 6.1 presents the indicators included in the survey.

## Results of research

The indicators considered to be important by respondents currently and those which will in their opinion be important over a short-term perspective are presented in Figure 6.3.

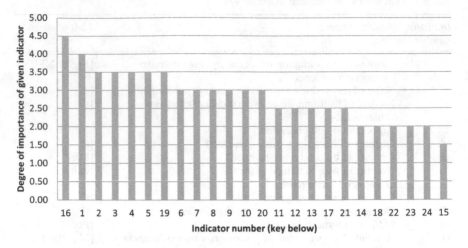

*Figure 6.3* Degree of importance of CE indicators identified by entities surveyed

The respondents surveyed considered seven indicators to be important. These include above all those related to the costs of management and the quantity of waste generated. Those concerning $CO_2$ emissions were however considered to be less important, which is probably related to a lack of a holistic approach, as it is not the producing companies which are responsible for emissions related to the consumption of energy in buildings, but rather residents, as well as a low awareness of the impact of materials consumed on $CO_2$. Such an approach also confirms the low level of interest in environmental product declarations (EPDs) or proprietary certification systems. This is probably an area to be targeted for direct support both in the context of education and finances in Poland, especially when compared to other EU countries, for example in Germany, where an EPD programme is proposed at the Institut Bauen und Umwelt e.V. (IBU). As a result, companies and associations have the ability to create environmental product declarations for their products. In addition to the individual manufacturer and the IBU as EPD programme operator, programme participants include the IBU Advisory Board, independent auditors, and interested individuals from the public. This guarantees objectivity and transparency (https://ibu-epd.com/en/epd-pro-gramme/). Another interesting certificate is the EMISSION CLASSIFICATION OF BUILDING MATERIALS (M1). The aim of this classification is to enhance the development and use of low-emitting building materials so that material emissions do not increase the requirement for ventilation. The classification presents requirements for the materials used in ordinary work spaces and residences. For air-handling components, there is a separate Cleanliness Classification of Air-handling Components (www.rakennustieto.fi/index/english.html). Another certification, this time related to the CE, is ISCC PLUS Certification for the Circular Economy and Bioeconomy (www.iscc-system.org/certificates/all-certificates/).

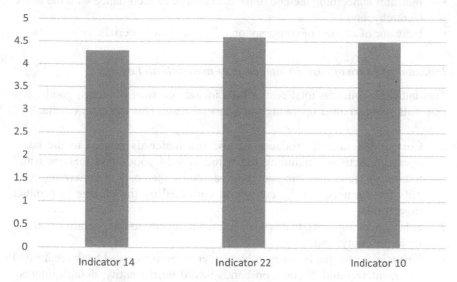

*Figure 6.4* Degree of importance of CE indicators over the short term (5 years)

In the next step, representatives of the entities surveyed were asked to pick CE indicators, which in their opinion is important for monitoring the CE in the building sector over the short term (5 years). The selected answers of respondents are shown in Figure 6.4. They considered three indicators to be important for the building sector in the future, and they are:

- The carbon footprint indicator;
- The indicator concerning the number of certificates held, for example EMAS, environmental declarations; and
- The indicator concerning the LCA environmental footprint.

## Discussion of results of research

It results from the research conducted that representatives of the building sector in Poland are more and more knowledgeable about the possibility of implementing the CE in their company and are able to indicate areas that are important for the development of companies over the short term. Respondents considered that the following indicators are important in their sector:

- Indicator of share of costs in total costs of materials and energy;
- Indicator of amount of consumption of secondary raw materials/amount of revenue;
- Indicator of share of renewable energy in total energy consumption;

- Indicator concerning method of handling waste in accordance with the waste hierarchy; and
- Indicator of amount of consumption of critical raw materials.

### *Indicator of share of costs in total costs of materials and energy*

This indicator concerns total costs which include costs of products, goods, and materials sold and other operating and financial costs. This indicator covers:

1    Cost of the sale of products, goods, and materials related to the basic operating activity, including the value of sold goods and materials and total costs (total operating cost) decreased by the costs of generating benefits for the needs of the entity and corrected by the change in product inventories;
2    Other operating costs, that is costs related indirectly to the operational activity of the entity; and
3    Financial costs, that is among others, interest from received bank credits and loans, interest and discount on bonds issued by the entity, default interest, loss on the sale of investments, write-offs updating the value of investment, the surplus of foreign exchange losses over gains (Statistics Poland [GUS], Methodological report. Non-financial enterprises surveys).

### *Indicator of amount of consumption of secondary raw materials/amount of revenue*

In Polish and EU law, there is no unequivocal definition of secondary raw materials, but the Polish National Waste Management Plan 2022 points out how important it is to stimulate the development of the market for secondary raw materials and products containing secondary raw materials, as well to replace primary raw materials with secondary raw materials. For the effective implementation and monitoring of the CE, it is thus important to separate and monitor not only recycled materials, but also secondary raw materials, as materials for further economic use. In this context it is important to organize monitoring procedures concerning the classification of substances or materials as by-products or raw materials, which have ceased to be waste.

According to the Polish Classification of Goods and Services (*Polska Klasyfikacja Wyrobów i Usług – PKWiU*), it is possible to indicate potential secondary raw materials (commonly used term) or, in other words, production waste or used products suitable for reprocessing. They undergo processes of recycling or recovery to obtain materials (semi-finished products) for primary or other use. The detailed and fairly comprehensive Polish Combined Commodity Nomenclature for Foreign Trade (PCN) is used by both suppliers and recipients of secondary raw materials, enabling a very precise definition of the type of secondary raw material. The concept of secondary raw material also appears in publications. In the 1990s, secondary raw materials were still considered to correspond to two types of products: waste (post-production scrap) generated at various stages of production and used directly on site and post-consumption waste (post-consumption scrap). Yet another classification is provided in the "Materials management" publication from Statistics Poland [GUS] (Statistics Poland, 2020), which divides raw materials into:

- Natural raw materials (mineral, plant, or animal);
- Raw materials obtained from processing (e.g. cement);
- Secondary raw materials (waste), which in turn are divided into post-production waste, generated in production processes and used products, that is post-consumption waste. The latter may be used by another user after appropriate preparation, as a replacement for primary raw material (Pietrzyk-Sokulska, Radwanek-Bąk, & Kulczycka, 2018).

Ultimately, the discussed indicator could be used as a measure of the share of use of secondary raw materials in the economy, which is already monitored in the green economy by Statistics Poland (GUS).

## Indicator of share of renewable energy in total energy consumption

This indicator is calculated as the share of gross final energy consumption from renewable sources in gross final energy consumption from all sources. Energy from renewable sources is the energy derived from natural, repetitive environmental processes, obtained from renewable non-fossil energy sources (water, wind, solar heat, waves, sea currents, and tides). Geothermal energy, energy produced from solid biofuels, biogas and liquid biofuels, as well as ambient energy – from the natural environment – is used by heat pumps (GUS).

## Indicator concerning method of handling waste in accordance with the waste hierarchy

The quantity of waste generated, in accordance with the waste hierarchy, should be minimised. The generation of waste should be prevented in the first place, but if it is generated, it should be prepared for reuse, recycling, and ultimately for safe disposal. Objectives in the handling of waste were set out in Poland in the National Waste Management Plan 2022. They take legislative proposals into consideration, recommended by the European Commission within the framework of the CE package, promoting an increasing emphasis on waste prevention and reduction. This allows society to obtain maximum value from resources and to adapt consumption to real needs and enables entrepreneurs to maximise profit. Application of the waste hierarchy also allows the highest economic value of the product to be maintained.

## Indicator of amount of consumption of critical raw materials

This indicator concerns raw materials of fundamental significance for the correct operation of the economy and fulfilment of the material needs of society and thus those of which a continuous supply must be guaranteed. These are also raw materials, of which there is a large domestic resource base and the use of which is fundamental to industrial activity, as well as important scarce resources. The EC carries out a criticality assessment at EU level on a wide range of non-energy and non-agricultural raw materials. The 2020 EU list contains 30 materials as compared to 14 materials in 2011 and 20 materials in 2014 and 27 materials in 2017 (https://ec.europa.eu/growth/sectors/raw-materials/specific-interest/critical_en). These are raw materials of important economic significance, characterised by a high risk of supply shortage or of a lack of supply, resulting from a limited quantity of sources from which they are derived and the low degree of use of secondary sources and substitution. Most of the raw materials included in this group are necessary for the development of new technologies.

Representatives of the companies surveyed assessed the importance of CE-monitoring indicators over a short-term perspective and, as a result, identified the following CE monitoring indicators:

- Indicator concerning carbon footprint;
- LCA indicator; and
- Indicator of number of environment certificates held.

### Indicator concerning carbon footprint

This indicator determines the total quantity of greenhouse gases emitted over the life cycle of the product by an organisation, event, or a given person. It is also defined as the total amount of greenhouse emissions emitted over the life cycle. The method of calculating carbon footprint was described in the ISO 14067:2014 standard and is based on the principles of LCA set out in ISO 14040:2006 ( ITB, 2010) It thus considers the product life cycle, analysing emissions which are direct and those which are indirect or, in other words, those originating from the supply chain.

### Life Cycle Assessment (LCA) indicator

This indicator is mentioned in many EU and national documents as a tool for the assessment of environmental impact. Its official definition and method of calculation have been described in the ISO 14040 group of standards, as well as in national publications. LCA is a process of compilation and evaluation of inputs, outputs, and potential environmental impacts throughout the life cycle (concerning the production, use, and disposal of products). LCA analysis is used as a basis to prepare the data necessary to obtain an environmental certificate, which is a tool supporting the development of modern construction.

### Indicator of number of environment certificates held

This indicator is defined as the number of organisations registered in EMAS or holding environmental certificates, for example ISO 14001, EDP, ETV, or other, recognised as national or international certificates. Within the scope of EMAS, data is reported by Statistics Poland (GUS) in its green economy indicators. In 2020, in Poland, according to data from the General Directorate for Environmental Protection, there were 69 organisations on the EMAS register, representing an increase of 6.2% compared to 2017.

From the research conducted, it can be seen that the Polish building sector has a great deal of potential, but, due to the specifics of the sector, an increase in competitiveness is only possible by engaging in new innovative activities. The CE may have a significant impact in terms of shaping and opening up prospects for the development in the sector, while stimulating business activity, contributing to an improvement in competitive position on global markets. During a period of rapid and dynamic development, construction is an economic sector which undergoes intensive development, consuming huge amounts of raw materials and energy. According to a report by the International Energy Agency (IEA, 2015), over 30% of global energy production and nearly 50% of the weight of processed materials are used in the building sector (Energy Efficiency Market Report, 2015; International Energy Agency, Paris, 2015). An appropriate analysis of the impact of the extraction of raw materials, as well as their transport and processing on the natural environment allows factors conflicting with the principles of Sustainable Development and the CE to be identified and limited. A tool that enables the impacts of building products to be verified based on an analysis of their life cycle are type III Environmental Product Declaration (EPDs), developed in accordance with the PN-EN 15804+A2:2020–03 standard (Piasecki, 2012). The research conducted clearly shows that Polish entrepreneurs see a great deal of potential and opportunities for the development in the sector thanks to holding the appropriate environmental certificates as a mark of responsibility for the natural environment.

## Environmental declarations as a tool supporting Sustainable Development in the building sector

Adopted in 2015 by all UN member states, the 2030 Agenda for Sustainable Development obliges the private sector to contribute to Sustainable Development. Environmental declarations are a tool supporting firms in conducting their activity in a more sustainable way and guaranteeing sustainability throughout the entire supply chain.

Environmental declarations are defined as voluntary statements by manufacturers that reflect the environmental nature of the product offered. Their purpose is to distinguish products of exceptional environmental quality or, in other words, those which have less of an impact on the natural environment compared to competing products on the market. So that environmental declarations can fulfil the functions entrusted to them, it is necessary to use established standards guaranteeing that the information communicated is reliable and unambiguous.

In accordance with the PN-EN ISO 14025:2010 standard, type III environmental labels contain quantified data concerning the environmental impact of the given product, taking its whole life cycle into consideration. This data is compiled on the basis of the results of studies conducted in accordance with procedures set out in the ISO 14040 series of standards concerning LCA analyses. In accordance with guidelines for the ISO 14040 group of standards, the term life cycle is defined as "consecutive and interlinked stages of a product system, from raw material acquisition or generation from natural resources to final disposal". In practice, this means that, in contrast to the energy labels used, for example, for household appliances and light sources, a type III environmental declaration refers to the whole life cycle of products (or at least to its most important stages) and not just one aspect. In accordance with PN-EN ISO 14040:2006 and ISO 14044:2006, product is understood as "any goods or service" and is used consistently in relation to various issues described in those standards – for example product system, co-product, intermediate product, and product flow. Standards take not only the stages of production or usage into consideration, but also all processes occurring in the supply chain and in the post-production stages. This means that it is also necessary to obtain information from suppliers. In addition, information related to distribution, consumption/use, and end-of-life waste management also fall within the scope of the analysis. The ISO 14025:2006 standard presents the principles and requirements relating to the preparation of type III environmental declarations (Environmental Product Declarations – EPDs). Moreover, for given product categories, within the framework of existing EPD systems, Product Category Rules (PCR) are determined. An example of this type of system is the International EPD® System (www.environdec.com/), which is a global environmental product declaration programme based on the ISO 14025 and EN 15804 standards.

In accordance with the guidelines provided in the PN-EN ISO 14040:2009/ A1:2021–03 standards, environmental declarations may be used to compare products in a given product category. The basic scope of data necessary to prepare a type III environmental declaration usually relates to:

- Determination of consumption of non-energy resources (renewable and non-renewable);
- Determination of consumption of energy resources including electrical energy, heat, process steam and fuels, as well as other energy products (renewable and non-renewable); and
- Determination of the quantity of polluting emissions and waste generated.

Calculated on the basis of the aforementioned data, environmental impacts are presented per functional unit, which, according to the definition (based on the PN-EN ISO 14020: 2002 standard), reflects the quantified performance of a product system. It determines the qualitative and quantitative aspects of the function or the service which the product being assessed provides. It contains answers to the questions: "what?", "how much?", "how well?", and "for how long?". The functional unit may be a physical unit, for example weight, volume, or another quantity, a function reflecting the analysed product. The prepared type III environmental declarations are subject to a process of verification and critical review in order to check whether the LCA analysis, which forms the basis for determining the ecological characteristics of a given product, is compliant with the requirements of ISO 1404x standards

Table 6.2 Selected products and companies holding an implemented type III environmental declaration

| Company/Product | Tubolit DG plus flexible polythene insulation produced at Armacell's plant in Środa Śląska |
| --- | --- |
| Company/Product | Gypsum stone and anhydrite, Gypsum plasterboard RIGIPS PRO and RIGIPS 4PRO™, Casoprano CASOBIANCA, CASOSTAR, CASOROC ceiling tiles, RIGIROC gypsum blocks, RIGIPS set of products for installing partition walls |
| Company/Product | Aluprof aluminium profiles |

Source: Own elaboration based on websites.

(EC, 2013). Selected products and companies holding an implemented type III environmental declaration are shown in Table 6.2.

The most important benefits of holding environment certificates include:

- Implementing helps investors to obtain green certificates, such as BREEAM or LEED, for their buildings and structures;
- Holding declarations may help to obtain additional points in other certifications thanks to the fact that products are 100% recycled;
- Buildings and structures, which have obtained a LEED or BREEAM certificate, are highly energy-efficient in the areas of energy, water, and materials and have a low impact on health and the environment;
- Raising of environmental production standards – reduction of environmental impacts and of costs of environmental use; and
- Providing investors with reliable information – publication of credible and confirmed information about products.

Based on an analysis of environmental characteristics, environmental declarations are considered to be a source of important information on the impact of individual aspects related to production processes on the natural environment. The widespread availability of environmental declarations for products on the market can be a tool that is strongly supportive of the formation of best practices in the field of Sustainable Development, both among consumers and producers. Information on impacts, expressed in a quantified manner, can provide consumers with support in the decision-making process, encouraging them to use products which are less harmful to the natural environment, while, at the same time, conclusions of the LCA analysis may serve as a basis for looking for new solutions, which bring measurable economic and environmental benefits.

## Summary

Environmental certificates held by companies may be a valuable source of guidance for designers, investors, and users, as to how to build and live in a sustainable and thus energy-efficient and ecological way. Building in a way that is compliant with the concept of the CE means building in a way that consumes less materials and energy and minimises waste. In the ideal sustainable building, there is nothing that is superfluous and nothing that is lacking (Panek, 2005). Such an approach shows how important it is to treat issues of Sustainable Development and the idea of the CE in a way which takes all stages of the life cycle into consideration from appropriate eco-design of the product to sustainable production and waste management. Environmental Product Declarations (EPDs) based on LCA are developed for products in different countries of the EU. In addition to this, the Product Environmental Footprint (PEF) initiated by the European Commission aims to provide benchmarking for products based on LCA. Though EPDs are well-developed in Europe, their application is difficult due to a lack of knowledge by key entities on the subject of the application of EPDs (Adibi, Mousavi, Escobar, Glachant, & Adibi, 2019). One of the tools for the implementation of the principles of Sustainable Development in the building sector is the introduction of the system of type III environmental declarations for building products. The most important part of environmental product declarations are the energy and ecological (environmental) characteristics, determined using the full LCA methodology. These characteristics provide the basic information necessary for the assessment of products from the point of view of their environmental impact. The use of new environmental solutions and tools can lead to more sustainable use of resources and bring long-term benefits not only for companies in the building sector, but also for the environment and society as a whole. The basic way in which the building sector should be modified in order optimise the benefits from construction, while limiting the problems related to it, is to make the transformation towards the CE.

# References

Adibi, N., Mousavi, M., Escobar, R. M., Glachant, M., & Adibi, A. (2019). *Mainstream use of EPDs in buildings: Lessons learned from Europe* (pp. 137–145). Dallas: ISBS.

Akerman, E. (2016). *Development of circular economy core indicators for natural resources – analysis of existing sustainability indicators as a baseline for developing circular economy indicators* (Master of Science thesis), Royal Institute, Stockholm.

Bukowski, H., & Sznyk, A. (2019). *Metodologia dopasowania cyrkularnych modeli bizneso wych do priorytetowych sektorów wdrażania gospodarki o obiegu zamkniętym w Polsce*. Retrieved from https://circulareconomy.europa.eu/platform/sites/default/files/the_circular_economy _in_policy_and_scientific_research.pdf

Deloitte. (2020). *Polskie Spółki Budowlane*. Retrieved from https://www2.deloitte.com/pl/pl/pages/real-estate0/articles/raport-polskie-spolki-budowlane-2020.html

Directive 2010/31/EU of May 19, 2010 on energy performance of buildings and its further updates.

Directive 2012/27/EU of October 25, 2012 on energy efficiency.

EC. (2013). 2013/179/EU: Commission Recommendation of 9 April 2013 on the use of common methods to measure and  communicate the life cycle environmental performance of products and organisations. Retrieved from https://eur-lex.europa.eu/legal-content/EN/TXT/?uri=CELEX%3A32013H0179

EC (European Commission). (2014). Communication from the Commission to the European Parliament, the Council, the European Economic and Social Committee and the Committee of the Regions on resource efficiency opportunities in the building sector; Communication COM(2014) 445.

EC (European Commission). (2018). *Communication from the commission to the European Parliament, the council, the European economic and social committee and the committee of the regions on a monitoring framework for the circular economy*. Retrieved from https://ec.europa.eu/environment/circular-economy/pdf/monitoring-framework_staff-workingdocument

EESC (European Economic and Social Committee). (2018). *Monitoring framework for the circular economy (communication)*. Retrieved from www.eesc.europa.eu/en/our work/opinionsinformation-reports/opinions/monitoring-framework-circular-economy-communication

Energy Efficiency Market Report. (2015). *Energy efficiency market report*. Paris: International Energy Agency.

IRP. (2020). *Resource efficiency and climate change: Material efficiency strategies for a low-carbon future* (E. Hertwich, R. Lifset, S. Pauliuk, & N. Heeren, Eds.). A report of the International Resource Panel. Nairobi, Kenya: United Nations Environment Programme.

ITB. (2010). Etykiety i deklaracje środowiskowe według norm ISO. Published on Zrównoważone Budownictwo. Retrieved from www.zb.itb.pl

Kulczycka, J., Bączyk, A., & Nowaczek, A. (2020). Monitorowanie transformacji gospodarki o obiegu zamkniętym w dokumentach strategicznych Polski i UE. In J. Kulczycka (Ed.), *Wskaźniki monitorowania gospodarki o obiegu zamkniętym*. Kraków: IGSMiE PAN.

Liu, X. (2014). Research on circular economy and industrial clusters. *Management & Engineering, 15*, 1838–5745.

Nowaczek, A., Kulczycka, J., & Bączyk, A. (2020). Postulowane mierniki monitorowania transformacji w kierunku gospodarki o obiegu zamkniętym. In J. Kulczycka (Ed.), *Wskaźniki monitorowania gospodarki o obiegu zamkniętym*. Kraków: IGSMiE PAN.

OECD (Organisation for Economic Cooperation and Development). (2002). *Indicators to me asure decoupling of environmental pressure from economic growth*. Retrieved October 15, 2019, from www.oecd.org/officialdocuments/publicdisplaydocumentpdf/?docla nguage=en&cote=sg/sd(2002)1/final

Panek, A. (2005). *Holistyczna metoda oceny oddziaływania obiektów budowlanych na środowisko naturalne uwzględniająca zasady rozwoju zrównoważonego*. Warsaw: PL.

Piasecki, M. (2012). *Deklaracje środowiskow e wyrobów budowlanych, typ III (EPD) – norma PN-EN 15804*. Zrównoważone budownictwo. Retrieved from www.itb.pl/zrownowazone-budownictwo1.html

Pietrzyk-Sokulska E., Radwanek-Bąk, B., & Kulczycka, J. (2018). Mineralne surowce wtórne – problemy polskiego nazewnictwa i klasyfikacji w związku z realizacją gospodarki o obiegu zamkniętym. *Przegląd Geologiczny, 66*(3), 160–165.

Statistics Pol and. (2020). *Business tendency in manufacturing, construction, trade and services 2000–2020*. Retrieved from www.stat.gov.en

Walker, S., Coleman, N., Hodgson, P., Collins, N., & Brimacombe, L. (2018). Evaluating the environmental dimension of material efficiency strategies relating to the circular economy. *Sustainability, 10*(666).

Wang, L. (2014). Construction on cluster green supply chain based on circular economy. *Contemporary Logistics, 16*, 78–82.

# 7 Enablers and barriers in the transition to circular business models

## Investigating the critical success factors for the tipping and break-even point

*Marek Ćwiklicki and Linda O'Riordan*

## Introduction

In this chapter, we provide an overview of the current key observations on the transition towards circular business model (CBM) covered by recent research. The study builds on our previous publication in this field aimed at describing the break-even or tipping-point of transition, a threshold we explore, whereby after its crossing, a CBM becomes successfully operational. We do so by demonstrating the balance of critical success factors required for enabling and even nudging towards a CBM. The chapter is based on a review of research articles identifying enablers and barriers involved in making this transition. The study does not merely limit itself to re-listing the enablers and barriers in the transition towards a CBM. While we do present an updated review of the related opportunities, constraints, and challenges in making this move, our investigation goes one significant step further. By way of clarification, it should be pointed out that we will treat the terms "constraints" and "barriers" and, to some degree, the negative impacts of "challenges" as synonyms in this chapter. Our analysis is based on the assumption that the degree of change in current business models to achieve circularity cannot result merely from the effect of one factor or merely from the micro-level of the organisation but from the arrangement of many enablers of varying strengths connected also with the broader macro-level network. Based on this assumption, we present our point of view by matching the identified enablers and barriers to particular elements of the CBM framework proposed by Antikainen and Valkokari (2016). We chose this framework for its merit in facilitating holistic reference to three levels of analysis: the macro-level – expressed by broad expectations with respect to sustainability requirements and behaviour regarding social, political, economic, and technological factors; the meso-level – by the business ecosystem (industry) level; and the micro-level – referring to the internal corporate focus in form of the classic business model canvas.

The chapter consists of three main parts. We first present the adopted CBM and discuss its main parts and role in making the transition to some notion of

DOI: 10.4324/9781003179788-7

a sustainable Circular Economy. This establishes the framework of reference. The second part explains the transition process from linear to Circular Business Models including a review of the related key concerns. The third part presents the authors' alignment of the identified factors influencing the shift in business model. We conclude with an evaluation of the significance of each of the identified opposing enablers and barriers. Ultimately, this evaluation allows us to identify the break-even point of transition which identifies when the linear *status quo* is broken, and the system begins to lean towards circularity.

## Underpinning concepts for transitioning to circularity

According to the Ellen MacArthur Foundation (2021), "the current system is no longer working for businesses, people, or the environment". From a socio-ecological perspective, we are taking resources from the planet to make products, which we use, and, when we no longer want them, throw them away. We call this current "take-make-waste" system a Linear Economy (LE). The underlying rationale of any linear system, given its inherent starting and finishing point, is predestined, by definition, to end. If, however, we wish the planet to continue to support life, we need to change that current logic. O'Riordan and Hampden-Turner (2021) described the challenge we face as follows:

> We have to realise that the effort to save the planet is an emergency akin to World War, save that our planet in its adverse reactions will do the killing. We need emergency powers and only governments with democratic mandates can do this.

In contrast with the LE, the Circular Economy (CE) advocates a new way to design, make, and use those things we need within the capability of the earth's resources to sustain those actions. The circularity of a CE system aspires to an approach based on the principles of architecting out waste and pollution, keeping products and materials in use, and regenerating natural systems.

The CE concept thereby inherently holds the potential to achieve sustainability, as defined by the World Commission on Environment and Development (1987). The opportunity to re-design the system to ensure that future generations can enjoy resources and satisfy their needs as we do today (or better) fosters the creation of sustainable value. This results in the generation of sustainable prosperity, wealth, and well-being within a system designed to ensure its fair distribution. The CE approach thereby poses a more equitable, viable, and bearable alternative to the single-minded emphasis on profits for the holders of equity inherent in the LE approach. Triple bottom line (TBL) and triple top line (TTL) objectives in the interests of people, planet, and profits are thus achieved (Elkington, 1997; McDonough & Braungart, 2002).

To transition to this new approach, we must transform all the elements of the take–make–waste system: how we manage resources, how we make and use products, and what we do with the materials afterwards. Only when we succeed in designing out waste and pollution, keeping products and materials in use, and regenerating natural systems can we reinvent everything, and only then can we create a thriving economy that can benefit everyone within the limits of our planet (Ellen MacArthur Foundation, 2021). An economic approach which is mindful of its impact on both ecological and social principles, aligning the Sustainable Development Goals 2030 (UN, 2015) with commercial objectives, can thereby serve to ensure a permanently peaceful and healthy, global well-being for mankind.

In the quest to accomplish the transformation to a CE, organisations play a crucial role. Commercial organisations can be interpreted as the platform via which solutions are generated in the form of products and services as outputs from invested resources (inputs). Business strategies which are aligned with CE principles are architected with regard to ensuring a "liveable" world for future generations. Given their significant resources and reach, at corporate level, leaders hold enormous potential to create sustainable stakeholder value by consciously designing strategic and operative goals to generate wealth for society via their business purpose and their business models. By considering the interaction between the social and environmental interests of the corporation's various stakeholder groups, they consciously adopt strategic corporate responsibility.

Organisations can undoubtedly deploy powerful solutions to address persistent issues, including global warming, climate change, human rights abuses, and poverty, among others, and make addressing such issues their corporate purpose. However, how precisely, when, and how well they achieve this shift remains to be seen. In their aim to align their business models with the CE concepts described before, how exactly must organisations in their diverse contexts change to achieve the goals of sustainable management? How will decision-makers re-organise their factors of production inputs (land, labour, capital, and enterprise) in a sustainable way?

## Circular business models as a theoretical framework

Currently, there is no common definition of CBM. Our previous study provided a list of nine definitions in the articles about the transition towards a CBM. From this list, we will choose two to be the most representative in terms of underlining the key issues which later are to be embedded into a CBM framework. According to Bocken et al., CBMs

> Are about slowing, closing and narrowing resource loops: strategies to provide products that last and support product life extension (slowing); strategies to close material loops through recycling (closing); and strategies to use less material and energy per product (narrowing loops).
>
> (Bocken, Strupeit, Whalen, & Nußholz, 2019, p. 241)

In this definition, we can see a focus on strategy defining what circularity means and to which business functions we can assign it (supply, sales, and marketing).

*Table 7.1* Key parts of a circular business model

| Level | Key dimensions | Parts |
|-------|----------------|-------|
| Macro | Sustainability impact | Environmental, social and business sustainability requirements and benefits |
| Meso | Business model ecosystems | Trends and drivers, stakeholder involvements |
| Micro | Business model canvas | Key partners, key resources, key activities, cost structure, value proposition, customer relationship, channels, revenue streams, customers and stakeholders |

Source: Own elaboration based on Antikainen and Valkokari (2016).

The second definition, proposed by Nußholz et al., explains that CBMs: "aim to utilise embedded economic and environmental value in products and materials for as long as possible, for instance through substituting primary materials with secondary materials" (Nußholz, Nygaard Rasmussen, & Milios, 2019, p. 309). Here, we can observe an underlying theme of value creation. Both definitions refer to the company level, thus limiting the point of view regarding enablers and barriers to those which refer to the micro-level.

As we mentioned in the introduction, a broader perspective is offered by including macro- and meso-levels of analysis. This expands the analysis to include the characteristics inherent in the sustainable Circular Business Model innovation framework proposed by Antikainen and Valkokari (2016). Table 7.1 shows the main parts of such an analysis.

## Transition towards a CBM

We understand transition to be a change, a modification, or a shift from an existing model to a new one. In the context of this chapter, we use this to refer to the shift from a linear business model (LBM) to a CBM already described. In order to describe the state of the LBM, we will identify its main features. The LBM is perceived as a "traditional business model of production of take-make-use-dispose" (Bocken, de Pauw, Bakker, & van der Grinten, 2016, p. 308). It is described as a model in which raw materials are extracted and moved to manufacturers, where goods are produced and shipped to retailers, sold, used, and discharged (Batista, Bourlakis, Smart, & Maull, 2019; Braungart, McDonough, & Bollinger, 2007). The opposite of linearity, also described as an open-loop, one-way, "cradle-to grave" model, is circularity, meaning a closed-loop, "cradle-to-cradle" model (Johannsdottir, 2014).

Transition means that the business model changes its features, redefines its components, and gains new qualities. Table 7.2 shows this transition from the LBM to a CBM at all levels. We noted that business models refer not only to the ecosystem at organisational level (supply chain and consumers). They are also linked with the broader macro-level.

*Table 7.2* Characteristics of linear and circular business models

| Level | Dimension | Linear business model | Circular business model |
|---|---|---|---|
| Macro | Sustainability requirements | None | Regarding used material, product discharge |
| | Sustainability impact | Mainly about discharging product | Stressing resource usage, circularity |
| Meso | Ecosystem drivers | Price | Protecting the natural environment and resources |
| | Stakeholder involvement | Basic customer–supply relation | Cooperation with suppliers within the supply chain and with customers |
| Micro | Resources | Raw materials | Recycled material |
| | Activities | One-way direction of business activities | Closed-loop activities |
| | Value proposition | Traditional product, transfer of ownership | Value retention, circular product, ownership of products are not key here |
| | Customer relationship | After-sales customer service | Cooperation in product usage and product disposal |
| | Channels | One-way, with several suppliers offering similar products, traditional sales channels delivering product | Circular, with less suppliers requiring close cooperation; sales channels requiring consumer engagement |
| | Revenue streams | Simple model based on sales | Extended due to added new additions to product-like services |

Source: Authors' elaboration.

## Factors in the transition towards circular business models: research results

### Methodological assumptions of the research

In this chapter, we identify the factors (enablers and barriers) which appear during the transition from an LBM towards a CBM. The initial list of barriers and enablers comes from previous research undertaken by the authors based on a systematic review of the literature Ćwiklicki and O'Riordan (in press) carried out in August 2019. As more than 1 year has passed since that date, an additional query was run covering the year 2020 in the two main databases (Scopus, Web of Science) using the same search term "circular business model" and "transition". In this update to the review, new research was identified, including among others: (Ferasso, Beliaeva, Kraus, Clauss, & Ribeiro-Soriano, 2020; Guldmann & Huulgaard, 2020; Urbinati, Franzò, & Chiaroni, 2021). The framework in Figure 7.2 provides a graphic illustration of this transition, highlighting the role of enablers in shifting companies towards a CBM and conversely, and that of barriers, which

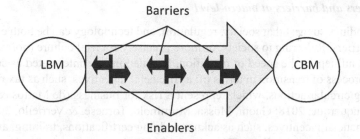

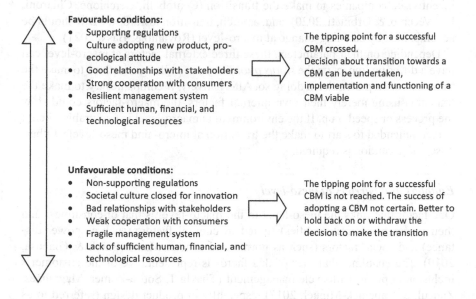

*Figure 7.1* The framework for transition from a linear to a circular business model

**Favourable conditions:**
- Supporting regulations
- Culture adopting new product, pro-ecological attitude
- Good relationships with stakeholders
- Strong cooperation with consumers
- Resilient management system
- Sufficient human, financial, and technological resources

The tipping point for a successful CBM crossed.
Decision about transition towards a CBM can be undertaken, implementation and functioning of a CBM viable

**Unfavourable conditions:**
- Non-supporting regulations
- Societal culture closed for innovation
- Bad relationships with stakeholders
- Weak cooperation with consumers
- Fragile management system
- Lack of sufficient human, financial, and technological resources

The tipping point for a successful CBM is not reached. The success of adopting a CBM not certain. Better to hold back on or withdraw the decision to make the transition

*Figure 7.2* The contingency theory for transition to a circular business model

constrain companies to remain stuck in their existing LBM. In the next step, the references to enablers and barriers were each classified according to key elements of the CBM framework using a three-level approach: macro, meso, and micro. In that way, we received, updated, and expanded upon the results compared to our previous work, arranging them in a different order. In the next sections, we will present the enablers and barriers identified according to a breakdown into the three levels mentioned, followed by a discussion indicating which factors constitute essential tipping-point elements, that is after reaching them, a CBM can successfully operate.

### Enablers and barriers at macro-level

The findings suggest that society, regulation, and technology can be both enablers and barriers. Referring to society, or more precisely societal culture (norms, fashion), it influences the speed of transition. Regulations are interpreted as supporting a process of transition in terms of: a) financial incentives, such as tax relief or funding circular actions, which represent a risk for business (de Mattos & Meira de Albuquerque, 2018; Gnoni, Mossa, Mummolo, Tornese, & Verriello, 2017); b) non-financial incentives, such as acknowledging certifications, training, and supportive procurement policies (Rizos et al., 2016). A proper level of technological infrastructure is additionally described as an enabler due to its positive role in easing collaboration within the value chain. However, a lack of IT infrastructure needed for smooth communication constitutes a barrier. These contextual factors incentivised companies to make the transition (Centobelli, Cerchione, Chiaroni, Del Vecchio, & Urbinati, 2020), and, as such, transition towards a CBM should be accompanied by enabling change at macro-level (Rovanto & Bask, 2021).

Depending on their parameters, these three external factors at macro-level can have a differing impact on the *status quo*. They can pull organisations to make the transition or stop them from doing so. Alternatively, if companies try to make the transition using merely their own internal factors, the macro-factors could slow the process or speed it up. If the environment at macro-level is favourable, then it is recommended to start to make the transition at micro- and meso-levels. Otherwise, more caution is required.

### Enablers and barriers at meso-level

One factor that links the meso-level to the macro-level is that of consumers and their preferences (irrationalities) rooted in cultural factors (such as power distance) and social factors (such as norms and consumerism) (Singh & Giacosa, 2019). The enabling character of this factor is representative of the customer's inclusion in product lifecycle management (Thayla T. Sousa-Zomer, Magalhães, Zancul, & Cauchick-Miguel, 2017), especially in product design (referred to as "user-centred eco-design" (Heyes, Sharmina, Mendoza, Gallego-Schmid, & Azapagic, 2018). The special case for a product-service system is a product with digital features. In the case of more digitally advanced products, consumers should be more digitally confident (Tunn, van den Hende, Bocken, & Schoormans, 2020). The double-headed factor identified is that of geographical proximity. The closer companies are located to each other, the more likely it is that there will be cooperation between them. This factor is related to the next one: the nature of relationships between partners (inter-organisational collaboration) (Sousa-Zomer, Magalhães, Zancul, & Cauchick-Miguel, 2018). Conversely, barriers include unsure and weak relationships among business partners. This is characterised by lack of trust, weak channel control, high confidentiality, or dependency on partners. Active communication with stakeholders and their engagement in companies' activities are regarded as factors favourable to transition as well as perceived mutual benefits.

The state of these factors is crucial. A close environment (ecosystem) is a key factor in determining whether it is possible to evolve towards a CBM, even if the macro-level is supportive and the business model is ready to shift.

### Enablers and barriers at micro-level

This level is more broadly discussed in the literature. The first factor assessed by us as being double-headed is that of human resources. Associated with this, the necessity of possessing knowledge, as well as an organisational culture, will enable or hamper transition. Next, strong investment power will allow the organisation to cope with the investment costs of making the transition towards the CE, while a weak capacity for investment will be a barrier. The third factor is the nature of the product. Products which are complex and difficult to re-design as circular make the transition troublesome, but products which make it easy to offer a new form of value can push companies towards a CBM. This also relates to another key activity: That of product design (Sumter, Bakker, & Balkenende, 2018). The fourth factor is internal technological infrastructure enabling communication and cooperation within companies, while the lack of such infrastructure can constitute a barrier. This is also related to the possibility of embedding changes in manufacturing and creating a product-service system (Bressanelli, Adrodegari, Perona, & Saccani, 2018). The next factor is that of the management system or governance, as it deals with the administrative burden and organisational barriers which will arise if such a system is poorly designed or malfunctioning. Otherwise, a properly functioning management system should support transition.

This level sets the minimum requirements for transition to a CBM. The value of these factors can differ according to context and thus may require individual assessment, as there is no unified set of standards or guidelines for making the required transition.

## Passing through the break-even point/tipping point

In total, it is possible to distinguish 11 key factors: 3 at macro- and meso-levels and 5 at micro-level (Table 7.3). The state of all of them influences the willingness to make the transition and has an impact on the speed of transition towards a CBM. According to our observations, each identified factor can both favour or

*Table 7.3* Key factors to consider during transition to a CBM

| Macro-level | Meso-level | Micro-level |
|---|---|---|
| Technology | Consumers (Market) | Human resources (employee level) |
| Regulations | Business Partners | Investment power |
| Society | Other Stakeholders | Product nature |
| | | Technological infrastructure |
| | | Management system |

prevent such change. The characteristics listed before provide an indication as to what values they should embrace in order to enable or hinder transition. A similar classification for the automotive industry has been proposed by Urbinati et al. (2021).

We thus propose a contingency theory of transition to a CBM, which allows us to build on the presented results. It explains how the existence of favourable conditions could encourage managers to change their business models. However, the recognition of disadvantageous factors which appear could cause managers to hold back on making the decision to proceed with the transition. Figure 7.2 shows this logic.

Based on the identification of the factors and the classifications shown in Table 7.3 and Figure 7.1, we can now discuss the salience of the different levels (macro/meso/micro) and which is most crucial for passing through the break-even point and thereby reaching the tipping point for transition towards a CBM.

Over the past 10 years, the use of the term 'tipping point' has significantly increased in the scientific literature. It is frequently loosely employed as a meta-phor to describe the phenomenon where, beyond a certain threshold, critical mass is achieved. Rapid replication, often compared with the spread of a virus, triggers runaway change thereby propelling a system to a new state. The term has been applied to any process in which, beyond a certain point, the rate of the progress increases dramatically (van Nes et al., 2016). The tipping point concept suggests that change is not an erratic, unpredictable, random occurrence. Instead, it arises at an inflection point where critical mass is achieved. The logic behind this notion has been applied in many fields.[1] It is interpreted by the authors as an interesting theoretical representation that helps to understand how and when a linear, exclu-sively profit-driven economic system might transition to a broader stakeholder-oriented Circular Economy construct. If this can be understood, it would be an insight that could conceivably be leveraged to purposefully stimulate change.

Could empirical exploration clarify whether it is possible to understand the critical factors accompanying certain phenomena which trigger swift transition? Assuming it is possible to identify the critical mass or tipping point factors in social—ecological systems, might it be conceivable to identify factors which help tip the system towards circularity or trigger the shift to a sustainable business model and when this might occur? At a macro-level, could such a study help to speed up the necessary processes of change in the field of climate change, as well as help to accelerate change in sustainable consumer and investor habits? At the micro-level of the firm itself, could it be possible to identify the factors and time when a project becomes self-sustaining or economically viable, and possibly even value-creating in the sense of achieving utility, such as some return-on-investment or profitability (break-even)?

The research presented indicates how three groups of factors (macro, meso, micro) appear to have different influencing powers on decision-making in the process of triggering a shift when making the transition towards a CBM. Macro-level factors define the context in which companies operate. It is possible that company leaders can initiate change in spite of the existence of unfavourable external factors at macro-level. As a result, this indirect impact is not decisive as long as other factors at the meso- and micro-level are favourable to transition.

The successful implementation of a CBM requires cooperation within certain value chains which cross organisational boundaries. We reason that meso-factors linked with internal processes responsible for cooperation are key to a successful change of business model. This is particularly relevant in the case of consumers for whom ethical purchase intention is impacted by "awareness, perceived value, and attitude" (Mostaghel & Chirumalla, 2021). Further indications of the salient role of customers are evident in previous studies, which demonstrated that the minimum level of readiness to change is to possess a proper value proposition for customers, along with the necessary resources, processes (cf. Johnson, Christensen, & Kagermann, 2008), as well as "management environmental awareness" (Urbinati et al., 2021). One recent suggestion from the literature is the beneficial impact of digital technology on CBM transformation (Ranta, Aarikka-Stenroos, & Väisänen, 2021), thus highlighting the requirement for digital competences. Additionally, for small- and medium-sized enterprises, the organisational learning process associated with knowledge management is perceived as being key in the shift to a CBM (Scipioni, Russ, & Niccolini, 2021). This is crucial, if organisations do not possess adequate resources, as then they are not capable of change, and transition is perceived as complex, highly uncertain, and radical (Hofmann & Jaeger-Erben, 2020). In such cases, different tools for the evaluation of CBMs, such as Life Cycle Assessment (LCA) (see Chapter 3) and Life Cycle Costing (LCC) (see Chapter 8), can be useful (Chen, Hung, & Ma, 2020).

Based on the reasoning given before, we identify the tipping point for the transition to a CBM to be at the meso-level. Being internally ready for transition is not sufficient. A closely cooperating environment, involving business partners and consumers, needs to reach a level favourable to transition. It is worth noting that these factors can be influenced by the companies themselves. Therefore, if a company wishes to initiate the change, it is possible to do so, on the condition that the relevant associated (stakeholder) parties are also ready to make the shift. Furthermore, the transition can be expedited when the identified enablers at macro-level are also present.

## Conclusion

Our analysis reveals that factors which are enablers and barriers can be related to the same issue. We observed that their types are similarly distinguished in the literature. The factors can be both enablers and barriers depending on the context and timing. Nevertheless, by making some generalisations, it is possible to arrive at a unified set of factors. The classification of opposing states for each factor allowed us to propose a contingency theory-based approach to the transition towards a CBM (Table 7.3). Figure 7.2 shows the key conclusions associated with some strongly practice-oriented recommendations which we list later. When presenting these recommendations, we do so, however, with the proviso that setting a transition pathway for each company is not an easy task. It requires taking a diverse range of data into consideration, which in some cases might not be fully available (due to factors such as trends, willingness of partners to adapt, etc.). Accordingly, it is only after the decision-makers at the company have recognised the organisation's own state of readiness to make the transition that an orchestration strategy should be chosen (Palmié et al., 2021).

According to our findings, the decision to make the transition towards a CBM should include the existence of conditional factors at meso- and micro-level. It is recommended that major decisions should only be made about starting or proceeding with the shift or, alternatively, restraining from action, based on the contingency theory guidelines presented in this chapter. Business practitioners should devote special attention to the capabilities internal to their own company and should carefully check their company's readiness to make the transition by applying evaluation tools, such as life cycle analysis. Our final key recommendation is to analyse the quality of stakeholder cooperation in terms of the possibility of fostering change towards a CBM. Our observations generally align with the findings of van Loon and Van Wassenhove (2020, p. 3420) that "transition to the CE requires realism" and the force to go beyond the comfort zone of the established orchestration strategy, including the necessity of resource modification and actor interaction (Palmié et al., 2021).

In this section, we highlight some further salient economic concepts which are related to, yet go beyond the scope of this chapter, but which could potentially be explored in separate subsequent research. The first one is the identification of factors for tipping phenomena: relating not just to business models but also to other stakeholder relationships in the entire system/network. The second one is the topic of Opportunity Cost: That is, considering the foregone benefit that would have been derived by taking an option not chosen, that is consideration of the costs and benefits of alternative choices under scenario planning assumptions. The third one is the law of diminishing marginal utility: Does the value or return on the invested resource decrease over time as supply increases – Can we over tip? Is there a point at which this can be identified? The next one is the search for a golden formula: Can we identify the mathematical point (of inflection or even the break-even point, or whatever other term one may wish to employ) at some "divine proportion" at which logical convergence/harmony would occur among the identified factors under certain conditions? The final topic refers to statistical probability: Is there some theory (e.g. possibly similar to the approach taken in the Dvoretzky–Kiefer–Wolfowitz inequality calculation), which could highlight a path for the empirically determined distribution function to forecast tipping point scenarios in circularity/sustainability contexts?

While the restorative CE approach can be considered to present many of the qualities needed to address a range of current disconnections in the LE mechanisms which are perilously threatening to endanger mankind's very future survival on this planet, the successful transition to achieve the needed transformation requires new approaches and mindsets to improve the sustaining linkages between business and society. Freeman (1984) shows that wealth is created by a broad alliance of stakeholders. Bringing about a shift in the system involves everyone and everything: businesses, governments, and individuals; our cities, our products, and our jobs.

In the circular approach, the rewards received by stakeholders must be commensurate to their contributions. Instead of overlooking those who do the actual work, the CE system thereby advocates a long-term strategic orientation and positioning of organisations within a network of stakeholder relationships. Profits ensue as the logical consequence of shared efforts, but are not the purpose of them (Grant, 2006, p. 41).

Sustainable Stakeholder Value Creation (O'Riordan, 2017) via this shift to a CE system therefore requires comprehensively, inclusively, and purposefully implementing integrated processes via partnerships to promote synergic sustainable effects for people by people to ensure long-term prosperity, including ultimately sustainable profits as a consequence (O'Riordan & Hampden-Turner, 2021).

Leaders can demonstrate their responsibility by taking the lead in addressing the issues and consciously designing their business processes to achieve sustainable solutions in a CE context through transitioning to CBMs.

## Note

1 The term "tipping point" has been applied in many fields. It can be compared to phase transition in physics or to the propagation of populations in certain ecosystems. In popular culture, it is frequently adopted by journalists and academics when referring to dramatic changes previously considered unforeseeable, such as political changes in China or the Arab Spring. The concept has been employed to explain a diverse range of events spanning from viral cat videos to why changing habits is so hard.

## References

Antikainen, M., & Valkokari, K. (2016). A framework for sustainable circular business model innovation. *Technology Innovation Management Review*, 6(7), 5–12. doi:10.22215/timreview/1000

Batista, L., Bourlakis, M., Smart, P., & Maull, R. (2019). Business models in the circular economy and the enabling role of circular supply chains. In L. de Boer & P. Houman Andersen (Eds.), *Operations management and sustainability* (pp. 105–134). Cham: Springer International Publishing. doi:10.1007/978-3-319-93212-5_7

Bocken, N., de Pauw, I., Bakker, C., & van der Grinten, B. (2016). Product design and business model strategies for a circular economy. *Journal of Industrial and Production Engineering*, 33(5), 308–320. doi:10.1080/21681015.2016.1172124

Bocken, N., Strupeit, L., Whalen, K., & Nußholz, J. (2019). A review and evaluation of circular business model innovation tools. *Sustainability*, 11(8), Scopus. doi:10.3390/su11082210

Braungart, M., McDonough, W., & Bollinger, A. (2007). Cradle-to-cradle design: Creating healthy emissions – a strategy for eco-effective product and system design. *Journal of Cleaner Production*, 15(13–14), 1337–1348. doi:10.1016/j.jclepro.2006.08.003

Bressanelli, G., Adrodegari, F., Perona, M., & Saccani, N. (2018). The role of digital technologies to overcome circular economy challenges in PSS business models: An exploratory case study. *Procedia CIRP*, 73, 216–221. doi:10.1016/j.procir.2018.03.322

Centobelli, P., Cerchione, R., Chiaroni, D., Del Vecchio, P., & Urbinati, A. (2020). Designing business models in circular economy: A systematic literature review and research agenda. *Business Strategy and the Environment, 29*(4), 1734–1749. doi:10.1002/bse.2466

Chen, L., Hung, P., & Ma, H. (2020). Integrating circular business models and development tools in the circular economy transition process: A firm-level framework. *Business Strategy and the Environment, 29*(5), 1887–1898. doi:10.1002/bse.2477

Ćwiklicki, M., & O'Riordan, L. (in press). Modes and factors in the transition towards a circular business model. In H. Lundberg, M. Ramirez-Pasillas, & V. Ratten (Eds.), *Entering the territory of the unknown: Sustainability through circularity, digitalization and exploration.* London: Taylor & Francis, Routledge.

de Mattos, C. A., & Meira de Albuquerque, T. L. (2018). Enabling factors and strategies for the transition toward a circular economy (CE). *Sustainability, 10*(12). doi:10.3390/su10124628

Elkington, J. (1997). *Cannibals with forks: The triple bottom line of 21st century business* Oxford: Capstone Publishing Ltd.

Ellen MacArthur Foundation. (2021). *Circular economy.* Retrieved from www.ellenmacarthurfoundation.org/circular-economy

Ferasso, M., Beliaeva, T., Kraus, S., Clauss, T., & Ribeiro-Soriano, D. (2020). Circular economy business models: The state of research and avenues ahead. *Business Strategy and the Environment, 29*(8), 3006–3024. doi:10.1002/bse.2554

Freeman, R. E. (1984). *Strategic management: A stakeholder approach.* Boston: Pitman.

Gnoni, M. G., Mossa, G., Mummolo, G., Tornese, F., & Verriello, R. (2017). Supporting circular economy through use-based business models: The washing machines case. *Procedia CIRP, 64*, 49–54. doi:10.1016/j.procir.2017.03.018

Grant, R. M. (2006). *Contemporary strategy analysis.* Malden: Blackwell Publishing.

Guldmann, E., & Huulgaard, R. D. (2020). Barriers to circular business model innovation: A multiple-case study. *Journal of Cleaner Production, 243*, 118160. doi:10.1016/j.jclepro.2019.118160

Heyes, G., Sharmina, M., Mendoza, J. M. F., Gallego-Schmid, A., & Azapagic, A. (2018). Developing and implementing circular economy business models in service-oriented technology companies. *Journal of Cleaner Production, 177*, 621–632. doi:10.1016/j.jclepro.2017.12.168

Hofmann, F., & Jaeger-Erben, M. (2020). Organizational transition management of circular business model innovations. *Business Strategy and the Environment, 29*(6), 2770–2788. doi:10.1002/bse.2542

Johannsdottir, L. (2014). Transforming the linear insurance business model to a closed-loop insurance model: A case study of Nordic non-life insurers. *Journal of Cleaner Production, 83*, 341–355. doi:10.1016/j.jclepro.2014.07.010

Johnson, M. W., Christensen, C. C., & Kagermann, H. (2008). Reinventing your business model. *Harvard Business Review, 86*(12).

McDonough, W., & Braungart, M. (2002). Design for the triple top line: New tools for sustainable commerce. *Corporate Environmental Strategy, 9*(3), 251–258. doi:10.1016/S1066-7938(02)00069-6

Mostaghel, R., & Chirumalla, K. (2021). Role of customers in circular business models. *Journal of Business Research, 127*, 35–44. doi:10.1016/j.jbusres.2020.12.053

Nußholz, J. L. K., Nygaard Rasmussen, F., & Milios, L. (2019). Circular building materials: Carbon saving potential and the role of business model innovation and public policy. *Resources, Conservation and Recycling*, 308–316, Scopus. doi:10.1016/j. resconrec.2018.10.036

O'Riordan, L. (2017). *Managing sustainable stakeholder relationships: Corporate approaches to responsible management* (S. Idowu & R. Schmidpeter, Eds.). Cham: Springer Publishing.

O'Riordan, L., & Hampden-Turner, C. (2021). CSR in Germany. In S. Idowu (Ed.), *Current global practices of corporate social responsibility: In the era sustainable development goals*. Cham: Springer International Publishing.

Palmié, M., Boehm, J., Lekkas, C. K., Parida, V., Wincent, J., & Gassmann, O. (2021). Circular business model implementation: Design choices, orchestration strategies, and transition pathways for resource-sharing solutions. *Journal of Cleaner Production, 280*, 124399. doi:10.1016/j.jclepro.2020.124399

Ranta, V., Aarikka-Stenroos, L., & Väisänen, J. M. (2021). Digital technologies catalyzing business model innovation for circular economy – multiple case study. *Resources, Conservation and Recycling, 164*, 105155. doi:10.1016/j.resconrec.2020.105155

Rizos, V., Behrens, A., van der Gaast, W., Hofman, E., Ioannou, A., Kafyeke, T., . . . Topi, C. (2016). Implementation of circular economy business models by small and medium-sized enterprises (SMEs): Barriers and enablers. *Sustainability, 8*(11), 1212. doi:10.3390/su8111212

Rovanto, I. K., & Bask, A. (2021). Systemic circular business model application at the company, supply chain and society levels – a view into Circular economy native and adopter companies. *Business Strategy and the Environment, 30*(2), 1153–1173. doi:10.1002/bse.2677

Scipioni, S., Russ, M., & Niccolini, F. (2021). From barriers to enablers: The role of organizational learning in transitioning SMEs into the circular economy. *Sustainability, 13*(3), 1021. doi:10.3390/su13031021

Singh, P., & Giacosa, E. (2019). Cognitive biases of consumers as barriers in transition towards circular economy. *Management Decision, 57*(4), 921–936, Scopus. doi:10.1108/MD-08-2018-0951

Sousa-Zomer, T. T., Magalhães, L., Zancul, E., & Cauchick-Miguel, P. A. (2017). Lifecycle management of product-service systems: A preliminary investigation of a white goods manufacturer. *Procedia CIRP, 64*, 31–36. doi:10.1016/j.procir.2017.03.041

Sousa-Zomer, T. T., Magalhães, L., Zancul, E., & Cauchick-Miguel, P. A. (2018). Exploring the challenges for circular business implementation in manufacturing companies: An empirical investigation of a pay-per-use service provider. *Resources, Conservation and Recycling, 135*, 3–13, Scopus. doi:10.1016/j.resconrec.2017.10.033

Sumter, D., Bakker, C., & Balkenende, R. (2018). The role of product design in creating circular business models: A case study on the lease and refurbishment of baby strollers. *Sustainability, 10*(7), Scopus. doi:10.3390/su10072415

Tunn, V. S. C., van den Hende, E. A., Bocken, N. M. P., & Schoormans, J. P. L. (2020). Digitalised product-service systems: Effects on consumers' attitudes and experiences. *Resources, Conservation and Recycling, 162*, 105045. doi:10.1016/j.resconrec. 2020.105045

UN. (2015). About the sustainable development goals. Şustainable Development Goals. Retrieved from www.un.org/sustainabledevelopment/sustainable-development-goals/

Urbinati, A., Franzò, S., & Chiaroni, D. (2021). Enablers and barriers for circular business models: An empirical analysis in the Italian automotive industry. *Sustainable Production and Consumption, 27*, 551–566. doi:10.1016/j.spc.2021.01.022

van Loon, P., & Van Wassenhove, L. N. (2020). Transition to the circular economy: The story of four case companies. *International Journal of Production Research*, *58*(11), 3415–3422. doi:10.1080/00207543.2020.1748907

van Nes, E. H., Arani, B. M. S., Staal, A., van der Bolt, B., Flores, B. M., Bathiany, S., & Scheffer, M. (2016). What do you mean, "tipping point"? *Trends in Ecology & Evolution*, *31*(12), 902–904. doi:10.1016/j.tree.2016.09.011

World Commission on Environment and Development (Ed.). (1987). *Our common future*. Oxford: Oxford University Press.

# 8 Costs and benefits of transition towards a circular business model

*Urszula Balon, Anna Prusak, and Marek Jabłoński*

## Introduction

Based on the subject literature, the following stages of the product life cycle in the CE were distinguished: design, production, distribution, consumption, collection of raw materials, and recycling. This division of the product life cycle into phases is a matter of convention, and their length and course depend on the specifics of the sector and the development strategy, as well as on environmental and economic conditions. Individual phases of the life cycle differ in terms of the type and amount of both the costs incurred and revenue generated. This is why the cumulative costs and revenues arising at all phases of the product life cycle should be taken into consideration when assessing and estimating the profitability of the product life cycle. It is also worth considering it in relation to the benefits that organisations can achieve by making the transition to or operating in the Circular Economy.

The purpose of this chapter is to develop a cost–benefit model for companies making the transition to or operating in the CE. In order to develop the model, the costs and benefits that a company can encounter in the context of the CE were identified. The identified costs and benefits were assigned to individual stages of the product life cycle. On this basis, a cost–benefit model was developed, which can be used by companies operating in the CE, or planning the implementation of such solutions, to diagnose the situation and to make decisions concerning concentrating activities on specific areas of the CE. The developed model, together with the applied Analytic Hierarchy Process (AHP) method, was implemented in two Polish companies, allowing its correctness to be verified. The purpose of the developed model is to aid decision-making and assist in identifying areas of costs and benefits, which are to be considered to be of the highest and lowest priority when engaging in activities related to the CE. Based on the developed cost–benefit model, companies can estimate the priorities and rankings of individual activities which are elements of the developed model.

## Cost–benefit model for companies operating in the CE

In the developed model, each identified step of the product life cycle was first analysed from the point of view of the costs (Figure 8.1) incurred by the organisation while conducting its basic activities as well as activities oriented towards the CE.

DOI: 10.4324/9781003179788-8

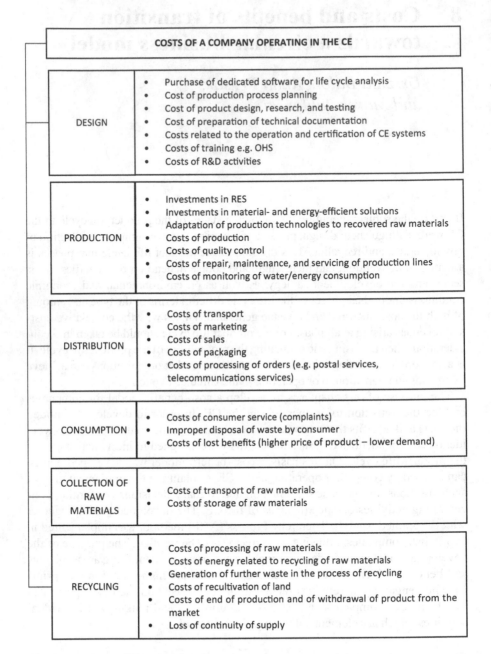

**COSTS OF A COMPANY OPERATING IN THE CE**

**DESIGN**
- Purchase of dedicated software for life cycle analysis
- Cost of production process planning
- Cost of product design, research, and testing
- Cost of preparation of technical documentation
- Costs related to the operation and certification of CE systems
- Costs of training e.g. OHS
- Costs of R&D activities

**PRODUCTION**
- Investments in RES
- Investments in material- and energy-efficient solutions
- Adaptation of production technologies to recovered raw materials
- Costs of production
- Costs of quality control
- Costs of repair, maintenance, and servicing of production lines
- Costs of monitoring of water/energy consumption

**DISTRIBUTION**
- Costs of transport
- Costs of marketing
- Costs of sales
- Costs of packaging
- Costs of processing of orders (e.g. postal services, telecommunications services)

**CONSUMPTION**
- Costs of consumer service (complaints)
- Improper disposal of waste by consumer
- Costs of lost benefits (higher price of product – lower demand)

**COLLECTION OF RAW MATERIALS**
- Costs of transport of raw materials
- Costs of storage of raw materials

**RECYCLING**
- Costs of processing of raw materials
- Costs of energy related to recycling of raw materials
- Generation of further waste in the process of recycling
- Costs of recultivation of land
- Costs of end of production and of withdrawal of product from the market
- Loss of continuity of supply

*Figure 8.1* Model of costs in the Circular Economy

The first phase – product design – is of key importance for the Circular Economy. It is at the product design stage that the product and its whole life cycle are shaped to be compliant with the concept and standards of the CE. At this stage, the following methods and tools for transition towards the CE may be used: resource savings, eco-design, prevention of product obsolescence, Life Cycle Assessment, and eco-labelling (Jaworski & Grochowska, 2017). There are costs of product design, research, and testing, as well as costs of production process planning and costs of preparation of technical documentation associated with these activities. Though these costs are incurred and declared at the design stage, the consequences of these decisions are visible in the production phase. Costs relating to future periods incurred in the pre-production phase should be deferred through accruals and amortisation (Ciechan-Kujawa & Sychta, 2018). In order to manage product life-cycle costs correctly, companies bear the costs of implementation of the requirements of standards and norms related to the CE. In order to certify the application of the requirements of these standards, companies can seek to obtain certification of these systems, which also has its associated costs. Costs are also incurred for the correct maintenance and operation of those systems. To aid with decision-making and life cycle analysis, companies may decide to purchase dedicated software (Dhillon, 2009).

The production stage is something that above all concerns the industrial sector, where resources are extracted and processed and waste is generated on a large scale. Particular attention must be paid to closing the loop in the production phase, however, especially in large companies, as it can be quite problematic. This applies, among other things, to the inefficient use of resources in production processes, which may lead to lost business opportunities and significant waste generation. Special attention must be paid to hazardous waste, chemicals, etc. (Communication EU, 2015). In this phase, it is important to apply the standards and good practices for each sector of industry to improve the efficiency of processes and introduce innovative solutions (Jaworski & Grochowska, 2017).

In the production phase, companies are mainly interested in management of the costs of consumption of materials and energy, costs of servicing of machines, as well as remuneration of production employees and costs of production. To a lesser degree, they are interested in management of the costs of quality control (Ciechan-Kujawa & Sychta, 2018). In order to adapt to CE standards, companies are investing in RES, as well as in material- and energy-efficient solutions. They should also adapt production technology and processes to make use of recovered raw materials (Elsayed, 2014).

The aim of activities in the distribution phase is to supply buyers with the products of the right quality, and in the right place at the right time, at the lowest possible cost. In the developed model, costs related to the preparation of the product for transport, as well as costs of packaging and of processing of orders from recipients, and costs of sales were distinguished in this phase. In many sectors, costs of marketing are one of the basic groups of expenses incurred by companies operating on the market. The aim of these activities is to increase sales and recognition

of the company, brand, and product. This group of costs includes costs of organisation of events, the handing out of free samples and gifts, advertising signs, and
sponsorship agreements (Wakabayashi, 2017; Ciechan-Kujawa & Sychta, 2018).

After the distribution phase, the product goes onto the market and to the customer, where the produced goods are then consumed. At this stage, the product
should be used in the most efficient way possible, and excessive consumption
should be limited, mainly by making consumers aware of ecological issues and
shaping their attitudes towards the environment (Jaworski & Grochowska, 2017).
At this stage, companies incur costs related to servicing, repairs under warranty,
as well as costs related to the loss of customers and the loss of benefits and revenues, as a result, among other things, of higher prices of products and lower
demand. An additional cost for the company is improper disposal of waste by the
consumer, which results in them not being able to be reused, or additional costs of
their disposal (Jansen, van Stijn, Gruis, & van Bortel, 2020; Ciechan-Kujawa &
Sychta, 2018).

To make rational use of used products and manage waste, companies incur
costs related to the collection of secondary raw materials as well as their transport
and storage. Thanks to such activities, organisations comply with the principles
in force in the CE and in the EU, namely, by preventing the generation of waste
and reusing used products (Gregson, Crang, Fuller, & Holmes, 2015; Winans,
Kendall, & Deng, 2017). Conscious companies and those operating in the CE
engage in activities in this direction already at the earlier stages of the product
life cycle and especially at the stage of design and preparation for production.
Another activity undertaken by organisations is recycling or, in other words, a
type of recovery, thanks to which waste, which has undergone prior processing,
can be put back into the loop, for example as a raw material in another cycle. If the
waste for various reasons cannot be recycled, it should undergo other processes
of recovery, which are, to a large extent, related to the recovery of energy from
waste, and these include processes of incineration and co-incineration of waste,
with a high level of energy recovery, and the processing of waste into solid, liquid,
or gaseous fuels. The least desirable form of waste management is its disposal,
as it is associated with wastage and does not fit with the concept of the Circular
Economy. Processes of disposal above all include the sending of waste to landfill
and the thermal treatment of waste without significant energy recovery (Jaworski & Grochowska, 2017). Where it is not possible to reuse products or their parts,
and where it is necessary to close the production line or plant, additional costs are
also incurred related to the necessity of ending production and of withdrawing the
product from the market, as well as the costs of recultivation of land. Wastage, as
well as the impossibility of making use of used and unnecessary products, leads
to a loss of continuity of supply of secondary raw materials, which may lead to
stoppages and even greater wastage of production input materials (Öner, Franssen, Kiesmuller, & Houtum, 2007; Ciechan-Kujawa & Sychta, 2018).

Implementation of the Circular Economy leads not only to costs, but also to
economic, social, and environmental benefits and to the construction of a system,
in which the economy, society, and the environment are coordinated in the area of

achievement of Sustainable Development goals. In the developed model, the benefits obtained by companies were also analysed according to stages of the product life cycle based on the principles of the CE (Figure 8.2).

Knowledge of the binding principles of the CE brings benefits already at the design stage. Activities undertaken with a view to closing the loop, as well as pro-environmental activities, contribute to a better understanding of environmental policy and thus to better management of the company and greater, informed engagement of employees. Another benefit for the company is the possibility of applying financial support in the form of subsidies and grants. There are many existing possibilities in the market, which have been created to increase investments in projects in accordance with CE principles (Öner et al., 2007).

In addition, company with the shared goal of applying CE principles makes new business contacts as part of cooperation, which reduces the risk of lack of continuity of supply of secondary raw materials for production.

At the production stage, organisations do incur costs related to the introduction of new solutions or to the improvement of existing ones, but, thanks to those activities, they achieve benefits in the form of reduced fees for the use of the eco-friendly materials and consumption of energy from external sources. They also avoid paying penalties for failure to comply with environmental regulations. Innovative solutions introduced at the production stage include material- and energy-efficient technologies, and solutions using renewable energy sources (RES), which contribute to a reduction in production costs, as well as a reduction in greenhouse gas emissions, are also employed. Another benefit thanks to the changes introduced is a reduction in the amount of waste and thus also in the costs related to their disposal.

In order to bring products to market and to the customer, companies may cooperate with each other, making new business contacts in the area of distribution. Due to the fact that the CE is, to a large extent, based on local markets, this leads to a shortening of the supply chain and a focussing of activities on local markets. This contributes not only to the development of the local economy, but also and most importantly from the company's point of view, to a reduction in costs of distribution.

CE-oriented activities are currently viewed in a positive light by the surrounding environment and society. Thanks to such solutions, and also to innovative projects, the company gains a positive corporate, brand, or product image, which may contribute to an increase in demand and also to gaining a competitive advantage. This results in an increase in sales, the winning of new markets, and thus of new group of customers too, who are interested in products offering new, improved solutions. The development of firms, along with the implementation of solutions, requirements, and standards dedicated to the CE, contributes to the extension of product lifetime. Companies offer their customers servicing and repair, which enable them to continue to make use of the product, while also generating positive relations with the surrounding environment, including by creating new jobs.

Companies operating in accordance with the concept of the Circular Economy may cooperate with other organisations on the basis of an industrial symbiosis.

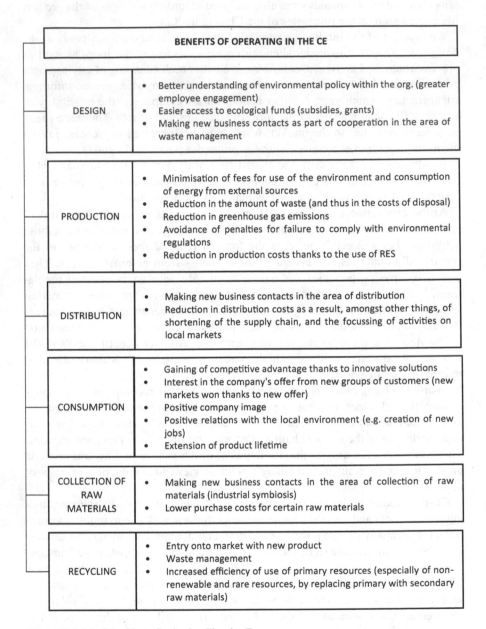

*Figure 8.2* Model of benefits in the Circular Economy

This is based on the physical exchange of materials, energy, water, by-products or the sharing of logistics resources. Such activities bring benefits to all participants in the symbiosis both in terms of its commercial and environmental aspects. Thanks to industrial symbiosis, companies make new business contacts in the area of collection of raw materials. The benefits at the raw materials collection phase also include lower purchase costs for certain raw materials.

The last stage included in the model is that of recycling, where the following benefits were identified: waste management, entry onto the market with a new product, as well as increased efficiency of use of primary resources, especially of non-renewable and rare resources, and replacing them with secondary raw materials.

Based on the costs and benefits identified, the developed model, in line with the author's assumptions, is of a general and universal nature. This means that it is constructed in such a way that it is able to be applied by all organisations regardless of the sector. To use the model, it is necessary to first adjust its elements to the company's own business, mode of organisation, and activities in the area of the CE and then conduct an analysis of the costs and benefits. Depending on the business activity conducted and the sector too, certain items may not apply. In that case, they should just be skipped and not used in the analysis.

## Use of Analytic Hierarchy Process (AHP) to analyse the costs and benefits of the CE for cleaning companies (CCs) and water and sewage companies (WSCs) (case study)

### Analytic Hierarchy Process (AHP)

The Analytic Hierarchy Process (AHP) is one of a group of methods for aiding decision-making, which use pairwise comparisons. Because this is a method which is very popular and readily used both in science and in industry, there are many references to the topic in the global literature (e.g. Kułakowski, 2021; Mu & Pereyra-Rojas, 2017; Brunelli, 2015). The creator of the AHP method, T.L. Saaty (e.g. Saaty & Vargas, 2012; Saaty, 2000, 1990) deserves a special mention in this respect. The AHP method usually consists of the following steps:

1  Development of the hierarchy model together with the questionnaire;
2  Making of judgements by means of pair wise comparisons using a ten-point scale of comparison (i.e. from "1" – the compared elements are equally important to "9" – one element has complete priority over the other);
3  Entering of the results of comparisons into the matrix and calculation of weighting coefficients (priorities), including local and global priorities;
4  Analysis of consistency of matrix based on consistency ratios (CRs) and a further analysis of the matrix for ratios of CR>0.10;
5  If the analysis is conducted by more than one person, aggregation of group results;
6  Analysis of sensitivity.

In these studies, an incomplete AHP model was used or, in other words, one which does not take decisional variants (alternatives) into consideration. The AHP analysis was thus used as a tool for the prioritisation of CE criteria. The resulting ranking may be used both to diagnose the situation and to make decisions concentrating activities on specific areas of the CE.

## Research methodology

General cost–benefit models, developed and discussed in the previous part of this chapter, had to be adapted to the specifics of the companies studied. The research study was conducted in two municipal companies of a metropolitan agglomeration in Poland. The first of these is a water and sewage company (referred to as *WSC*), and the other is a cleaning company (referred to as *CC*), which are local service providers. These municipal companies, as autonomous organisations owned by municipalities which do not play a role in local bureaucracy, have tariffs and commercial revenues and produce and deliver local public services (Voorn, van Genugten, & van Thiel, 2017, p. 820). They also respect following principles: accessibility, adaptability, conflict resolution, continuity, equality, participation, transparency, and universality (Marques, 2010). The specifics of the companies selected for the research strictly match the determining characteristics of the Circular Economy. These companies are responsible for collecting waste in metropolitan areas (i.e. refuse – CC and sewage – WSC), which they treat, clean, and transform for reuse by final recipients. Not all of the elements of the model defined before (Figures 8.1 and 8.2) were relevant for this type of company. For example in the case of the analysis of costs for the CC, such elements as the Recycling, the *End of production and withdrawal of products from the market,* and the *Loss of continuity of supply* criteria were not taken into consideration, while these elements were taken into consideration in the case of the WSC. The analysis was prepared by experts associated with these companies, who were representatives their management boards, and pair wise comparisons were made on the basis of a consensus for the individual degrees of preference in all of the comparison matrices or, in other words, aggregation of individual judgements (AIJ) was used. An analysis of the results was conducted using *Super Decisions v3.2* software. One very important element of this analysis was the determination of the consistency ratios (CRs). For the majority of the matrices, the CR was not greater than $0.10$, so they were considered to be consistent. Matrices for which the $CR>0.10$ were analysed again. Except for one matrix with n = 7 (which required 21 pair wise comparisons), for which, at the outset, the value of the consistency was $CR = 0.687$, for the others, the CR did not have a value in excess of $0.250$. A second analysis of comparisons allowed the CR for these matrices to be reduced to a value of between $0.06$ and $0.20$. Due to the large number of elements in each model, a value of $CR = 0.20$ was taken to be an acceptable level of inconsistency (cf. Prusak, 2017). In the process, it was noted that there was not a single case in which the order of ranking of the elements changed as a result of correction of the ratios.

*Analysis of costs and benefits for the CC*

First, the experts were asked to compare the importance of the overall costs and benefits with regard to the aspect of the CE (*What is more important from the point of view of analysis of stages of the CE – benefits or costs?*). Both for the CC, and for the WSC, the priority for costs was *.833 (83.3%)*, and for benefits, it was *.167 (16.7%)*. This shows that an analysis of costs is in any case of much greater importance than an analysis of benefits. The results of the costs model are presented in Table 8.1 and those of the benefits model are presented in Table 8.2.

With reference to the costs criteria, the most important ones proved to be *Production (.407)* and *Distribution (.379)*, while for the benefits criteria, they were *Distribution (.389)* and *Design (.304)*. It can thus be seen that, in the case of this company, *Distribution* is an important element both in the analysis of costs and in that of benefits. For the individual elements (sub-criteria), global priorities were calculated. These are weighting coefficients showing the importance of individual elements in the cost and benefit models (Figures 8.3 and 8.4) and combined together in both models (Figure 8.5). Global priorities of individual sub-criteria were obtained by multiplying their local priorities by the priorities of their overriding criteria. The global priorities for costs and benefits were however obtained by multiplying the global priorities of sub-criteria by the priorities of *costs (.833)* and *benefits (.167)*. The priorities on the graphs are shown as percentages. In the case of the benefits and costs model (Figures 8.1 and 8.2), only those elements are shown, for which the weighting coefficients were greater than *.050 (5%)*, while for the combined model, those elements are shown, for which the value of priorities was greater than .020 (2%). The remaining elements were considered to be of minor significance.

As can be seen, the most important element (with the highest global priority) in the analysis of costs is *Transport (.316)*, while in the analysis of costs, it is *Making new business contacts in the area of distribution (.324)*. Based on the results from Tables 8.1 and 8.2, the elements not shown on the graphs given before have very low priorities. The lowest priorities in the costs model (*.002*, equivalent to *0.2%*) are ranked *ex aequo, Product design, research and tests; Generation of further waste in the process of recycling; Consumption of energy related to recycling of raw materials* and *Operation; and certification of CE systems*. The lowest priorities in the benefits model were obtained for *Interest in the company's offer from new groups of customers* (.001, equivalent to 0.1%) and *Positive relations with the local environment (.002*, equivalent to *0.2%*). As for the combined share of each of the elements in the combined analysis of benefits and costs, this share is shown in a different way (Figure 8.5). Priorities for the cost elements are shown on the graph in blue, while benefits are shown in orange.

In this case, the most important element in the analysis of benefits, namely *Making new business contacts in the area of distribution*, obtained a share of only 5.4% (.540) in the overall analysis of benefits and costs.

*Table 8.1* Analysis of costs for the CC

| Criterion | Priority of criteria | Sub-criterion | Local sub-criterion | Global sub-criterion | Global for costs (.833) |
|---|---|---|---|---|---|
| Design | 0.052 | Production process planning | 0.519 | 0.027 | 0.022 |
| | | Product design, research, and tests | 0.029 | 0.002 | 0.001 |
| | | Preparation of technical documentation | 0.304 | 0.016 | 0.013 |
| | | Operation and certification of CE systems | 0.044 | 0.002 | 0.002 |
| | | Training, for example occupational health & safety | 0.104 | 0.005 | 0.005 |
| Production | 0.407 | Investments in RES | 0.119 | 0.048 | 0.040 |
| | | Investments in material- and energy-efficient solutions | 0.204 | 0.083 | 0.069 |
| | | Current expenditure on production processes | 0.362 | 0.147 | 0.123 |
| | | Quality control | 0.028 | 0.011 | 0.009 |
| | | Repair, maintenance, and servicing of production lines | 0.078 | 0.032 | 0.026 |
| | | Monitoring of water and energy consumption | 0.016 | 0.007 | 0.005 |
| | | Adaptation of production technologies to recovered raw materials | 0.193 | 0.079 | 0.065 |
| Distribution | 0.379 | Transport | 0.833 | 0.316 | 0.263 |
| | | Packaging | 0.167 | 0.063 | 0.053 |
| Collection of raw materials | 0.123 | Transport of raw materials | 0.009 | 0.111 | 0.092 |
| | | Storage of raw materials | 0.100 | 0.012 | 0.010 |
| Recycling | 0.039 | Processing of raw materials | 0.669 | 0.026 | 0.022 |
| | | Consumption of energy related to recycling of raw materials | 0.055 | 0.002 | 0.002 |
| | | Generation of further waste in the process of recycling | 0.055 | 0.002 | 0.002 |
| | | Recultivation of land | 0.220 | 0.009 | 0.007 |

*Table 8.2* Analysis of benefits for the CC

| Criterion | Priority of criteria | Sub-criterion | Local sub-criterion | Global sub-criterion | Global for benefits (.167) |
|---|---|---|---|---|---|
| Design | 0.304 | Better understanding of environmental policy | 0.105 | 0.032 | 0.005 |
| | | Easier access to ecological funds | 0.637 | 0.194 | 0.032 |
| | | Making business contacts | 0.258 | 0.078 | 0.013 |
| Production | 0.155 | Minimisation of fees for use of the environment | 0.549 | 0.085 | 0.014 |
| | | Reduction in the amount of waste | 0.132 | 0.020 | 0.003 |
| | | Reduction in greenhouse gas emissions | 0.033 | 0.005 | 0.001 |
| | | Avoidance of penalties for failure to comply with regulations | 0.079 | 0.012 | 0.002 |
| | | Reduction in production costs thanks to the use of RES | 0.207 | 0.032 | 0.005 |
| Distribution | 0.389 | Making new business contacts in the area of distribution | 0.833 | 0.324 | 0.054 |
| | | Reduction in distribution costs | 0.167 | 0.065 | 0.011 |
| Consumption | 0.026 | Gaining competitive advantage | 0.248 | 0.006 | 0.001 |
| | | Interest in the company's offer from new groups of customers | 0.045 | 0.001 | 0.000 |
| | | Positive company image | 0.626 | 0.016 | 0.003 |
| | | Positive relations with the local environment | 0.081 | 0.002 | 0.000 |
| Collection of raw materials | 0.045 | Making new business contacts in the area of collection of raw materials | 0.833 | 0.037 | 0.006 |
| | | Lower purchase costs for certain raw materials | 0.167 | 0.008 | 0.001 |
| Recycling | 0.080 | Entry onto market with new product | 0.279 | 0.022 | 0.004 |
| | | Waste management | 0.649 | 0.052 | 0.009 |
| | | Increased efficiency of use of primary resources | 0.072 | 0.006 | 0.001 |

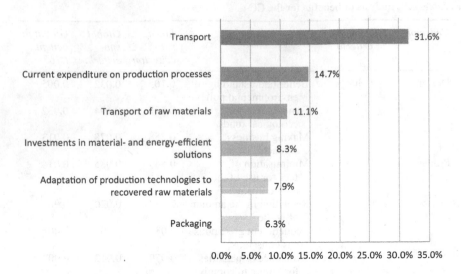

*Figure 8.3* Global priorities of sub-criteria in the costs model for the CC

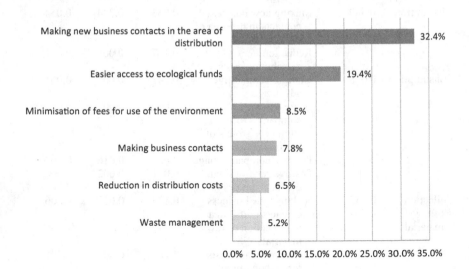

*Figure 8.4* Global priorities of sub-criteria in the benefits model for the CC

### Analysis of costs and benefits for the WSC

The analysis was conducted in a similar manner as for the CC. The results of the costs model are presented in Table 8.3 and those of the benefits model are presented in Table 8.4.

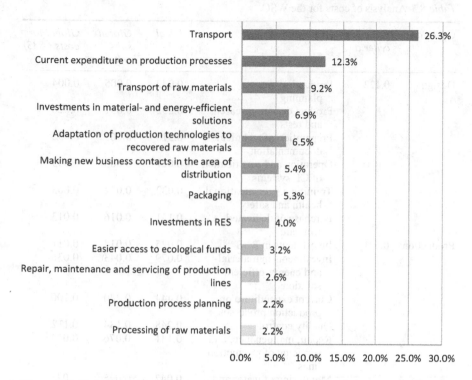

Transport — 26.3%
Current expenditure on production processes — 12.3%
Transport of raw materials — 9.2%
Investments in material- and energy-efficient solutions — 6.9%
Adaptation of production technologies to recovered raw materials — 6.5%
Making new business contacts in the area of distribution — 5.4%
Packaging — 5.3%
Investments in RES — 4.0%
Easier access to ecological funds — 3.2%
Repair, maintenance and servicing of production lines — 2.6%
Production process planning — 2.2%
Processing of raw materials — 2.2%

0.0%   5.0%   10.0%   15.0%   20.0%   25.0%   30.0%

*Figure 8.5* Global priorities of sub-criteria in the combined benefits and costs model for the CC

With reference to the costs criteria, the most important ones proved to be *Production* (*.541*) and *Distribution* (*.255*), while for the benefits criteria, they were *Distribution* (*.488*) and *Design* (*.238*). It can thus be seen that, like in the case of the WSC, for this company, *Distribution* is also an important element both in the analysis of costs and in that of benefits. For the individual elements (sub-criteria), global priorities were calculated, showing the share of individual priorities in the overall model of costs (Figure 8.6), benefits (Figure 8.7), and combined together in both models of benefits and costs (Figure 8.8).

The most important element (with the highest global priority) in the analysis of costs is *Current expenditure on production processes* (*.240*), while in the analysis of costs, it is *Making new business contacts in the area of distribution* (*.439*). The elements not shown on the graphs shown before have very low priorities. The lowest priorities in the costs model (*.001*, equivalent to *0.1%*) were obtained for and ranked *ex aequo, End of production and withdrawal of product from the market,* and *Recultivation of land*. The lowest priorities in the benefits model (*.002*, equivalent to *0.2%*) were obtained for *Entry onto market with new product* and *Extension of product lifetime*. As for the combined share of each of the elements

*Table 8.3* Analysis of costs for the WSC

| Criterion | Priority of criteria | Sub-criterion | Local sub-criterion | Global sub-criterion | Global for costs (.833) |
|---|---|---|---|---|---|
| Design | 0.122 | Production process planning | 0.042 | 0.005 | 0.004 |
| | | Product design, research, and tests | 0.252 | 0.031 | 0.026 |
| | | Preparation of technical documentation | 0.516 | 0.063 | 0.052 |
| | | Operation and certification of CE systems | 0.027 | 0.003 | 0.003 |
| | | Training e.g. occupational health and safety | 0.032 | 0.004 | 0.003 |
| | | Purchase of dedicated software | 0.131 | 0.016 | 0.013 |
| Production | 0.541 | Investments in RES | 0.025 | 0.014 | 0.011 |
| | | Investments in material- and energy-efficient solutions | 0.079 | 0.043 | 0.036 |
| | | Current expenditure on production processes | 0.444 | 0.240 | 0.200 |
| | | Quality control | 0.248 | 0.134 | 0.112 |
| | | Repair, maintenance, and servicing of production lines | 0.141 | 0.076 | 0.064 |
| | | Monitoring of water and energy consumption | 0.047 | 0.025 | 0.021 |
| | | Adaptation of production technologies to recovered raw materials | 0.017 | 0.009 | 0.008 |
| Distribution | 0.255 | Transport | 0.672 | 0.171 | 0.143 |
| | | Marketing | 0.037 | 0.009 | 0.008 |
| | | Sales | 0.167 | 0.043 | 0.035 |
| | | Processing of orders | 0.123 | 0.031 | 0.026 |
| Collection of raw materials | 0.052 | Transport of raw materials | 0.500 | 0.026 | 0.022 |
| | | Storage of raw materials | 0.500 | 0.026 | 0.022 |
| Recycling | 0.030 | Processing of raw materials | 0.283 | 0.008 | 0.007 |
| | | Consumption of energy related to recycling of raw materials | 0.142 | 0.004 | 0.004 |
| | | Generation of further waste in the process of recycling | 0.077 | 0.002 | 0.002 |
| | | Recultivation of land | 0.037 | 0.001 | 0.001 |
| | | End of production and withdrawal of product from the market | 0.025 | 0.001 | 0.001 |
| | | Loss of continuity of supply | 0.436 | 0.013 | 0.011 |

Table 8.4 Analysis of benefits for the WSC

| Criterion | Priority of criteria | Sub-criterion | Local sub-criterion | Global sub-criterion | Global for benefits (.167) |
|---|---|---|---|---|---|
| Design | 0.238 | Better understanding of environmental policy | 0.279 | 0.066 | 0.011 |
| | | Easier access to ecological funds | 0.072 | 0.017 | 0.003 |
| | | Making business contacts | 0.649 | 0.154 | 0.026 |
| Production | 0.133 | Minimisation of fees for use of the environment | 0.076 | 0.010 | 0.002 |
| | | Reduction in the amount of waste | 0.227 | 0.030 | 0.005 |
| | | Reduction in greenhouse gas emissions | 0.024 | 0.003 | 0.001 |
| | | Avoidance of penalties for failure to comply with regulations | 0.440 | 0.059 | 0.010 |
| | | Reduction in production costs thanks to the use of RES | 0.233 | 0.031 | 0.005 |
| Distribution | 0.488 | Making new business contacts in the area of distribution | 0.900 | 0.439 | 0.073 |
| | | Reduction in distribution costs | 0.100 | 0.049 | 0.008 |
| Consumption | 0.083 | Gaining competitive advantage | 0.544 | 0.045 | 0.008 |
| | | Interest in the company's offer from new groups of customers | 0.072 | 0.006 | 0.001 |
| | | Positive company image | 0.179 | 0.015 | 0.002 |
| | | Positive relations with the local environment | 0.179 | 0.015 | 0.002 |
| | | Extension of product lifetime | 0.026 | 0.002 | 0.000 |
| Collection of raw materials | 0.035 | Making new business contacts in the area of collection of raw materials | 0.500 | 0.018 | 0.003 |
| | | Lower purchase costs for certain raw materials | 0.500 | 0.018 | 0.003 |
| Recycling | 0.022 | Entry onto market with new product | 0.072 | 0.002 | 0.000 |
| | | Waste management | 0.279 | 0.006 | 0.001 |
| | | Increased efficiency of use of primary resources | 0.649 | 0.014 | 0.002 |

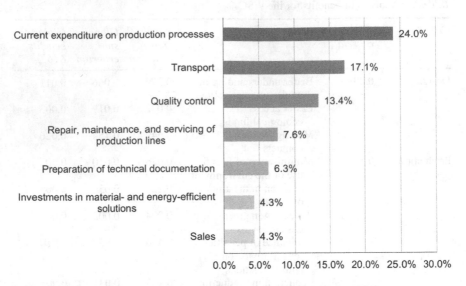

*Figure 8.6* Global priorities of sub-criteria in the costs model for the WSC

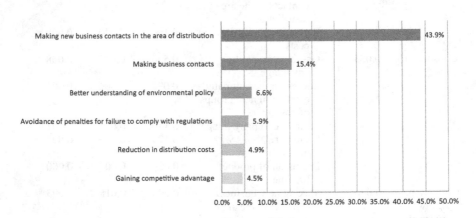

*Figure 8.7* Global priorities of sub-criteria in the benefits model for the WSC

in the combined analysis of benefits and costs, this share is shown in Figure 8.8. As given before, priorities for the cost elements are shown on the graph in blue, while benefits are shown in orange.

In the case of this company, the most important element in the analysis of benefits, namely *Making new business contacts in the area of distribution*, obtained a priority of 7.3% (.730) in the overall analysis of benefits and costs.

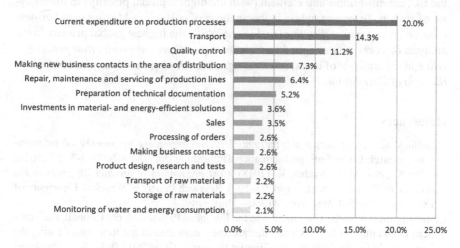

*Figure 8.8* Global priorities of sub-criteria in the combined benefits and costs model for the WSC

## Summary

Companies conducting activities oriented towards the Circular Economy not only incur costs, but also achieve benefits at the same time. One important aspect for companies to compare the costs incurred with the benefits that can be achieved once such activities have been introduced. To make such comparisons, a cost–benefit model was developed, thanks to which it is possible to estimate priorities and rankings of individual activities and also to compare the costs and benefits resulting from the implementation of, or operation in, the CE and determine which costs and benefits are most important for the company, and which of them are of the highest or lowest priority. The developed model is of a general and universal nature and may be used by various companies in different sectors.

To verify the correctness of the constructed model, it was tested in two Polish companies. The studies showed that the developed model, after slight adjustment of the costs and benefits to the business activity, structure, and legal and organisational form of the company, may be applied to determine the activities which are the most and least important for operation in the CE. These activities may vary depending on the company's sector and its level of pro-ecological awareness.

The studies conducted show that, for both companies, an analysis of costs is of greater importance than an analysis of benefits. In both companies, with reference to the costs criteria, the most important costs proved to be *Production* and *Distribution*, and for the benefits criteria, they were *Distribution* and *Design*. On this basis, it is possible to conclude that, for the companies studied, *Distribution* is an important element of the Circular Economy both in the analyses of costs and benefits. The priority activities for both companies were however different. For the CC, the most important element (with the highest global priority) in the analysis of costs is *Transport*, while in the analysis of costs, it is *Making new business contacts in the area of distribution*. For the WSC, the highest global priority in the analysis of costs was assigned to *Current expenditure on production processes*, while in the analysis of costs, it was assigned to *Making new business contacts in the area of distribution*.

## References

Brunelli, M. (2015). *Introduction to the analytic hierarchy process*. Springer Briefs in Operations Research. Cham: Springer International Publishing. doi:10.1007/978-3-319-12502-2

Ciechan-Kujawa, M., & Sychta, K. (2018). *Cost accounting of product life cycle in the practice of Polish enterprises* (pp. 95–107). Research Papers of Wrocław University of Economics, No. 514. Wrocław: Wrocław University.

Communication from the Commission to the European Parliament, the Council, the European Economic and Social Committee and the Committee of the Reg ions Closing the loop – An EU action plan for the Circular Economy, COM/2015/0614 final. Retrieved from https://eur-lex.europa.eu/legal-content/EN/TXT/?uri=CELEX%3A52015DC0614

Dhillon, B. S. (2009). *Life cycle costing for engineers*. London: CRC Press. doi:10.1201/9781439816899

Elsayed, E. A. (2014). *Life cycle costs and reliability engineering* (N. Balakrishnan, T. Colton, B. Everitt, W. Piegorsch, F. Ruggeri, & J. L. Teugels, Eds.). New York: Wiley. Statistics Reference Online. doi:10.1002/9781118445112.stat04150

Gregson, N., Crang, M., Fuller, S., & Holmes, H. (2015). Interrogating the circular economy: The moral economy of resource recovery in the EU. *Economy and Society, 44*(2), 218–243. doi:10.1080/03085147.2015.1013353

Jansen, B. W., van Stijn, A., Gruis, V., & van Bortel, G. (2020). A circular economy life cycle costing model (CE-LCC) for building components. *Resources, Conservation & Recycling, 161*, 104857. doi:10.1016/j.resconrec.2020.104857

Jaworski, T. J., & Grochowska, S. (2017). Circular economy – the criteria for achieving and the prospect of implementation in Poland. *Archives of Waste Management and Environmental Protection, 19*(4), 13–22.

Kułakowski, K. (2021). *Understanding the analytic hierarchy process*. Series in Operations Research. London: Chapman and Hall, CRC Press.

Marques, R. C. (2010). *Regulation of water and wastewater services. An international comparison*. London, UK: IWA Publishing.

Mu, E., Pereyra-Rojas, M. (2017). *Practical decision making [An introduction to the analytic hierarchy process (AHP) Using super decisions V2]*. Springer Briefs in Operations Research. Cham: Springer International Publishing. doi:10.1007/978-3-319-33861-3

Öner, K. B., Franssen, R., Kiesmuller, G. P., & Houtum, van, G. J. J. A. N. (2007). *Life cycle costs measurement of complex systems manufactured by an engineer-to-order company* (BETA Publication: Working Papers; Vol. 209). Eindhoven: Technische Universiteit Eindhoven.

Prusak, A. (2017). *Niespójność osądów w analitycznym procesie hierarchicznym*. Kraków: Wydawnictwo Uniwersytetu Ekonomicznego w Krakowie.

Saaty, T. L. (1990). *Decision making for leaders: The analytic hierarchy process for decisions in a complex world*. Pittsburgh: RWS Publications.

Saaty, T. L. (2000). *Fundamentals of decision making and priority theory with the analytic hierarchy process*. Pittsburgh: RWS Publications.

Saaty, T. L., & Vargas, L. G. (2012). *Models, methods, concepts & applications of the analytic hierarchy process*. International Series in Operations Research & Management Science. New York: Springer. doi:10.1007/978-1-4614-3597-6

Voorn, B., van Genugten, M. L., & van Thiel, S. (2017). The efficiency and effectiveness of municipally owned corporations: A systematic review. *Local Government Studies*, *43*(5), 820–841. https://doi.org/10.1080/03003930.2017.1319360.

Wakabayashi, Y., Peii, T., Tabata, T., & Saeki, T. (2017). Life cycle assessment and life cycle costs for pre-disaster waste management systems. *Waste Management*, *68*, 688–700. doi:10.1016/j.wasman.2017.06.014

Winans, K., Kendall, A., & Deng, H. (2017). The history and current applications of the circular economy concept. *Renewable Sustainable Energy Reviews*, *68*, 825–833. doi:10.1016/j.rser.2016.09.123

# 9 Utilisation of digitalisation in sustainable manufacturing and the Circular Economy

*Riccardo Beltramo, Enrica Vesce, and Stefano Duglio*

## Introduction

Since 1987, when the definition of Sustainable Development (World Commission on Environment and Development, 1987) was established, a debate has been ongoing among academics, policymakers, entrepreneurs, and representatives of public opinion in order to convert the theoretical concept into practical actions to be implemented by economic operators. In this context, the Circular Economy (CE) paradigm was defined as a model in which *"the value of products, materials and resources is maintained in the economy for as long as possible, and the generation of waste minimised"* (European Commission, 2015). This new circular production model is opposed to the linear one, which has been in force since the first industrial revolution and is to be held accountable for the main environmental concerns of today.

In parallel, giant leaps have been made in the area of digitalisation. The power of personal computers, mainframes, and Intranet networks has been augmented since 1992 by the Internet, which opened up new possibilities for global connectivity. The Open Source phenomenon has dramatically increased the speed of innovation, both in hardware and software. It has led to new business solutions, the development of which has been made possible due to favourable availability of data.

Today, we are in the midst of a new industrial revolution, known as Industry 4.0, boosted by the spread of the Internet of Things (IoT). Distributed networks microcomputers, linked to a wide range of low-cost sensors collecting data, allow companies to find their strategies on a wide knowledge base that is fed into artificial intelligence techniques to generate substantial added value. This makes it possible to trace the environmental burdens of production cycles and obtain up-to-date information in real time on the overall efficiency of economic organisations. Not only data on operations can be collected in digital form, but consumer perceptions of products and services can be also analysed using artificial vision techniques in order to gain a complete overview of the whole supply chain.

In the broadest sense, Industry 4.0, based on the digitalisation of processes, is at the service of sustainability. The CE leads to a search not only for ecological

DOI: 10.4324/9781003179788-9

advantages but also for economic and social ones, according to the logic of multidimensional sustainability. This chapter examines the fundamental role of digitalisation in boosting sustainability, considering that an essential condition for determining whether or not processes or products are sustainable is the availability of accurate datasets, enabling their environmental profile to be checked in real time, as well as over time.

Furthermore, we will analyse the tools available to companies which are attempting to apply sustainability principles and practices. This is an area which has benefited from considerable support from the European Commission and has been the subject of numerous research projects. Tools for application of the CE, which aim to make efficient use of resources, such as Environmental Management Systems and Ecolabels or other models for waste prevention and management such as eco-design, will therefore be analysed at the case study review level in order to discover profitable interactions with data collected by digital means.

As regards the product level and the Life Cycle Assessment (LCA) method, digital tools can be used to feed data into so-called "primary" databases (i.e. those containing information collected in the field), permitting an improvement in terms of data and accuracy of results. Similarly, digital information can be fed into the Initial Environmental Review of Environmental Management Systems at the process level.

## State of the art and conceptual definitions

When focussing on phenomena that are as complex and challenging as the ones introduced before, it is mandatory to define the right study boundaries for carrying out the analysis. Essentially, this consists of providing a clarification of the terms employed to illustrate some case studies relating to the digitalisation of sustainable production through the implementation of the CE paradigm.

An analysis of the state of the art and the breadth of studies conducted was not conducive to a structured literature review. The review of the scientific literature proved useful for definitions and for identification of CE applications that were facilitated by digitalisation.

In order to achieve the goal of the study, which is to highlight the importance of digital tools for achieving sustainability within companies based on the CE paradigm, three major topics are taken into consideration in this chapter: the CE, sustainable manufacturing, and digitalisation. There are multiple facets to each of these topics, and it is possible to approach them from numerous points of view: It is crucial thus to establish some definitions of system boundaries, before examining the interactions between them.

### *The Circular Economy*

At the European level, the guiding strategy for policies is clear: The CE action plans from 2015 (European Commission, 2015) and 2020 (European Commission,

2020) proposed practical actions first to establish a basis and then to explore some measures to stimulate the transition towards the CE. Nonetheless, at a scientific level, the concept of the CE still remains ambiguous.

In fact, the CE can be defined in different ways, as can Sustainable Development, which the CE helps to achieve (Schroeder, Anggraeni, & Weber, 2019).

Kirchherr, Reike, and Hekkert (2017) introduce 114 definitions of the CE, and Korhonen, Nuur, Feldmann, and Birkie (2018) arrived at the conclusion that the CE is an "essentially contested concept" (Bjørnbet, Skaar Ch. Fet, & Schulte, 2021).

According to Kirchherr et al. (2017), in this study, the term CE is understood to refer to a model in which the "end-of-life" concept is replaced by reducing, reusing, recycling, and recovering (4R) materials in production/distribution and consumption processes. In addition to this definition, it is necessary to bear the principles of application in mind *Preserve and enhance natural capital, Optimise resource yields,* and *Foster system effectiveness* (Ellen MacArthur Foundation, 2015) and to understand at which level ("macro", "meso", and "micro") the CE can be implemented. The "micro" level refers to products and companies, the "meso" to eco-industrial parks, and the "macro" to the city, regional and national level (Kirchherr et al., 2017; Ćwiklicki & Wojnarowska, 2020; Harris, Martin, & Diener, 2021; Saidani, Yannou, Leroy, Cluzel, & Kendall, 2019; Ghisellini, Cialani, & Ulgiati, 2016).

### *"Micro-level" for sustainable manufacturing*

This contribution highlights the use of digital tools to facilitate the achievement of CE objectives at the micro-level. As reported in Dopfer, Foster, and Potts (2004), the micro-perspective is one of a "complex structure of rules that constitute systems such as firms" (p. 267): in this case, a manufacturing firm that has to be sustainable. Due to the plurality of definitions of sustainability and due to the fact that it is not so clear how to apply the concept of sustainability at company level, this means that it is difficult to elucidate the notion of sustainable manufacturing too.

Moldavska and Welo (2017), who carried out a review on this topic as a starting point, stated that "Sustainable production is one of the Sustainable Development Goals (SDG) set by UN in 2015, which defines manufacturing as one of the measures toward Sustainable development".

Since sustainable manufacturing is considered to be key to achieving Sustainable Development (Chen & Zhang, 2009; Loglisci, Priarone, & Settineri, 2013), as the CE is being considered too, it was decided to include both in the discussion in this chapter.

### *Digitalisation*

At the end, digitalisation has to be understood from different perspectives too.

Although it has been widely discussed by scholars, there is still a lack of knowledge on how technologies have an impact the everyday activity of industrial businesses (Ranta, Aarikka-Stenroos, & Väisänen, 2021). Starting from a company viewpoint, in defining how digitalisation may help to introduce circularity into companies' business activities, this chapter will assume a more practical definition of digitalisation. According to Gartner (2021), digitalisation *"is the use of digital technologies to change a business model and provide new revenue and value-producing opportunities"* or, in other words, *"the process of moving to a digital business"*, specifically focussing on business model challenges. Applying digitalisation in the field of CE may thus enable the transition from linear business models to circular ones. Moving on from the general concept to practical tools, IoT and Big Data Analytics are often referred to as the tools that are the most adaptable and adoptable by firms which are aiming to achieve circular models (Kristoffersen et al., 2020; Ćwiklicki & Wojnarowska, 2020). These tools provide opportunities for companies to experiment with new business models (Baines et al., 2017) and, digitalisation being the basis of the new Industry 4.0 paradigm, the aforementioned digital tools may become a condition *sine qua non* for companies' survival (Canestrino, Ćwiklicki, Kafel, Wojnarowska, & Magliocca, 2020).

Considering that the literature has mainly been focussed on strategies rather than on case studies of actual and more practical implementation (Ingemarsdotter, Jamsin, & Balkenende, 2019), we have decided to dispense with a brainstorming on CE and digitalisation tools in order to present case studies of manufacturing companies in which these tools have been adopted. Before introducing the case studies, however, it may be useful for the reader to be provided with some context on the scope of the analysis in order to allow for a better understanding of the different topics and content.

## Conceptual framework

After the general definitions, it is useful to define the conceptual framework in which Sustainable Development, the CE, Sustainable Manufacturing (SM), and digitalisation are connected to each other at different levels. As shown in Figure 9.1, the first step was to identify the links between concepts at the micro-level, while the boundaries of the focus of this chapter are defined in the second step: in particular, the discussion regarding the relationship between digitalisation and the CE, which was accompanied by an analysis of the case studies.

## Sustainable Development and Circular Economy tools at the micro-level

Although great efforts have been made to clarify the points of contact between sustainability and CE, the link between the two is still rather confusing (Geissdoerfer, Savaget, Bocken, & Hultink, 2017).

Comparing the two concepts reveals some differences: Sustainability is seen as an approach that integrates environmental, social, and economic aspects; the CE leads to some benefits mainly for the economic actors implementing it, along with some environmental benefits and social benefits that are only implicit gains

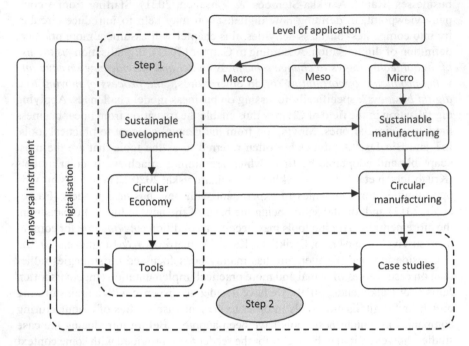

*Figure 9.1* Conceptual framework and steps

Source: Authors' elaboration.

(Geissdoerfer et al., 2017). Sustainability concepts have been developed according to a broader vision based on the general objectives to be pursued, while the CE aims to close loops and save resources: The latter thus seems to be more easily applied at the company level.

Sustainability seems to be developed at the macro-level with a top-down approach: It was first established at the international level (Brundtland Report) and only later applied at other levels. However, the first experiences of the CE implemented spontaneously at the micro-level (based on 4R strategies) made it possible to define this approach at the higher levels.

Based the aforementioned considerations, it can be observed that:

- Many authors point to the CE as the basis of SM (Geissdoerfer para. 4.2);
- Other authors, less numerous, do not agree with this point of view;
- There is still no consensus on SM (Moldavska & Welo, 2017); and
- Most importantly, the focus of the CE is on economic actors.

We thus decided to evaluate the tools for application recognised as being useful for the implementation of the CE at the company level including the SM point of view.

Moreover, the CE mainly leads to improvements in the environmental field and not to an increase in sustainability in the broader sense (Geissdoerfer et al., 2017).

According to the authors, to examine the CE at the micro-level, a distinction needs to be made. We would need to identify the strategies that, based on concrete action plans, can allow us to achieve the CE goals we have set and to improve the 4R policy. For example to increase the percentage of recycled products as input, manufacturers have to be certain of the technical usability of the new raw material. Or they will need to verify the availability of that new recycled raw material in the marketplace within an acceptable distance.

The latter is the real area of application of CE tools at the micro-level, and, being very different from each other, they need to be integrated with other business governance variables. In this sense, management systems are very useful, as they can monitor the steps towards the implementation of the CE in the company from the initial analysis towards the expected improvements.

In reality, today, there is a lot more discussion in academic circles about how best to measure the degree of achievement of circularity objectives than about concrete tools in the area of the CE.

Many researchers are working in this area by defining dedicated sets of indicators that represent the second aspect and that we can define as CE "tools".

An indicator measures one aspect of a more complex phenomenon and is useful to monitor the transition from an initial state to a better one or to compare two different realities for benchmarking purposes. Thus, indicators are not an application tool in the strict sense. Application tools and indicators become part of the framework that enables producers to become circular and, thus, sustainable.

When referring to sustainability, the EMSs – Environmental Management Systems – are normally considered to be a suitable tool for integrating ecological sustainability with managerial operations. As widely known, there are two main schemes for adopting an EMS at the company level: the ISO 14001 Standard, defined by the International Organization for Standardization, and the EMAS Regulation (Eco-Management and Audit Scheme), issued by the European Union.

Both of them are based on the Deming Cycle approach, recognising four main interconnected phases (Plan–Do–Check–Act), under the umbrella of the continual improvement of the company's environmental profile (Salomone, 2008). Even if the two schemes have similarities that make it possible for enterprises to adopt both of them simultaneously, there are several fundamental distinctions between the EMAS and ISO 14001, especially with regard to external communication. In the EMAS, a dedicated document is in fact required, that is the Environmental Declaration (Commission Regulation, 2017). This is the main reason why the motivations for adopting the two schemes may differ in terms of external forces or internal drivers (Neugebauer, 2012) as it is related to the willingness to share the company's environmental profile and the tools for assessing environmental performance with external stakeholders.

Although EMSs can be implemented in any kind of organisation, they were drafted to be adopted *in primis* by manufacturing industries (Beltramo, Duglio, Peira, & Gerbino, 2014), and, according to Hall and Wagner (2012), it is possible to identify a positive correlation in the integration of strategic decisions and the environmental management both in terms of product and process innovation.

This last consideration seems to be supported by the 2015 edition of the ISO 14001 in which some challenging concepts were introduced (ISO, 2015). In particular, the 2015 edition contains three structural innovations, that is the organisational context, the life-cycle perspective, and the risk-based perspective (Beltramo, Duglio, & Paolo, 2016). In particular, the life-cycle perspective creates a direct link with the LCA methodology, as will be presented later, and the risk-based perspective plays a fundamental role in considering the environmental component as a strategic driver. Indeed, not only has the risk management concept to be assessed in relation to the environment, but also in terms of the organisational risks, listed as business continuity, corporate image/reputation, asset integrity, legal liability, market response, and financial balance (Assolombarda, Università Commerciale Luigi Bocconi – IEFE, 2015).

In the light of this new paradigm, IoT and digitalisation can be useful as support tools in guaranteeing input of real-time environmental data for implementing risk-based management, providing a more holistic view of the relationships between companies and the environment.

In terms of indicators that not only measure the CE but also help with its implementation, we should also consider the LCA perspective.

This perspective is cited as being one of the tools that make it possible to achieve SM (Gundes, 2016), even if this is rarely included in its definition (Moldavska & Welo, 2017). Similarly, application to the CE is based on the same logic, since reuse, reduce, and recycling are concepts which are strongly connected to lengthening of the life cycle.

Understandably, impact assessment across the product life cycle is useful for all approaches that lead to the closure of cycles, as well as regeneration and design strategies for sustainability purposes. Besides, companies are indeed able to influence the entire life cycle of the product with their choices from design to end of life (Bjørnbet et al., 2021).

Life cycle sustainability assessment (LCSA) is a method that considers all the sustainability aspects of the product from cradle to grave throughout the life cycle (Gundes, 2016).

The most well-known methods are LCA which highlights the environmental impacts of the product; Life Cycle Costing (LCC), which refers to costs; and Social Life Cycle Assessment (SLCA) concerning social aspects.

## Digitalisation and sustainable manufacturing

As mentioned before, the digitalisation tools which are most often cited are to be found in the Internet of Things (IoT) and Big Data Analytics. As digitalisation is defined as *"the use of digital technologies to change a business model"* (Gartner, 2021), it can encompass a very wide range of digital technologies. When referring to the digitalisation tools for managing the company's environmental variables, however, there is a lack of a complete and integrated framework for digitalisation, as pointed out by Canestrino et al. (2020). In their study, which explores the adoption of digitalisation tools within Polish and Italian companies with an EMS, the authors identify four different groups of technological tools and usages: i) remote access to utilities; ii) releasing environmental data for external parties; iii) automated environmental monitoring; and iv) digital documents, mainly for internal uses and processes.

IoT can therefore play an important role in managing environmental data by providing support for real-time monitoring. Hardware technologies can be linked to the cloud to create a network for collecting environmental data in dedicated databases (Beltramo et al., 2016). Data can be read in real time, help managers in abnormal or emergency situations, and can be used to assess the company's environmental profile over time. The same data can be shared with external stakeholders in accordance with the company's policy.

With Big Data, information is more precise, and managers have data at their disposal that is more easily available and allows more accurate evaluations, yielding better results, as it was confirmed by recent research (Şerban Radu-Alexandru, 2017).

## Case studies: digitalisation of manufacturing in the transition towards the CE

To support the overview provided and to offer a practical clarification of the numerous interdependencies in manufacturing between the CE and digitalisation, some case studies will be presented later. Digitisation is closely linked to manufacturing, and, although the results of its implementation in production are, in practice, positive overall, there have also been some criticisms and difficulties. Regarding the interdependence between the manufacturing sector and sustainability, it is to be noted that, among the sectors that are most responsible for pollution and resource consumption, manufacturing plays an important role. To solve these environmental problems, the adoption of CE principles seems to be a correct solution because it aims both to reduce consumption of resources and to keep materials in the production cycle for longer (Acerbi & Taisch, 2020).

The link between digitisation and sustainability is a supportive one: Digital tools create many opportunities to foster sustainability (such as the allocation of resources and utilities) (Stock & Seliger, 2016).

Many theoretical aspects on the different models for applying the CE to manufacturing have been developed over the years. There have been fewer cases of such application and fewer suggestions regarding strategies as to how implement them successfully. Digitalisation has certainly been a useful transversal phenomenon to increase the applications of the CE (Hoosain, Paul, & Ramakrishna, 2020).

An analysis of the literature was conducted to recover case studies at micro-level, which have adopted digital techniques to move towards the CE.

Using "digitalisation", "circular economy", and "case study" as keywords, seven documents were found on Scopus and nine on the Web of Science. Given that four of these results were common and some were discarded for references to the non-micro-level, the possible examples for deriving indications were considerably reduced.

Table 9.1 presents an overview of the indications obtained from a review of selected articles with results that are of interest for our purposes.

Furthermore, we have decided to add a new case because of its specific reference to the advantages derived from digitalisation in large companies. The study by Hoosain et al. (2020) focussed on circular thinking and the impact of digital technologies on the achievement of Sustainable Development Goals, highlighting some large companies, which have started to implement CE paths and have made use of digital tools. For example:

- Apple Inc., which, by using a robot that disassembles the iPhone, recovers valuable materials for reuse;
- Michelin (Effifuel system), which, by using sensors, provides indications for more sustainable driving;
- EON (New York start-up) that makes the fashion world more sustainable through digital technologies by means of a Circular ID digital identity (useful for maintaining product information) (Hoosain, Paul, & Ramakrishna, 2020).

## Case study: digitalisation for implementation of the Circular Economy at the micro-level: LCA and chemical industry

In general, the operations of a manufacturing company are based on extensive data collection: The production process draws data upon this database and returns information at every moment in time. Digitalisation provides better quality data by recording changes in real time, which facilitates timely management (de Sousa et al., 2018). Even more so in recent years, the ability to manage and control strategic information concerning the company has taken on increasingly fundamental importance.

As anticipated, the transition towards the CE means adopting a life-cycle perspective from the first stage of product design in order to provide for correct end-of-life management and production that impact as little as possible on the surrounding environment.

In this context, the LCA tool plays an integral part in the environmental assessment of the life cycle of the product. This tool allows us to know which phases of a production cycle have the most impact in a delimited product system in relation to a functional unit. This tool allows us to determine the complete and objective impact, only for that specific product represented by the functional unit, throughout the entire supply chain from the cradle to the grave and, therefore, not to just shift pollution from one phase to another of the process or from one sector to another.

Since it is not possible to have all the specific data at our disposal relating to the impacts of each element to be analysed, we rely on databases that use what is referred to as the secondary data. This makes it possible to work on specific issues (how much pollution is caused by the steel used for the construction of the

Table 9.1 Examples of support for digitalisation in the CE

| Authors | Title | Year | Focus | Case studies |
|---|---|---|---|---|
| Ingemarsdotter, Kambanou, Jamsin, Sakao, and Balkenende | Challenges and solutions in condition-based maintenance implementation – A multiple case study | 2021 | Maintenance | 3 case studies at plant manufacturers (forklifts, industrial robots, heat pumps) in relation to challenges and solutions in the implementation of digital technologies |
| Ranta, Aarikka-Stenroos, and Väisänen | Digital technologies catalyzing business model innovation for circular economy – Multiple case study | 2021 | Digital solution for the CE | 4 case studies: Tools/construction, machinery, oil refinery, waste management (public enterprise). Technologies to improve information on the efficient use of resources, on the lengthening of the life cycle, and on the closure of flows. |
| Ingemarsdotter, Jamsin, and Balkenende | Opportunities and challenges in IoT-enabled circular business model implementation – A case study | 2020 | IoT in the CE | Case study of a lighting company (LED). Reference to the support that IoT can provide to the CE. |
| Kovacic, Honic, and Sreckovic | Digital platform for circular economy in AEC industry | 2020 | Digital technologies | Case study: use of digital technologies in the construction industry |
| Rocca, Rosa, Sassanelli, Fumagalli, and Terzi | Integrating virtual reality and digital twin in circular economy practices: A laboratory application case | 2020 | Support for digitalisation in the CE | Application of Industry 4.0 to a CE laboratory case reports, some interesting examples of IoT technologies applied to CE strategies (remanufacturing, maintenance, and disassembly) in companies. |
| Rossi, Bianchini, and Guarnieri | Circular economy model enhanced by intelligent assets from Industry 4.0: The proposition of an innovative tool to analyse case studies | 2020 | Intelligent assets for the CE | Review with case studies in different sectors about intelligent assets in the CE. |
| Ingemarsdotter, Ella, Gerd, and Balkenende | Circular strategies enabled by the Internet of Things – A framework and analysis of current practice | 2019 | IoT, circular models | Review. 40 cases of application of IoTs in the CE and, for each case, which IoT capacity has been exploited and which circular models. |

plant that produces the product we are analysing?), but at the same time affects the accuracy of the result. The more secondary data is used, the more approximate the resulting impact will be. However, a study of this type requires so much data that some initial assumptions have to be made.

The solution may be to make the acquisition of primary data automatic by using a sensor system (IoT) which, in this way, will capture useful data which can be fed into databases for making decisions in real time.

The reported case study concerns data on electricity consumption.

This variable was chosen because, as can be seen from various reports by the International Energy Agency, industry is still the main consumer of electricity, both in advanced and developing economies. Furthermore, these are important indications for the monitoring of sensitive variables for the achievement of the CE at the company level. This research has highlighted the importance of direct measurement above all to feed data into predictive models in the face of a general scarcity of primary data. In this case, reference was made to the feeding of an LCA study with data from sensors for two reasons: Not only to understand how much the secondary data can differ from the real data in general, but also to evaluate the differences in environmental impact that the LCA system returns to us depending on whether the inventory is set up with primary or secondary data. In fact, in the analysis carried out by Ingrao et al. (2021) in one case, the data was captured by sensors, while in another, it was calculated with a mathematical model based on licence plate data (less specific data), and then finally was calculated with a mathematical model based on licence plate data plus other data provided by the company.

To improve the acquisition of data relating to energy consumption, the application of sensors for the acquisition of primary data to a plant (chemical mixer) in a chemical product for the tannery industry was studied.

The study envisaged three measurement methods: The first obtained through the use of a theoretical model based on the characteristics of the machine, the second supplemented by more information derived from the type of product processed, and the third carried out by means of direct measurement. The results showed a large difference in results with an overestimation of 74% in the case of the first method compared to the third.

This is the first fundamental aspect to point out: Where the data was essential for making management decisions in real time, the importance of direct detection is immediately highlighted.

Furthermore, if this data is used to be fed into LCA assessment systems, the resulting analysis worsens the environmental picture, since the indirect measurement data are overestimated (Ingrao et al., 2021).

### Case study: digitalisation for implementing an Environmental Management System

Regardless of the implemented scheme (ISO 14001 or EMAS), an Environmental Management System needs core environmental data.

Environmental data is necessary to draft the Context Analysis, that is the initial document that constitutes the starting point for a company in the move towards a formalised management system, as well as for making it possible for the system to improve the company's environmental profile. Such data serves two main purposes: First, it can be used as a variable for defining the severity associated with an environmental aspect thanks to a dedicated procedure (Zobel et al., 2002; Darbra et al., 2005; Beltramo et al., 2014), and, most importantly, it is used for calculating both core and additional environmental indicators (Commission Regulation, 2017). These indicators aim to provide evidence of achievement of the company's objectives and targets and, ultimately, of the validity of the company's policy on environmental issues as a whole.

To do this, companies normally rely on registers (physical documents) that contain data on different environmental impacts, such as waste production and consumption of energy and water, depending on their business model and operations. Data collection and analysis, however, can be costly, time-consuming and work-intensive, and finally not completely error-free if they are done "by hand". Furthermore, this can increase the level of bureaucracy in the company.

In this context, digitalisation helps in two ways. The first level at which it helps is that of the digitalisation of EMS documents (Canestrino et al., 2020). In this case, digitalisation helps companies to decrease bureaucracy, store up-to-date data, and improve the effectiveness and efficiency of document management. Data input may still, however, remain a time-consuming process and one that is not error-free. A second and deeper level at which digitalisation can help is by creating a remote sensing network based on sensors that are able to automatically fill out the environmental registers from dedicated databases (Beltramo et al., 2016; Beltramo et al., 2018). Such a tool avoids the negative consequences associated with the manual completion of environmental data registers because it allows for real-time monitoring, enabling data to be collected directly to a server, where it is then stored. This data can then be fed into dedicated environmental registers to provide a snapshot of the company's environmental profile in real time and/or over time and to calculate both core and additional indicators.

## Conclusion

Digitalisation is an essential part of current developments in technology as it provides a pathway to sustainable production, understood both as sustainable in the narrow sense and as circular production.

Among the various tools that allow digitalisation at the micro-level, IoT and Big Data certainly have a particular role to play, precisely because of their capacity to collect and systematise data in real time at company level.

The latest trends in LCA and EMS represent this need very well: as regards LCA, there is an increasing amount of interest in dynamic LCA which could be useful to feed data into analyses, allowing them to precisely capture variations in flows over time. Similarly, the need to keep certain variables under control leads us to discuss the possibility of dashboards which, once again in real time, allow us to keep track of changes in the magnitudes of primary LCT data fed into databases or to monitor environmental analyses that move from initial to subsequent steps, with a view to continuous improvement.

Furthermore, an LCA path evaluates the impacts in absolute terms, that is it compares the environmental impacts of the functional unit considered with the load capacity of a system. In contrast to the context in which LCA was born, this innovative approach is even more based on a need for data, which, in terms of quantity and quality, can only be managed by sensor systems.

In general terms, the need to improve a company's performance, including that of sustainability, is based on the acquisition of different types of data to be managed in an integrated way. A general recommendation for practitioners is to adopt a holistic approach when implementing IoT tools in order to foster the synergic value which can be derived from different datasets that are able to provide an integrated picture of management of the company.

# References

Acerbi, F., & Taisch, M. (2020). A literature review on circular economy adoption in the manufacturing sector. *Journal of Cleaner Production, 273*, 123086. doi:10.1016/j.jclepro. 2020.123086

Assolombarda, Università Commerciale Luigi Bocconi – IEFE. (2015). *ISO 14001:2015. Le novità della norma e le linee guida per l'applicazione dei nuovi requisiti*. Milano: Assolombarda.

Baines, T., Bigdeli, A. Z., Bustinza, O. F., Shi, V. G., Baldwin, J., & Ridgway, K. (2017). Servitization: Revisiting the state-of-the-art and research priorities. *International Journal of Operations & Production Management, 37*(2), 256–278. doi:10.1108/IJOPM-06-2015-0312

Beltramo, R., Cantore, P., Vesce, E., Margarita, S., & De Bernardi, P. (2018). The internet of things for natural risk management (Inte.Ri.M.). *Perspectives on Risk, Assessment and Management Paradigms*. doi:10.5772/intechopen.81707

Beltramo, R., Duglio, S., & Paolo, C. (2016). SCATOL8®: A remote sensing network for risk assessment in the environmental management system. *Quality – Access to Success, 17*.

Beltramo, R., Duglio, S., Peira, G., & Gerbino, L. (2014). The environmental management system: A vector for the territorial development. The experience of the town of Giaveno (Italy). In T. Sikora & J. Dziadkowiec (Eds), *Commodity science in research and practice – towards quality – management systems and solutions* (pp. 19–29). Cracow: Polish Society of Commodity Science.

Bjørnbet, M. M., Skaar Ch. Fet, A. M., & Schulte, K. Ø. (2021). Circular economy in manufacturing companies: A review of case study literature. *Journal of Cleaner Production, 294*, 126268. doi:10.1016/j.jclepro.2021.126268

Canestrino, R., Ćwiklicki, M., Kafel, P., Wojnarowska, M., & Magliocca, P. (2020). The digitalization in EMAS-registered organizations: Evidences from Italy and Poland. *The TQM Journal, 32*(4), 673–695. doi:10.1108/TQM-12-2019-0301

Chen, M., & Zhang, F. (2009). End-of-life vehicle recovery in China: Consideration and innovation following the EU ELV directive. *JOM, 61*(3), 45–52. doi:10.1007/s11837-009-0040-8

Commission Regulation (EU) 2017/ 1505 – of 28 August 2017 – Amending Annexes I, II and III to Regulation (EC) No 1221 / 2009 of the European Parliament and of the Council on the Voluntary Participation by Organisations in a Community Eco-Management and Audit Scheme (EMAS)". s.d. 20. Retrieved from https://eur-lex.europa.eu/legal-content/EN/TXT/?uri=uriserv:OJ.L_.2017.222.01.0001.01.ENG

Ćwiklicki, M., & Wojnarowska, M. (2020). Circular economy and industry 4.0: One-way or two-way relationships? *Engineering Economics, 31*(4), 387–397. doi:10.5755/j01.ee.31.4.24565

Darbra, R. M., Ronza, A., Stojanovic, T. A., Wooldridge, C., & Casal, J. (2005). A procedure for identifying significant environmental aspects in sea ports. *Marine Pol lution Bulletin, 50*(8), 866–874. doi:10.1016/j.marpolbul.2005.04.037

Dopfer, K., Foster, J., & Potts, J. (2004). Micro-meso-macro. *Journal of Evolutionary Economics, 14*(3), 263–279. doi:10.1007/s00191-004-0193-0

European Commission. (2015). *Communication from the commission to the European Parliament, the Council, the European economic and social committee and the committee of the regions closing the loop – an eu action plan for the circular economy.* Retrieved from https://eur-lex.europa.eu/legal-content/EN/TXT/HTML/?uri=CELEX:52015DC0614&from=ES

European Commission. (2020). *Communication from the commission to the European Parliament, the Cou ncil, the European economic and social committee and the committee of the regions. Circular economy action plan.* Retrieved from https://ec.europa.eu/environment/strategy/circular-economy-action-plan_it

Gartner. (2021). Digitalization. (s.d.). *Gartner glossary.* Retrieved from www.gartner.com/en/information-technology/glossary/digitalization

Geissdoerfer, M., Savaget, P., Bocken, N., & Hultink, E. (2017). The circular economy – a new sustainability paradigm? *Journal of Cleaner Production, 143*, 757–768. doi:10.1016/j.jclepro.2016.12.048

Ghisellini, P., Cialani, C., & Ulgiati, S. (2016). A review on circular economy: The expected transition to a balanced interplay of environmental and economic systems. *Journal of Cleaner Production, Towards Post Fossil Carbon Societies: Regenerative and Preventative Eco-Industrial Development, 114*, 11–32. doi:10.1016/j.jclepro.2015.09.007

Gundes, S. (2016). The use of life cycle techniques in the assessment of sustainability. *Procedia – Social and Behavioral Sciences, Urban Planning and Architectural Design for Sustainable Development (UPADSD), 216*, 916–922. doi:10.1016/j.sbspro.2015.12.088

Harris, S., Martin, M., & Diener, D. (2021). Circularity for circularity's sake? Scoping review of Assessment methods for environmental performance in the circular economy. *Sustainable Production and Consumption, 26*, 172–186. doi:10.1016/j.spc.2020.09.018

Hoosain, M. S., Paul, B. S., & Ramakrishna, S. (2020). The impact of 4IR digital technologies and circular thinking on the united nations sustainable development goals. *Sustainability, 12*(23), 10143. doi:10.3390/su122310143

Ingemarsdotter, E., Jamsin, E., & Balkenende, R. (2020). Opportunities and challenges in IoT-enabled circular business model implementation – a case study. *Resources, Conservation and Recycling, 162*, 105047. doi:10.1016/j.resconrec.2020.105047

Ingemarsdotter, E., Jamsin, E., Kortuem, G., & Balkenende, R. (2019). Circular strategies enabled by the internet of things – a framework and analysis of current practice. *Sustainability, 11*(20), 5689. doi:10.3390/su11205689

Ingemarsdotter, E., Kambanou, M. L., Jamsin, E., Sakao, T., & Balkenende, R. (2021). Challenges and solutions in condition-based maintenance implementation – a multiple case study. *Journal of Cleaner Production, 296*, 126420. doi:10.1016/j.jclepro.2021.126420

Ingrao, C., Stella, E. R., Paolo, C., Paola De, B., Del Borghi, A., Vesce, E., & Beltramo, R. (2021). The contribution of sensor-based equipment to life cycle assessment through improvement of data collection in the industry. *Environmental Impact Assessment Review, 88.*

ISO, International Organization for Sta ndardization. (2015). *ISO 14001:2015 Environmental management systems – requirements with guidance for use.* ISO. Retrieved from https://www.iso.org/standard/60857.html

Hall, J., & Wagner, M. (2012). *Integrating sustainability into firms' processes: Performance effects and the moderating role of business models and innovation. Business strategy and the environment.* New York: Wiley Online Library.

Kirchherr, J., Reike, D., & Hekkert, M. (2017). Conceptualizing the circular economy: An analysis of 114 definitions. *Resources, Conservation and Recycling, 127,* 221–232. doi:10.1016/j.resconrec.2017.09.005

Korhonen, J., Nuur, C., Feldmann, A., & Birkie, S. E. (2018). Circular economy as an essentially contested concept. *Journal of Cleaner Production, 175,* 544–552. doi:10.1016/j.jclepro.2017.12.111

Kovacic, I., Honic, M., & Sreckovic, M. (2020). Digital platform for circular economy in AEC industry. *Engineering Project Organization Journal, 9.* Retrieved from https://publik.tuwien.ac.at/files/publik_290949.pdf

Kristoffersen, E., Blomsma, F., Mikalef, P., & Li, J. (2020). The smart circular economy: A digital-enabled circular strategies framework for manufacturing companies. *Journal of Business Research, 120,* 241–261. doi:10.1016/j.jbusres.2020.07.044

Loglisci, G., Priarone, P. C., & Settineri, L. (2013). Cutting tool manufacturing: A sustainability perspective. *11th Global Conference on Sustainable Manufacturing,* 275–280. doi:10.14279/depositonce-4781

Moldavska, A., & Welo, T. (2017). The concept of sustainable manufacturing and its definitions: A content-analysis based literature review. *Journal of Cleaner Production, 166,* 744–755. doi:10.1016/j.jclepro.2017.08.006

Neugebauer, F. (2012). EMAS and ISO 14001 in the German industry – complements or substitutes? *Journal of Cleaner Production, 37,* 249–256. doi:10.1016/j.jclepro.2012.07.021

Ranta, V., Aarikka-Stenroos, L., & Väisänen, J. H. (2021). Digital technologies catalyzing business model innovation for circular economy – multiple case study. *Resources, Conservation and Recycling, 164,* 105155. doi:10.1016/j.resconrec.2020.105155

Rocca, R., Rosa, P., Sassanelli, C., Fumagalli, L., & Terzi, S. (2020). Integrating virtual reality and digital twin in circular economy practices: A laboratory application case. *Sustainability, 12*(6), 2286. doi:10.3390/su12062286

Rossi, J., Bianchini, A., & Guarnieri, P. (2020). Circular economy model enhanced by intelligent assets from industry 4.0: The proposition of an innovative tool to analyze case studies. *Sustainability, 12*(17), 7147. doi:10.3390/su12177147

Saidani, M., Yannou, B., Leroy, Y., Cluzel, F., & Kendall, A. (2019). A taxonomy of circular economy indicators. *Journal of Cleaner Production, 207,* 542–559. doi:10.1016/j.jclepro.2018.10.014

Salomone, R. (2008). Integrated management systems: Experiences in Italian organization. *Journal of Cleaner Production, 16*(16), 1786–1806. doi:10.1016/j.jclepro.2007.12.003

Schroeder, P., Anggraeni, K., & Weber, U. (2019). The relevance of circular economy practices to the sustainable development goals. *Journal of Industrial Ecology, 23*(1), 77–95. doi:10.1111/jiec.12732

Şerban, R. A. (2017). The impact of big data, sustainability, and digitalization on company performance. *Studies in Business and Economics, 12*(3), 181–189.

Sousa Jabbour, A. B. L., Chiappetta, J. C. J., Foropon, C., & Filho, M. G. (2018). When Titans meet – can industry 4.0 revolutionise the environmentally-sustainable manufacturing wave? The role of critical success factors. *Technological Forecasting and Social Change, 132*, 18–25. doi:10.1016/j.techfore.2018.01.017

Stock, T., & Seliger, G. (2016). Opportunities of sustainable manufacturing in industry 4.0. *Procedia CIRP, 13th Global Conference on Sustainable Manufacturing – Decoupling Growth from Resource Use, 40*, 536–541. doi:10.1016/j.procir.2016.01.129

TCE, Ellen MacArthur Foundation. (2015). Retrieved from www.ellenmacarthurfoun dation.org/assets/downloads/publications/TCE_Ellen-MacArthur-Foundation_26-Nov-2015.pdf

World Commission on Environment and Development. (1987). *Our common future.* Oxford Paperbacks. Oxford: Author.

Zobel, T., Almroth, C., Bresky, J., & Burman, J. O. (2002). Identification and assessment of environmental aspects in an EMS context: An approach to a new reproducible method based on LCA methodology. *Journal of Cleaner Production, 10*(4), 381–396. doi:10.1016/S0959-6526(01)00054-3

# 10 Assessing sustainability across circular inter-firm networks

## Insights from academia and practice

*Anna M. Walker, Katelin Opferkuch,*
*Erik Roos Lindgreen, Alberto Simboli,*
*Walter J.V. Vermeulen, and Andrea Raggi*

## Introduction

Since companies using Circular Economy (CE) practices are usually embedded within regional or global networks of supply chains, it is essential to consider the comprehensive sustainability impact of these actors (Vegter, van Hillegersberg, & Olthaar, 2020). While the academic literature on sustainability assessment approaches for circular inter-firm networks (CIFN) has already been summarised by the authors (Walker, Vermeulen, Simboli, & Raggi, 2021c), research addressing the industry perspective on this topic is limited. Walker et al. (2021c) describe CIFNs as company networks which "consist of actors that are connected through open (intersectoral) and/or closed (intrasectoral) supply chains which are *de facto* circular" (p. 3).

This chapter aims to map out which sustainability assessment approaches are implemented by frontrunner companies from Italy and the Netherlands involved in CE practices. The 43 companies interviewed are assumed to be frontrunners, as they are members of national and international CE networks. Their insights can contribute to advancing the application of sustainability assessment across CIFNs, as CE practices mostly take place beyond company boundaries (Vegter et al., 2020). The second aim of this chapter is to benchmark the sustainability assessment approaches from academia and practice against sustainability assessment criteria identified from literature to further the development of sound assessment approaches.

The following section sets out the academic state of the art in sustainability assessment of CIFNs and then provides a brief literature review of the most pertinent criteria for analysing the efficacy of sustainability assessment approaches in CIFNs. The third section offers an overview of the methods employed, while the fourth section presents the empirical results. In the fifth section, the authors benchmark the identified approaches against criteria for sound sustainability assessment. The findings are discussed in section six and complemented with some recommendations for CE practitioners on how to conduct a meaningful sustainability assessment across supply chains.

DOI: 10.4324/9781003179788-10

## Theoretical background

### *The state of the art in academia*

Sustainability assessment approaches from industrial ecology (IE) and circular supply chain management (CSCM) are particularly well adapted to the realities of CIFNs (Walker et al., 2021c). IE provides mainly *ex-post* assessments focussing on the environmental dimension of sustainability, the most frequently applied approaches being life cycle thinking (LCT)-based methodologies. Meanwhile, the field of CSCM offers more *ex-ante* assessments, which sometimes cover all three sustainability dimensions traditionally addressed by companies: the social, environmental, and economic dimensions. Furthermore, CSCM literature mainly promotes the employment of mathematical programming for identifying the optimal constellation of a circular supply chain. However, these mathematical programming models are often based on singular indicators, which may obscure trade-offs and could be considered a reductionist approach (Gasparatos, El-Haram, & Horner, 2008). In line with other scholars, results from the literature review further revealed that assessment approaches for social sustainability should be further developed and integrated with assessment approaches for the environmental and economic dimensions (Roos Lindgreen, Salomone, & Reyes, 2020; Vegter et al., 2020; Walzberg et al., 2021). This discussion has been taken up by Walker et al. (2021b), documenting companies' understanding of the social dimension in the CE and its potential assessment, even across supply chains. Figure 10.1 presents the sustainability assessment framework for CIFNs developed based on the academic literature by Walker et al. (2021c). This framework is divided into two types of assessment approaches: first, evaluation approaches, with the goal of mapping sustainability impacts, and second, decision-supporting approaches, enabling informed decision-making based on the data observed.

### *Criteria for sound inter-firm sustainability assessment*

Sustainability assessments are conducted to identify (potential) sustainability impacts of CIFNs. These impacts either bring society closer to (if positive) or further away (if negative) from achievement of the UN Sustainable Development Goals (SDGs) (United Nations, 2015). Several requirements for meaningful sustainability assessments in CIFNs can be identified from the literature for various stages of assessment. While scholars have provided insights into the selection process of assessment approaches (Gasparatos & Scolobig, 2012; Zijp et al., 2017), such as being aware of the underlying value system of each approach (Gasparatos, 2010; Sala, Farioli, & Zamagni, 2013), the ontology of the approaches (Sala, Ciuffo, & Nijkamp, 2015), or their feasibility in a certain context (Schöggl, Fritz, & Baumgartner, 2016), the authors would like to present those criteria most pertinent to CIFNs. Therefore, the four criteria presented by Muñoz-Torres et al.

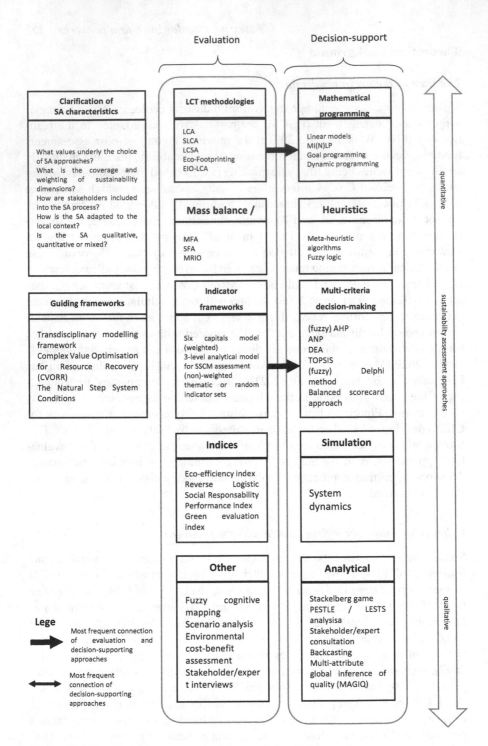

*Figure 10.1* Sustainability assessment framework for CIFNs

Source: Based on Walker et al. (2021c, p. 15).

(2018) are applied here in a slightly adapted version, selected due to their frequent mentioning in supply chain literature and their aptness to CE practices. The four criteria are *balance of the different sustainability dimensions aligned with the SDGs, consideration of the intergenerational nature of sustainability, stakeholder involvement, and LCT.*

The first criterion entails the question whether the assessment approach can sufficiently *cover and integrate the sustainability dimensions*, aligned with the SDGs (Valenzuela-Venegas, Salgado, & Díaz-Alvarado, 2016), and whether trade-offs between them are made transparent (Haffar & Searcy, 2017; Morrison-Saunders & Pope, 2013).

The second criterion evaluates if the assessment approach can take the *intergenerational nature of sustainability* into consideration or, in other words, whether implications across time can be captured (Gasparatos & Scolobig, 2012; Sala et al., 2013).

The third (*stakeholder involvement*), mentioned by several authors (Kühnen & Hahn, 2018; Sala et al., 2015; Silva, Nuzum, & Schaltegger, 2019), can help develop trust and empower stakeholders. This step is crucial for sustainability assessment, providing it with legitimacy and concurrently embedding it into its local context (Schöggl et al., 2016). The engagement of different types of stakeholders further allows for interdisciplinarity (across disciplines) or even transdisciplinarity (including practitioners), providing a more holistic perception of reality (Sala et al., 2015).

The fourth and final criterion is related to the circular nature of the inter-firm networks: *LCT*. As has been pointed out by several scholars (Peña et al., 2021; Sala et al., 2013), an assessment approach with a life-cycle perspective considers the sustainability impacts of a given product and the producing companies from the sourcing of material and production through to its final disposal or, in the case of circular networks, to the recovery of that product or its material/energy content (Vegter et al., 2020).

## Methods

To compare the assessment approaches proposed by academia with those implemented by frontrunner companies engaged in CE practices, the sustainability assessment framework for CIFNs (Figure 10.1) is to be populated with approaches found in practice.

The empirical data was collected via semi-structured interviews from a subset of respondents who had participated in a survey on the connection between the CE and sustainability (Walker et al., 2021a). The 43 interview participants were mostly in upper management positions within micro-companies with less than 10 employees (49%), while the rest were from sustainability departments in small and medium companies (26%) and large companies with more than 250 employees (25%). All firms were operating either in Italy (n = 20) or the Netherlands (n = 23) and were members of a CE network. For more details on the interview methodology, please refer to Walker et al. (2021a).

## Supply chain assessment in frontrunner companies engaged in CE practices

Over two-thirds of the interviewees acknowledged the importance of sustainability assessment in supply chains and confirmed that they performed some type of such an assessment. The main reason for this was that respondents could obtain a better oversight of the materials processed in their supply chains, use this knowledge to set internal performance targets, and strengthen their supplier relationships. Furthermore, the results of the assessment were also used for external communication at both a corporate and regional level. However, about a third of the interviewees did not conduct any sustainability assessments of their supply chains because they were either small companies with few supply chain partners, or larger companies, whose clients did not lend significant weight to sustainability criteria. Nevertheless, they still included information regarding their CE practices in their external communication.

### *Supply chain assessment approaches applied*

In Figure 10.2, all the applied approaches are presented in their respective assessment category. Companies which conducted sustainability assessments mostly did so in a qualitative way, together with their supply chain partners, or opted for a method based on LCT. The qualitative approach usually implied client and/or supplier meetings to jointly evaluate product solutions. External consultants were sometimes involved in this process to help establish the sustainability impacts of products, mainly focussing on the environmental domain. Another type of evaluation consisted of visiting supplier sites or, where not possible, requiring suppliers to be members of ethical supplier networks. Whereas the aforementioned qualitative assessment practices were more common among smaller companies, large companies conducted regular supplier audits in line with their corporate purchasing policies. Concerning more intricate assessments, the most sophisticated methodology used was Life Cycle Assessment (LCA), either conducted by the company itself or, more frequently, by external consultants. Related thereto, company supply chains also shared data on carbon emissions to determine their comprehensive carbon footprint, a process which was often initiated by clients. Given that no indices were mentioned by companies within the sample, the category "Indices" has been replaced by "Certification and labels required for suppliers" in Figure 10.2. Additionally, the category "Indicator frameworks" has been extended to specify "Indicator frameworks for suppliers".

Furthermore, there were several cases where companies monitored their CE practices, such as the number of items of used equipment handed down to other companies, the amount of demolition material used to make bricks, or the amount of used cooking oil utilised as input to produce lubricants or detergents. Yet, the sustainability impact of these CE practices was not necessarily calculated.

### *Reliance on supplier criteria (ex-ante assessment)*

While in the past the most important indicator for choosing a supplier has been the cost factor, the criteria have now widened considerably. Around three-quarters

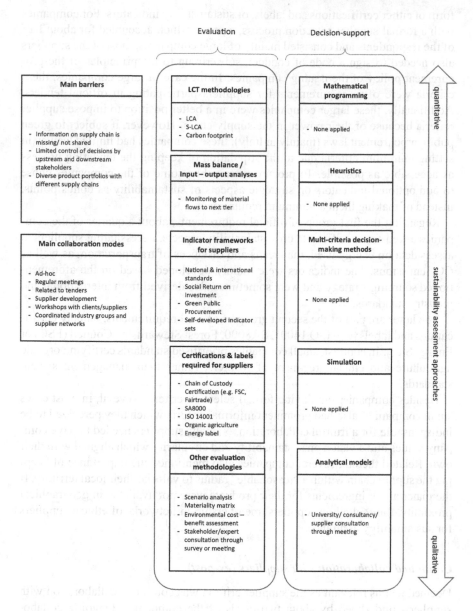

*Figure 10.2* Use of supply chain sustainability assessment approaches by sampled companies

of the companies have determined both formal and/or informal criteria for their supplier selection. Within larger companies, these criteria were often defined explicitly in a socially responsible purchasing policy or circular sourcing strategy. Though these were frequently developed internally, external consultants were sometimes also involved in the process. The supplier criteria usually took the form of either certifications and labels, or sustainability indicators. For companies with a formal supplier selection process, a group which accounted for about half of the respondents and consisted mainly of large companies, most of the suppliers also needed to sign a code of conduct, subscribing to the principles of the procurement policy of the issuing companies. In the case of large companies, these criteria were often requirements for suppliers participating in official tenders. Additionally, these larger companies were in a better position to impose supplier criteria because of their power in the supply chain. However, if subject to green public procurement laws (mainly in Italy), these companies had limited leeway in setting minimum criteria due to the stipulation of keeping the tendering process as accessible as possible. To procure sustainably, some of the companies chose to add optional indicators on strategic aspects of sustainability as bonus points, instead of making them a minimum requirement.

Regarding the first group of official requirements, about a quarter of the companies asked for specific data on sustainability aspects, necessitating suppliers to assess data on energy used, the means and distances of transportation, as well as $CO_2$ emissions. The indicators were mainly developed based on the aforementioned sourcing strategy and were sometimes also derived from international and industry standards.

The latter are part of the second group of official requirements, namely certifications and labels [e.g. ISO 14001, SA8000, Forest Stewardship Council (FSC) or Energy Star]. In contrast, small companies prioritised standards certifying organic agriculture and ethical treatment of workers, rather than management system standards.

Besides companies employing formal selection criteria, several, in most cases small companies, also had prominent informal criteria which they perceived to be indispensable for a fruitful collaboration. Namely, suppliers needed to have company values and a vision on sustainability and circularity which aligned with their own. Related thereto, several companies also mentioned the importance of keeping the supply chain within a "reasonable" radius to valorise their local territory. If the materials or ingredients for their products were not available in geographical proximity, they relied on suppliers' membership in networks of ethical suppliers for sustainability assurance.

### Continued collaboration with suppliers (ex-post)

Deemed just as relevant as the supplier criteria was continued collaboration with suppliers, underlined by about two-thirds of the companies. Ongoing collaboration was seen, on the one hand, as a driver for CE, while, on the other hand, some companies also saw the novel CE paradigm as a catalyst for more collaboration. Especially in the case of smaller companies, collaboration with suppliers was often based on long-term relationships with partners that pursued similar values, establishing trust. For larger companies with more formal supplier protocols ensuring competitiveness, this kind of relationship was less frequent; but taking a stance on circularity still had a positive impact on supply chain collaboration.

In addition to collaborating with key suppliers, these companies also conducted educational workshops on sustainability and CE with their suppliers. On this collaborative basis, it was possible to evaluate the production process for a joint product, either by means of an official assessment or by deliberating the different production options as a first step. Some companies mentioned that the level of collaboration with their suppliers has also increased due to the supply chain sustainability assessment, because it required them to actively reach out to their suppliers and to find out how their supply chain was organised. While, in some cases, incumbent companies came together to develop new product solutions and services, often facilitated by smaller, more agile companies, others did not necessarily want to change the product. Instead, they wanted to better understand their suppliers' actions and the impacts on the product. At the end, collaboration did not only happen within supply chains, but also across industries and supply chains, for example in industry working groups such as the CE100, the Factor10 of the World Business Council for Sustainable Development, or industry consortia, and the aforementioned ethical supplier networks.

Depending on the type of collaboration, Figure 10.2 shows that the meetings with suppliers and clients were regular, *ad-hoc,* and, in some cases, facilitated by consultancies or a coordination team, mainly as part of cross-industry initiatives. These external parties were often also essential to initiate collaboration by, for example, finding suitable buyers for by-products of companies, while ensuring the financial competitiveness of these transactions. Besides the economic aspects, geographical proximity was also seen as a driver of collaboration both within and across sectors.

### Barriers to implementing supply chain sustainability assessment

The respondents mentioned several reasons why their assessment did not go further in detail and scope, which are summarised in Figure 10.2 in the top left. The main issue was that information on the supply chain was often simply missing, given it would have to be collected from the lower tiers. This was particularly relevant for companies working with secondary materials and products, given these materials had already gone through at least one life cycle or were complex products such as electronics. In other cases, the suppliers did not directly want to share information for competitive reasons, though this was less frequently mentioned. Where companies were themselves suppliers for larger companies, they also did not necessarily have the leverage to ask their downstream partners for further information. This is closely related to the second main barrier, namely the limited leverage over the decision-making and behaviour of upstream and downstream partners, as well as that of consumers. Several respondents mentioned that their clients were often informed about the sustainability impacts of certain product and material choices. Yet, the decision on whether to opt for the potentially more (expensive) sustainable option or not lay beyond the interviewees' sphere of influence. There were several instances where clients did not demand a sustainability assessment, especially when informed that it might increase costs. Another barrier, faced mainly by large companies with diverse product portfolios, was the high number and diversity of supply chains, so the assessments were kept to the most important partnerships and impact categories such as global warming.

## Benchmarking approaches from academia and practice

Based on the four criteria of sound sustainability assessment, approaches from academia and practice are compared in the following subsections. An overview of the comparison is provided in Table 10.1.

### *Balancing sustainability dimensions aligned with SDGs*

Both academics[1] and practitioners use LCT methodologies, but the assessment is mainly put into operational practice for the environmental dimension of sustainability. While LCA covers a wide range of environmental impacts, the integration of these impacts with social and socio-economic factors is still limited across supply chains. Regarding mass balance and input–output analyses, the social dimension is not covered in the literature and in practice. In contrast, the identified indicator frameworks are more inclusive and contain a more holistic set of indicators. Yet, it was found that both in practice and in the literature, social indicators are often limited to job creation, sometimes focussing on people disadvantaged in the labour market. Indices found in the literature encompass either an environmental or a social dimension but do not necessarily address the balance among them. In the newly created category of certifications and labels for suppliers, it was noted that companies require both environmental and social certifications when selecting suppliers. The other evaluation approaches such as scenario analysis and stakeholder consultation can also be considered holistic in terms of sustainability assessment for both academia and practice, whereas the environmental cost–benefit analysis is focussed only on the environmental and economic factors. With regards to decision-supporting approaches, none of the interviewed companies applied mathematical programming, heuristics, multi-criteria decision-making, and simulation approaches. In the academic literature, there are instances of mathematical programming models covering all sustainability dimensions, but they often consider only a limited number of indicators per dimension. Also, costs are the most common indicator to be optimised, while the social (mostly job creation) and environmental indicators (mostly greenhouse gas emissions) are generally modelled as constraints, rather than simultaneous optimisation goals. Regarding multi-criteria decision-making models proposed by academia, they can indeed include all sustainability dimensions, depending on the underlying indicators chosen in the evaluation step. Similarly, simulation can also model the effects on all sustainability dimensions, if the cause-and-effect relations are known. At the end, concerning analytical models, the Political, Economic, Sociological, Technological, Legal and Environmental (PESTLE/LESTS) analysis, stakeholder/expert consultation, as well as backcasting usually include the sustainability dimensions considered to be important by the decision-makers, both in practice and academia.

Table 10.1 Benchmarking literature and practice according to criteria for sound sustainability assessment

| | Assessment approach categories | Balance of sustainability dimensions | | Addressing intergenerational nature of sustainability | | Stakeholder involvement | | Life-cycle thinking | |
|---|---|---|---|---|---|---|---|---|---|
| | | Literature[a] | Practice[b] | Literature[a] | Practice[b] | Literature[a] | Practice[b] | Literature[a] | Practice[b] |
| **Evaluation** | LCT methodologies | Possible | Not implemented | By default | Implemented | Possible | Implemented | By default | Implemented |
| | Mass balance and input–output analyses | Not possible | Not implemented | Possible | Not implemented | Possible | Implemented | Possible | Not implemented |
| | Indicator frameworks (for suppliers) | Possible | Implemented | Possible | Not implemented | By default | Implemented | By default | Implemented |
| | Indices/ Certifications and labels for suppliers | Possible | Implemented | Not possible/ found | Implemented | Possible | Implemented | Possible | Not implemented |
| | Other evaluation methods | Possible | Implemented* | Possible | Not implemented | Possible | Implemented | Possible | Implemented |
| **Decision-support** | Mathematical programming and heuristics | Possible | N/A | Not possible/ found | N/A | Possible | N/A | Possible | N/A |
| | Multi-criteria decision-making methods | Possible | N/A | Possible | N/A | By default | N/A | Possible | N/A |
| | Simulation | Possible | N/A | Possible | N/A | Possible | N/A | Possible | N/A |
| | Analytical models | Possible | Implemented* | Possible | Not implemented | Possible | Implemented | Possible | Implemented |

[a] For details, refer to Walker et al. (2021c).
[b] Based on interviews with companies.
* Social dimension mostly left out.

However, it has been found that the companies within this sample rarely asked external parties (e.g. consultancies) for social assessments.

### Addressing the intergenerational nature of sustainability

With regards to intertemporal aspects, the LCT methodologies are well suited, because they consider the long-term impacts of resource use and depletion beyond generations (Sala et al., 2013). However, besides those respondents who conducted a full LCA, several companies only examined single-impact categories, such as the carbon footprint, providing a limited picture of the long-term impacts. In the category of the mass balance and input–output analyses, the temporal scope of such analyses can extend over long time periods, especially when covering large regions. Yet, the monitoring of materials as described by the interviewees and in literature does not explicitly take a long-term perspective. Regarding the indicator frameworks in the literature, they do not necessarily specify the importance of long-term availability of resources, though indicators related to the R-hierarchy (Reike, Vermeulen, & Witjes, 2018), for example, do point in this direction. In practice, companies have started to consider introducing indicators related to the CE, such as recyclability of products or amount of recycled content, into their purchasing requirements, but in most cases have not yet done so. When it comes to indices proposed in the literature, scholars do not include any long-term indices. In contrast, some of the certifications and labels for suppliers contain inter-temporal aspects, such as the FSC and the organic agriculture label, which stand for sustainable and regenerative forestry and agricultural practices. Regarding the other evaluation methods, several can depict different timeframes, in particular, scenario analysis. For the decision-supporting approaches, temporal aspects can be included in mathematical programming models, also combined with heuristics. However, these models are usually not optimised for time spans across generations and often are modelled with infinite stocks of resources. Concerning the multi-criteria decision-making models, they have the option to include indicators based on the R-hierarchy, yet the indicators' importance depends on their weighting. Similarly, a simulation can model any time span required if the dynamics within the system are known. Yet, fast-changing CIFNs usually have time horizons of less than one generation. In contrast, analytical models such as backcasting are well suited to address the intergenerational aspects of sustainability, depending on the timeframe set. Regarding the other methods such as stakeholder consultation and the PESTLE/LESTS analysis, these might, both in theory and practice, focus more on short-term impacts, as they are meant to find solutions for current situations, and the needs of future stakeholders are not necessarily considered (Wannags & Gold, 2020; Wu & Pagell, 2011).

### Stakeholder involvement

In literature, stakeholder involvement has been considered to a rather limited degree for LCT methodologies, besides data collection from suppliers (Sala et al.,

2013). In particular, the social LCA addresses various stakeholder categories beyond suppliers, though their consultation does mostly not go beyond data collection. Whereas the interview respondents report LCA information throughout supply chains, the actual assessments are usually done by external consultancies. Involvement is also limited for mass balance and input–output analyses as described by scholars. Yet, when considering the monitoring of material flows in practice, there is considerable communication between supply chain partners, but not necessarily beyond this stakeholder category. Regarding indicator frameworks, stakeholder involvement is seen as being critical for the validation and selection of indicators. In practice, the indicator frameworks for suppliers are often developed based on clients' needs, legal requirements imposed by public stakeholders, or indicators communicating corporate information to civil society stakeholders. Similar considerations are also true for the required standards and labels. In addition to this, labels are sometimes promoted to sell products within industry networks. When looking at the indices proposed by scholars, it becomes evident that stakeholder involvement is limited, except for indices which also include the social dimension. As already anticipated previously, stakeholder involvement is often used as a first step to identify relevant indicators. Stakeholder involvement can further be used to inform scenario analysis, as well as cost–benefit calculations. In practice, consultation of stakeholders is essential to both set strategic priorities and legitimise CE practices. This is often carried out in the form of a (materiality) survey or stakeholder meetings. Scenario analysis has also been informed by potential partners of a regional industrial symbiosis project for evaluating its feasibility. Moving on to the decision-supporting approaches, mathematical programming models and heuristics have traditionally not been considered participative. However, both Stindt, Sahamie, Nuss, and Tuma (2016) and Voinov et al. (2016) report that initial involvement of (non-academic) stakeholders before the actual modelling is fundamental for the applicability of the models. Furthermore, a considerable number of authors propose the stakeholder involvement when applying multi-criteria decision-making methods, though expert involvement is slightly more common. Simulation approaches could also be informed by stakeholder involvement, though this is not necessarily the case in literature. At the end, analytical models in literature involve stakeholders to help with decision-making and to inform the PESTLE/LESTS analysis. In practice, stakeholder consultation for decision-making is frequent, primarily limited to supply chain partners and consultancies through meetings and workshops. With regards to clients, companies often provide them with sustainability evaluation results, while leaving the final decision up to them.

## Life-cycle thinking

As the name already anticipates, LCT methodologies are rooted in LCT. Nevertheless, it needs to be underlined that, in practice, LCAs do not always cover the whole life cycle of a product. Likewise, a mass balance or input–output-based method can be used to depict a whole life cycle or just a part of it. In practice, it was found that the monitoring of material flows does often not include the whole life cycle of materials but just the respective next tiers of the supply chain. Regarding indicator frameworks and indices in the literature, neither of these is necessarily LCT-oriented, as some are based on direct impacts, particularly the social ones. Yet, a number of LCT-based indicators are included. The same is true for companies using the indicator frameworks and certification and labels intended for suppliers. As regards the other evaluation methodologies, scenario analysis is often combined with LCT methodologies in the literature and thus is well suited for the integration of LCT; something which is also done in practice. The same is true for stakeholder or expert interviews, though whether LCT is included or not depends on the content of the discussion. In the environmental cost–benefit analysis, it is also possible to cover the whole life cycle of a product. Similarly, the respondents underlined that, through their CE practices and collaboration with their partners, they have taken an LCT approach to organising and assessing their supply chain network. When looking at the decision-supporting approaches, LCT-based indicators are frequently included in the mathematical programming models. In contrast, multi-criteria decision-making methods do not necessarily employ LCT-based indicators, though they would be able to integrate them. The same is true for system dynamics, which can model whole life cycles of products within a system. Concluding with the analytical models, they are not based on LCT per se, if this is not explicitly required – for example through expert or stakeholder consultation, or backcasting. Yet, in practice, it was found that companies engaged in CE practices attribute a high degree of importance to inclusion of the whole product life cycle, making LCT essential in decision-making.

## Discussion and conclusion

When comparing the approaches proposed in the literature with the ones actually applied by frontrunner companies engaged in CE practices, it is noticeable that most of the approaches used were qualitative. There is however one important exception which are the LCT methodologies, such as LCA and the carbon footprint. These two approaches were principally established in larger companies and smaller specialised firms. Otherwise, the approaches primarily consisted of indicator frameworks for supplier selection and other methodologies such as stakeholder consultation, scenario analysis, or cost–benefit calculation. Furthermore, while some companies monitored their material flows across supply chains, they did not do so in a comprehensive material flow analysis but rather collected data on the volume of material flows between two tiers only. Another interesting observation was that most of the approaches used in practice were evaluation approaches, while quantitative decision-supporting approaches were absent. The main decision-supporting approach applied was expert consultation, namely through consultancies. This finding is in contrast with the literature, where the most frequently proposed approach was mathematical programming, often in connection with LCT-based indicators or heuristics. Furthermore, the combination of indicator frameworks and multi-criteria decision-making models also prevalent in the literature was not employed in practice. The low application of more quantitative decision-supporting approaches might partly be explained by the high share of micro-companies in the sample. The only similarity between the findings from literature and practice was that both proposed employing LCT-based methodologies, though mainly for larger companies. Schöggl et al. (2020) have arrived at a similar finding in their review of CE research. Swarr et al. (2015) also pointed out increased collaboration with suppliers and customers and cross-functional integration within companies as benefits of LCT approaches for SMEs. When Galindro et al. (2020) asked practitioners why they conducted LCAs, companies reported that they mostly responded to clients' requests. The lack of these requests was seen as one of the main barriers to assessment by the interviewees in this study.

Regarding the fulfilment of the criteria for sound sustainability assessment, it was generally found that LCT methodologies strongly favoured the environmental dimension. Another notable result from literature and practice was that the requirement of addressing the intergenerational nature of sustainability was least fulfilled. It further emerged that stakeholder engagement was already a best practice for most of the interviewed companies, while, in the literature on LCT methodologies as well as mass balance and input–output analyses, this was documented only sporadically. Concerning LCT, this criterion was also broadly applied in practice by the respondents, though sometimes practitioners did not involve the whole life cycle but only the next tiers of the supply chain.

A third main finding of the comparison is that the answers to the question of what approaches were suitable for assessing sustainability across CIFNs were provided on different levels. While literature offered plenty of assessment approaches, thus providing methodological support, practitioners underlined the importance of relationships when initiating CE practices and assessing them. It thus became clear that rather than relying on quantitative assessment results for decision-making, high importance was attributed to frequent exchange of best practices, trust, and supply chain management tools such as supplier selection criteria, codes of conduct, audits, and certifications and product labels. As already anticipated by Qian, Seuring, and Wagner (2020) and Brown, Bocken, and Balkenende (2019), collaboration based on trust was seen as a critical starting point both for initiating CE practices and evaluating them jointly. It was also found that the connection between collaboration and assessment was reciprocal; the more collaboration there was, the more likely an assessment would be conducted. Simultaneously, the need for assessing a joint product led to more collaboration. Scholars have also noted that collaboration had positive effects on performance, especially when sharing knowledge along the supply chain (Qian et al., 2020). This was put into practice by several interviewees, who conducted workshops for and with suppliers to share information on sustainability aspects and CE practices. Frontrunner companies engaged in CE practices with short supply chains also mentioned that the geographical proximity to their suppliers was an essential driver for their collaboration.

*Table 10.2* Recommendations for developing and applying assessment procedures

| Development steps | Recommendation |
| --- | --- |
| 1 Identifying partners | • When reaching out to companies within your supply chain, be clear and vocal about the overall vision you are pursuing as a company. |
| | • When contemplating about whom to include in the assessment, aim to involve actors across the whole value chain of a product, including those recovering the materials at the end of life. |
| 2 Building trust | • When planning to do a supply chain sustainability assessment, ensure that the partnership with the supply chain actors is sufficiently strong for sharing information and inform them why you are doing the assessment. |
| 3 Conceptualising the assessment | • When conceptualising the supply chain assessment, ensure that all three sustainability dimensions are sufficiently covered to make trade-offs transparent. |
| | • When selecting indicators, ensure the fulfilment of the criteria for sound sustainability assessment for CIFNs, either through the indicators themselves or the way they are chosen (e.g. the criteria of stakeholder involvement can be covered by including stakeholders in setting the priorities of the assessment or a corporate strategy). |
| | • When setting supplier selection criteria, award the inclusion of LCT-based indicators and other sustainability indicators in line with your CSR strategy with higher scores, if you do not want to be too restrictive for suppliers. |
| 4 Continuous learning | • Join an industry initiative or retailer network related to CE or sustainable sourcing to stay up to date regarding innovative best practices in your sector. |

Besides these informal ties between supply chain actors, almost half of the interviewees had additionally established formal supplier selection criteria, either in the form of sustainability performance indicators, often based on LCA data, or certifications and product labels. The importance of balancing formal supplier criteria with trust when managing the corporate social responsibility (CSR) performance was also underlined by Hyder, Chowdhury, and Sundström (2017).

The authors would like to bring this chapter to a close with the recommendations for practitioners presented in Table 10.2, based on best practices from the interviews and literature.

## Acknowledgements

This research constitutes part of the research project CRESTING (Circular Economy: Sustainability implications and guiding progress), funded by the European Union's Horizon 2020 research and innovation programme under the Marie Skłodowska-Curie grant agreement number 765198.

## Note

1 The findings from the literature are taken from Walker et al. (2021c), where further information can be found on specific articles.

## References

Brown, P., Bocken, N., & Balkenende, R. (2019). Why do companies pursue collaborative circular oriented innovation? *Sustainability*, *11*, 635.

Galindro, B. M., Welling, S., Bey, N., Olsen, S. I., Soares, S. R., & Ryding, S. O. (2020). Making use of life cycle assessment and environmental product declarations: A survey with practitioners. *Journal of Industrial Ecology*, *24*, 965–975.

Gasparatos, A. (2010). Embedded value systems in sustainability assessment tools and their implications. *Journal of Environmental Management*, *91*, 1613–1622.

Gasparatos, A., El-Haram, M., & Horner, M. (2008). A critical review of reductionist approaches for assessing the progress towards sustainability. *Environmental Impact Assessment Review*, *28*, 286–311.

Gasparatos, A., & Scolobig, A. (2012). Choosing the most appropriate sustainability assessment tool. *Ecological Economics*, *80*, 1–7.

Haffar, M., & Searcy, C. (2017). Classification of trade-offs encountered in the practice of corporate sustainability. *Journal of Business Ethics*, *140*, 495–522.

Hyder, A. S., Chowdhury, E. H., & Sundström, A. (2017). Balancing control and trust to manage CSR compliance in supply chains. *International Journal of Supply Chain Management*, *6*, 1–14.

Kühnen, M., & Hahn, R. (2018). Systemic social performance measurement: Systematic literature review and explanations on the academic status quo from a product life-cycle perspective. *Journal of Cleaner Production*, *205*, 690–705.

Morrison-Saunders, A., & Pope, J. (2013). Conceptualising and managing trade-offs in sustainability assessment. *Environmental Impact Assessment Review*, *38*, 54–63.

Muñoz-Torres, M. J., Fernández-Izquierdo, M. Á., Rivera-Lirio, J. M., Ferrero-Ferrero, I., Escrig-Olmedo, E., Gisbert-Navarro, J. V., & Marullo, M. C. (2018). An assessment tool to integrate sustainability principles into the global supply chain. *Sustainability*, *10*, 535.

Peña, C., Civit, B., Gallego-Schmid, A., Druckman, A., Pires, A. C., Weidema, B., & Motta, W. (2021). Using life cycle assessment to achieve a circular economy. *International Journal of Life Cycle Assessment*, *26*, 215–220.

Qian, C., Seuring, S., & Wagner, R. (2020). Reviewing interfirm relationship quality from a supply chain management perspective. *Management Review Quarterly*, *71*, 625–650.

Reike, D., Vermeulen, W. J. V., & Witjes, S. (2018). The circular economy: New or refurbished as CE 3.0? – Exploring controversies in the conceptualization of the circular economy through a focus on history and resource value retention options. *Resources, Conservation and Recycling*, *135*, 246–264.

Roos Lindgreen, E., Salomone, R., & Reyes, T. (2020). A critical review of academic approaches, methods and tools to assess circular economy at the micro level. *Sustainability*, *12*, 4973.

Sala, S., Ciuffo, B., & Nijkamp, P. (2015). A systemic framework for sustainability assessment. *Ecological Economics*, *119*, 314–325.

Sala, S., Farioli, F., & Zamagni, A. (2013). Progress in sustainability science: Lessons learnt from current methodologies for sustainability assessment: Part 1. *International Journal of Life Cycle Assessment*, *18*, 1653–1672.

Schöggl, J. P., Fritz, M. M. C., & Baumgartner, R. J. (2016). Toward supply chain-wide sustainability assessment: A conceptual framework and an aggregation method to assess supply chain performance. *Journal of Cleaner Production, 131*, 822–835.

Schöggl, J. P., Stumpf, L., & Baumgartner, R. J. (2020). The narrative of sustainability and circular economy – a longitudinal review of two decades of research. *Resources, Conservation and Recycling, 163*, 105073.

Silva, S., Nuzum, A. K., & Schaltegger, S. (2019). Stakeholder expectations on sustainability performance measurement and assessment: A systematic literature review. *Journal of Cleaner Production, 217*, 204–215.

Stindt, D., Sahamie, R., Nuss, C., & Tuma, A. (2016). How transdisciplinarity can help to improve operations research on sustainable supply chains – a transdisciplinary modeling framework. *Journal of Business Logistics, 37*, 113–131.

Swarr, T. E., Asselin, A. C., Milà i Canals, L., Datta, A., Fisher, A., Flanagan, W., & Rasteiro, M. G. (2015). Building organizational capability for life cycle management. In G. Sonnemann & M. Margni (Eds.), *Life cycle management, LCA compendium* (pp. 239–256). Dordrecht: Springer.

United Nations. (2015). *Transforming our world: The 2030 agenda for sustainable development [WWW Document]: Sustainable development knowledge platform*. Retrieved January 12, 2012, from https://sustainabledevelopment.un.org/post2015/transformin gourworld

Valenzuela-Venegas, G., Salgado, J. C., & Díaz-Alvarado, F. A. (2016). Sustainability indicators for the assessment of eco-industrial parks: Classification and criteria for selection. *Journal of Cleaner Production, 133*, 99–116.

Vegter, D., van Hillegersberg, J., & Olthaar, M. (2020). Supply chains in circular business models: Processes and performance objectives. *Resources, Conservation and Recycling, 162*, 105046.

Voinov, A., Kolagani, N., McCall, M. K., Glynn, P. D., Kragt, M. E., Ostermann, F. O., & Ramu, P. (2016). Modelling with stakeholders – next generation. *Environmental Modelling & Software, 77*, 196–220.

Walker, A. M., Opferkuch, K., Roos Lindgreen, E., Raggi, A., Simboli, A., Vermeulen, W. J. V., Caeiro, S., & Salomone, R. (2021a). What is the relation between circular economy and sustainability? Answers from frontrunner companies engaged with circular economy practices. *Circular Economy and Sustainability*. doi:10.1007/s43615-021-00064-7

Walker, A. M., Opferkuch, K., Roos Lindgreen, E., Simboli, A., Vermeulen, W. J. V., & Raggi, A. (2021b). Assessing the social sustainability of circular economy practices: Industry perspectives from Italy and the Netherlands. *Sustainable Production and Consumption, 27*, 831–844.

Walker, A. M., Vermeulen, W. J. V., Simboli, A., & Raggi, A. (2021c). Sustainability assessment in circular inter-firm networks: An integrated framework of industrial ecology and circular supply chain management approaches. *Journal of Cleaner Production, 286*, 125457.

Walzberg, J., Lonca, G., Hanes, R. J., Eberle, A. L., Carpenter, A., & Heath, G. A. (2021). Do we need a new sustainability assessment method for the circular economy? A critical literature review. *Frontiers in Sustainability, 1*, 620047.

Wannags, L. L., & Gold, S. (2020). Assessing tensions in corporate sustainability transition: From a review of the literature towards an actor-oriented management approach. *Journal of Cleaner Production, 264*, 121662.

Wu, Z., & Pagell, M. (2011). Balancing priorities: Decision-making in sustainable supply chain management. *Journal of Operations Management, 29*, 577–590.

Zijp, M. C., Waaijers-van der Loop, S. L., Heijungs, R., Broeren, M. L. M., Peeters, R., Van Nieuwenhuijzen, A., & Posthuma, L. (2017). Method selection for sustainability assessments: The case of recovery of resources from waste water. *Journal of Environmental Management, 197*, 221–230.

# 11 Deliberation as a tool in cooperation with stakeholders in companies deploying the Circular Economy based on the example of Unimetal Recycling Sp. z o.o.

*Olga Janikowska, Joanna Kulczycka, and Agnieszka Nowaczek*

## Introduction

It has been clear for some time now that it is necessary for us to change the model implemented to ensure the fulfilment of our needs, including the model used for the management of resources. The idea of the Circular Economy (CE) is part of the trend related to the practical implementation of the concept of Sustainable Development. Development should be made sustainable on many levels, taking ecological, social, and economic aspects into account. The role of social participation in decision-making processes is emphasised by involving as many social groups as possible in processes guiding change. In promoting the rational management of resources and the minimisation of generated waste, the CE is consistent in many areas with the implementation of the principles of Sustainable Development, climate protection, and the deployment of eco-innovation (Kulczycka et al., 2016).

From the point of view of this chapter, business models and support for the cooperation of stakeholders in the value chain in the practical implementation of the principles of the CE seem to be of key importance. In this chapter, it is indicated that deliberation should become an essential tool in cooperation with stakeholders in entities deploying the CE. The chapter presents the process for the mapping of company stakeholders based on the example of Unimetal Recycling Sp. z o.o. (UMR), selected activities of the company for the benefit of its stakeholders, and also deliberation by the company together with stakeholders, so that they are included in processes of implementing the CE, because the opening up of decision-making processes to a broad spectrum of social actors is one of the key principles both of Sustainable Development and of the CE which has grown up around it. The company presented in the chapter may be considered to be the one that is implementing the CE model, though a critical analysis of its activity indicates that it is not a company that is implementing the principles of the CE completely. This results from the fact that the elements recovered from the catalytic monolith are processed further after being transported outside of the European Union. The company is already working on a solution to this problem

DOI: 10.4324/9781003179788-11

and, as one of its next activities, is planning to build a processing plan, which will make it possible to fully close the loop.

## Deliberation as a key component in the socialisation of decision-making processes, the building of social capital and support for investment in the CE

In a time of economic crisis, the reasons for which many theorists attribute to weakness in the system of representation, deliberative democracy seems to be an alternative that may allow this impasse to be overcome. The inclusion of citizens in the decision-making process opens up many new possibilities and may fulfil a range of functions in assisting with democratic processes, as well as processes of social education. Many theorists (Gutmann, 1998; Frith, 2008; Dryzek, 2001; Gagnon, 2011; Dahl, 2012; Held, 2006; Gutman & Thompson, 2002; Šardecka-Nowak, 2008; Janikowska & Słodczyk, 2016) consider that the essence of deliberative democracy comes down to the belief that the decision-making process should be based on a non-violent and non-coercive process of argumentation. This brings clear benefits as a result of the deliberation process, as the company's stakeholders will feel that they are not only the addressees of the changes, but also their co-creators. This is especially valuable in the context of dissemination of the principles of the CE as well as its implementation by cooperating entities. Deliberative democracy emphasises the greater importance of dialogue, discussion, and debate. Its proponents argue that debate improves the quality of collective decisions and increases the chances of their acceptance. Thus, the development of new solutions as part of either the practical implementation of Sustainable Development or the CE can be achieved through social discourse and with social acceptance of the changes taking place. It is important to identify and involve relevant entities, that is people, groups, non-profit organisations, entrepreneurs, experts, and educational institutions, which create a social forum of stakeholders and also contribute to generating a sense of community (Janikowska & Słodczyk, 2016).

A model process of deliberation with company stakeholders should take the following elements into account; all stakeholders should have a real and equal opportunity to express their preferences. Stakeholders should be able to have an influence over the setting of agendas and the presentation of their reasons for the preferred outcome. When determining the result, all expressed preferences should be taken into account.

The introduction of a model of deliberation with stakeholders into the investment process builds social capital, which is indispensable for social participation in decision-making processes, and is thus a boundary condition for social participation in implementation of the CE. This requires both an understanding of the essence of the CE, as well as assessment and monitoring of the risks and benefits associated with the planned investment for local communities and development of the economy. In this respect, it is important to identify stakeholders in the CE and their role across the entire value chain. This is a new challenge as the studies conducted to date have been for individual entities.

The notion of stakeholder was first used by H.I. Ansoff in the context of the definition of company goals. However, the term was formulated and defined in the mid-1980s in an article by R.E. Freeman entitled "Strategic Management – a Stakeholder Approach". However, there is still no one generally accepted interpretation of the concept of stakeholder in the literature, and Freeman and Reed distinguished between a narrower and a broader interpretation:

• The narrower one – assumes that the company cannot meet the expectations of all stakeholders, so it should focus on achieving the goals of a limited group of them, having a real impact on the company's operations and the expected benefits from this.
• the broader one – includes anyone as stakeholders who can have an influence on the organisation or is under its influence or, in other words, groups that are in any way involved in the company's interests or make demands on it.

This approach to stakeholders becomes even more complicated when we assume broad economic cooperation, which is the basis of the CE model. Identifying just one of the models, for example economic symbiosis, which consists of cooperation between companies with the aim of exchanging surplus energy, materials, and the implementation of joint ventures whose purpose is not only to facilitate organisational and technical activities, but also to reduce costs, we must take into account the stakeholders in individual as well as common entities (Figure 11.1).

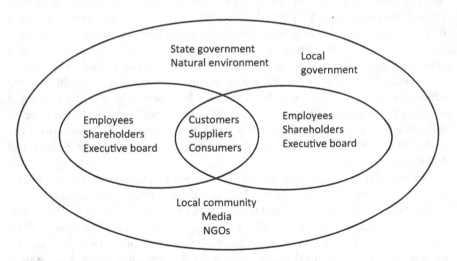

*Figure 11.1* Map of stakeholders of cooperating companies broken down into common, internal, and external shareholders forming an economic symbiosis

That is why it is so important to identify and rank stakeholders. D. Walker, A. Shelley, and L. Bourne draw attention to the fact that there are many different tools for use in the process of identification of stakeholders, for example brainstorming, media, mind mapping, and benchmarking. In the case of the CE, the map is created for the whole chain of suppliers and consumers, which means not only it is more time-consuming, but also makes it possible to see more dependencies and potential benefits, especially for local markets. This, however, requires transparent communication, as well as active and substantive engagement in discussion.

## Social communication and the ethics of discourse as indispensable elements of stakeholder participation in the process of deployment of the CE for entities in the value chain

From the point of view of involvement of stakeholders in the process of deployment of the CE, an important element is social deliberation with the stakeholders grouped around cooperating entities. As shown in Figure 11.1, there are situations where they may have common stakeholders in their closer and more distant surroundings. The input condition for such deliberation is thus communication, which should start at the level of the cooperating companies. The procedure to follow in such a situation should begin with the mapping of common stakeholders, both those from the closer and more distant surroundings.

Social communication, which is understood as a process of a bilateral or multilateral flow of information and experiences between people, social groups, the authorities, and the economic environment, is of immeasurable importance for subsequent deliberation with stakeholders. As J. Habermas (1999) writes, the *world of life* opens up only through a subject who makes use of his linguistic competence and capacity for action. Therefore, the first and most important condition for participation or co-participation in the *world of life* is the use of linguistic competences. In view of this, we assume that the most important element in decision-making by many decision-makers should be mutual communication between them.

The document regulating matters related to access to information, public participation in decision-making, and access to justice in environmental matters is the Aarhus Convention adopted on 25 June 1998 in Denmark at the 4th Pan-European Conference of Environment Ministers. In Poland, it was adopted by the Act of 21 June 2001 on Ratification of the Convention on Access to Information, Public Participation in Decision-Making and Access to Justice in Environmental Matters (Journal of Laws of 2001 No. 89 item 970). Further legal regulations and high EU standards introduced in the field of environmental protection allow freedom of access to environmental data and also require the consent of stakeholders to be sought in the case of new investments, which is usually a long-term process.

It should be emphasised that social acceptance of changes is extremely important for the company's operations. Often, eco-innovations introduced as part of the CE are related to the re-management of waste, which may be a source of social concern. This applies not only to landfills or waste incineration plants, but also to recycling plants. One example of an investment that aroused controversy among the local community, and thus with one of the company's potential stakeholders, is the planned battery and catalytic converter recycling plant in the Kosówka district of Libiąż. Seven hundred and fifty inhabitants of Libiąż and Chełmek signed up to a protest against this investment. This opposition resulted mainly from social concerns about the nuisance associated with this type of activity. The investment was planned in the area of the former branch of the Ruch II "Janina" Coal Mine. There have also been protests and disputes ending in court judgements concerning the expansion of plants already in operation, for example the company Krynicki Recykling which planned to purchase additional equipment for the sorting of a mixture of glass cullet and to increase the efficiency of the plant. In some cases, the granting of approval for investments for recycling plants has been the subject of parliamentary interventions, for example in the case of the plant in Łosice.

In Poland, public participation in decision-making processes is regulated, *inter alia*, by Article 58.1 of The Constitution of the Republic of Poland which guarantees the freedom of association to everyone, as well as by the Law on Foundations of April 6, 1984 and the Law on Associations of May 31, 2001. In the context of processes of deliberation, attention should be paid to the European Charter of Local Self-Government. Its establishment was the next step in the building of a democratic system. The Charter speaks of the right of citizens to participate in the conduct of public affairs, which is one of the democratic principles of member states of the Council of Europe and the need to equip society with real powers enabling effective administration. That is why it is so important to create appropriate instruments allowing for the process not only of deliberation, but also that of education of the company's stakeholders. It is thus also important to indicate that investments carried out within the CE will be able to bring long-term social, environmental, and economic benefits. In the course of deliberation with stakeholders, the company should be guided by the following principles:

1    Participation in such a deliberation should be based on the norm of both equality and symmetry, which means that all its participants, regardless of whether they are in the company's closer or more distant surroundings, should have the right conditions for initiating acts, asking questions, proposing conclusions, opening discussion, and questioning the designated topics of discussion.

2    All participants in deliberation, regardless of whether they are in the company's closer or more distant surroundings, should have the right to dispute the principles of the procedure of discourse and the forms of their application or execution. Thus, in this model, there is no *prima facie* principle limiting the subjects of deliberation or determining the identity of participants, as long as each excluded person or group is able to prove that they are in some way subject to the norm that is being deliberated.

Deliberation as a special form of cooperation between the company's stakeholders is based on the following principles:

- Difficult choices – this means that often the issue that is the subject of deliberation will focus on problems that will require stakeholders to make decisions between competing values. For example the local community may have legitimate concerns about the processing in their area of waste, which will become useful as soon as it has been processed. The resolution of these concerns requires some difficult decisions and choices to be made. These decisions should be determined by way of compromise – related to the situation where several well-founded propositions (values) diverge in opposite directions.

- Democratisation of decision-making within and outside the company – deliberation assumes that activities in favour of the CE understood to fall within the scope of the company's activities require involvement and coordination on the part of stakeholders.

- Inclusion and equality – the processes of deliberation with the company's stakeholders must be based on the principle of inclusion and equality. Here, a balance is to be stuck between the importance of the opinions and judgements of the participants in deliberation. Deliberation requires genuine diversity of thinking and perspectives offered by stakeholders from various walks of life and cultural groups, from different educational or professional backgrounds, etc.

One of the most important advantages of deliberation is the fact that, when making decisions, all stakeholders have a real and equal opportunity to express their preferences and can influence the setting of the agenda, presenting their arguments in favour of the preferred solution to the problem. Moreover, in the decisive stage of the decision-making process, each of the stakeholders has the same opportunity to express their preferences, which must be considered to be just as valid as the preferences of other stakeholders. The outcome should however be determined by taking all the presented positions into account.

## Use of mapping of company stakeholders in the CE based on the example of Unimetal Recycling Sp. z o.o. (UMR)

A detailed analysis of the scientific literature, of strategic documents developed at the EU forum, and of countries and regions regarding the CE shows that many methods and indicators have been proposed for monitoring activities related to the CE. It focuses in real terms on the use of resources (raw materials), seeking to maximise their value and increase the efficiency of management of resources and the durability of products, and on methods of preventing and minimising waste, which are appropriate to and in line with the waste hierarchy. Such activities may be carried out not only by means of eco-innovative technological but also organisational solutions, taking maximisation into account in the value chain or, in other words, from designer to consumer. The related monitoring of the transformation towards the CE concerns not only the technological aspects and methods of resource and waste handling, but also its economic, social, and environmental aspects. It is important to make an assessment that takes the whole value chain into account, that is from the design phase, production, consumption, repair, and regeneration through to waste management and the recovery of secondary raw materials that can be put back into circulation in the economy. For each of these steps, it is important to identify key stakeholders, so that they can be involved in the process of implementation of the CE at every step in the company's activities, because the opening up of decision-making processes to a broad spectrum of social actors is not only a key principle of the CE, but above all an excellent tool for obtaining social acceptance of the changes taking place.

In this respect, the activity of the company UMR, which consists of the processing of automotive catalytic converters, may be a good example of a company whose activity is in line with the principles of the CE. The company has many years of experience in the purchase and processing of automotive catalytic converters in Poland and around the world. UMR purchases automotive catalytic converters with a ceramic and metal core, as well as ceramic monolith in pieces or in powder form. Importantly, the company pays the actual price for them – based on a specified amount of precious metals by introducing an IT solution proposed by the WasteMaster start-up (Urban, 2020).

The company is based on a CE model in line with the Polish National Smart Specialisation (*Krajowa Inteligentna Specjalizacja – KIS 7* – water, mining raw materials, waste). The process of valuation of catalytic converters is based on two different techniques, one of which consists of drilling into the monolith, taking samples for laboratory analysis and determining the content of platinum, palladium, rhodium (and other components that have an influence on the result). However, the second, more precise technique consists of cutting up the catalytic converters, removing the monolith, and crushing it into smaller pieces in a vessel, before then being placed in a mixer where the monolith is then ground up by blades. The samples are subsequently sent to the laboratory for examination under a spectrometer. The company's activity is of particular significance from the point of view of one of the fundamental messages of Sustainable Development, namely the sustainable management of raw materials. It is also important from the point of view of the founding principle that any material that has been taken from the environment once should be re-used several times in cycles of a closed loop.

A critical analysis of the company's activity however shows that it cannot be considered to be a company that is implementing the CE model completely, as, after recovering the platinum, palladium, and rhodium from the catalytic monolith, the company sends these elements outside the European Union. Full implementation of the CE model would require the creation of the last fraction, namely a facility for the recovery of pure metal. One interesting solution could also be to create an industrial symbiosis with entities specialising in this area or with start-ups.

In the case of platinum metals, recycling makes it possible to obtain platinum much more economically than the traditional method of extraction. To recover 1 kilogram of platinum, it is necessary to process 1 metric ton of catalytic converters (1 kg of platinum can be produced from 300 catalytic converters), while the extraction of 1 kg platinum in the traditional way requires the processing of 150 metric tonnes of ore and 400 metric tons of waste rock down to a depth of 1,000 metres. The process of recycling of catalytic converters in accordance with the technology used in the company consists of the following steps:

1   Sorting of catalytic converters according to type of monolith core (ceramic, metal);
2   Tearing off or cutting up of the metal cover;
3   Removal of insulating material;
4   Extraction and grinding down of the ceramic monolith into a fine powder in ball mills;
5   Refining.

*Internal stakeholders*

The added value of UMR is its personnel, who have the opportunity to improve their qualifications on an ongoing basis thanks to the fact that company guarantees them access to training. The company has a network of suppliers and contractors not only in Poland, but also in other European countries, including in Germany, Norway, Italy, and Ireland. It purchases catalytic converters in 16 voivodeships of Poland. The company has its own fleet of vehicles, allowing it to pick up catalytic converters directly from the customer. The company operates in a transparent manner: An initial valuation of the catalytic converter can be obtained over the phone or by e-mail. Catalytic converters are classified according to transparent and understandable criteria, and the company has all permits required by law. UMR is one of the few companies in Poland with permission to collect and process waste with the code 16 08 01 and 16 08 03. The beginning of the code 16 08 refers to the used catalytic converters, while the ending 01 specifies products containing gold, silver, rhenium, rhodium, palladium, iridium, or platinum (excluding those that have been contaminated with hazardous substances). The ending (03) designates products containing transition metals or their non-hazardous compounds. According to data for 2019, the company purchased 3,000 catalytic converters every day which it then recycled. As a result, UMR has 50–60% share of the market for the recycling or automotive catalytic converters in Poland. In 2019, it obtained around 300 metric tons of the 500–550 metric tons of monolith available on the market originating from 514 thousand registered vehicles handed over to disassembly stations (Kapczyńska, 2020).

However, such activities require not only modern technological solutions, but also the construction of an entire supply system for used catalytic converters and the definition of transparent rules of cooperation. A map of stakeholders is presented in Figure 11.2.

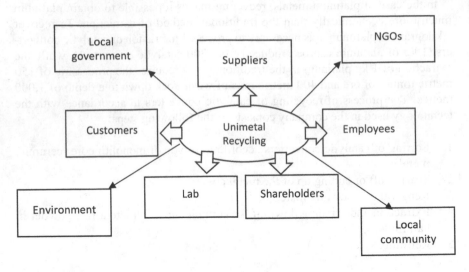

*Figure 11.2* Map of UMR stakeholders

### *Relations of the company UMR with external stakeholders*

A company implementing the principles of the CE should place particular emphasis not only on relations with stakeholders, but also on open dialogue with them. Under conditions where prices of recovered precious metals change daily and where there is a wide range of catalytic converters purchased and destined to be recycled, it is necessary to develop solutions that allow all stakeholders to be involved in changes. This requires permanent contact with your stakeholders, conducting deliberation with them that underlines not just the economic aspects, but the ecological aspects too. This should be made easier to the extent that most companies today apply a model of Sustainable Development, which means that the company focusses its activities not only on improvement of the entity's situation in terms of finances and assets, but also on the improvement of the conditions under which its employees operate and their quality of life, and making efforts to improve standards for the protection of the natural environment. The consolidation of activities upon the achievement of predetermined strategic goals must not only take conditions of operation into account, but also be an expression of the necessity to care about the economic entity's future. For the company's essential operations, this means that, in addition to the achievement of economic goals, such as profit, financial liquidity, and competitiveness, social responsibility becomes an extremely important or even indispensable part of their image, thus necessitating a continuous process of communication with its stakeholders.

UMR, as the largest recycler of catalytic converters in Poland, is a company which implements a CE model in its activities known as the waste value model. This is a model which concentrates on the recovery of used resources by means of processes of recycling that make waste from one production process usable as input material for another production process. Activities such as recycling and upcycling play a key role here. Recovery and recycling allow products to be recovered at the end of their life cycle and new value to be created from them.

An actively updated list of stakeholders and partners providing professional customer service in the purchase of catalytic converters by voivodeship within Poland and on international markets is available on the company's website and contains detailed information, including address, opening hours, and name of person with mobile phone contact number, making it possible to get in touch quickly.

Figure 11.3 presents a detailed and more complex map of UMR's stakeholders. It is possible to clearly distinguish entities cooperating closely with the company, which share stakeholders with it. In the case of companies deploying the CE, it is very often the case that their activities extend beyond the scope of activities for their own needs and become to some degree intertwined with companies from the closer or more distant surroundings, an example of which may be industrial parks, which are an inseparable element of industrial symbiosis. Such close cooperation between companies leads to an aggregation of stakeholders. The procedure to follow in the case of companies which remain in close cooperation, the effect of

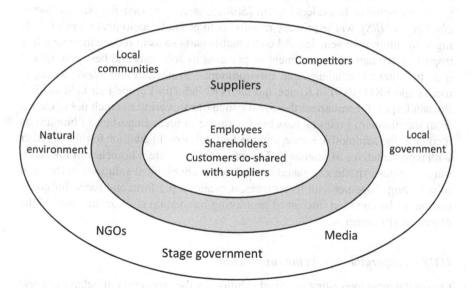

*Figure 11.3* Map of stakeholders for entities deploying the CE based on the example of UMR

which is to bring together some of the stakeholders, should be based on the following points:

- Selection of leading companies, which will play a coordinating role in the stakeholder mapping process;
- Internal consultation within cooperating companies, with the aim of drawing up an initial mapping of stakeholders common to the companies;
- Process of proper mapping of stakeholders common to companies.

Importantly, in its activities, UMR clearly demonstrates that it is capable of fulfilling the role of a leader among the companies around it, while maintaining good relations with all its stakeholders, and the activities carried out by the company for the benefit of stakeholders include active participation in research programs and conferences, as well as ecological and educational campaigns. In 2020, UMR started the implementation of a project co-financed under the Polish Operational Programme for Smart Growth [*Program Operacyjny Inteligentny Rozwój – POIR*] entitled "*Opracowanie innowacyjnej w pełni zautomatyzowanej i mobilnej technologii recyklingu katalizatorów*" [Development of an innovative, fully automated and mobile catalytic converter recycling technology] worth: 14,736,427.72 PLN. These activities also include analyses and research conducted in its own laboratory, including a catalogue that has already been put together containing information on more than 20 thousand types/models of catalytic converter (Unimetal, https://unimetalrecycling.pl/en/offer/). The company is a member of the Polish Automotive Recyclers Forum [*Stowarzyszenia Forum Recyklingu Samochodów – FORS*], whose primary mission is to promote ecological ways of dealing with automotive vehicles and consumable parts so as to reduce their negative impact on the natural environment. It presents its solutions and best practices at industry shows (including 22nd Environmental Protection and Waste Management Expo EKOTECH in Kielce and the ITM Industry Europe Fair held at MTP Poznań Expo). It emphasises the significance of recovering as much noble metals from the platinum group as possible for re-use in areas important to humans as medicine, the automotive sector, and the fuel industry. In addition to this, together with representatives of science and business from different branches of the economy, it shares experience related to the latest technological solutions in the field of recycling. Together with its partners, it creates a platform and space for cooperation in the common interest of promoting innovation and for the good of the natural environment.

### UMR – cooperation across the supply chain

One of the most interesting practical solutions is the promotion of behaviours and practices that foster mutual trust as well as caring about mutual benefits and a sense of security when concluding transactions. To do this, UMR has developed the "Healthy Business Relations" programme, the purpose of which is to promote the idea of ethical behaviour towards customers among entrepreneurs and to support those companies to give them greater opportunities for development. The programme is also dedicated to people who are just planning to set up business in the field of the collection of catalytic converters. To this end, UMR has developed its own certificate. UMR organises a wide range of social actions such as joint cleanups of woodland areas and cemeteries to remove bottles, cans, glass, and other waste, which also serve to bring the local community together. Such events are often combined with educational workshops with the aim of raising awareness in the field of waste segregation and recycling and improving knowledge of best practices in environmental protection. Moreover, UMR is willing to take part in activities to shape pro-ecological attitudes among young children and to encourage them to care for the natural environment by using methods for the creative transfer of knowledge in the field of ecology. In this area, the company cooperates with local schools and foundations, pointing to the need to protect the health of the people and keep the air clean, among other things, thanks to the use of catalytic converters and their proper recycling.

UMR also provides financial support for a variety of social initiatives, which aim to promote a healthy life style, sport, and safe behaviours on the road (*"Bezpieczny Powiat Chrzanowski"* [Safe District of Chrzanów]). Another important area is the cooperation with charity organisations and those providing help to those in need and the disabled, for which the company received the *"Serce za serce"* ["Heart for Heart"] award for its support for the Trzebinia branch of the *Towarzystwo Przyjaciół Dzieci* [Polish Society for the Friends of Children]. The company takes part in an action to collect toys and sweets, including for children at the Care and Educational Centre in Miękinia, as well as in auctions of various types held at local events. The company also encourages the development of young talent, for example by providing financial support to the 23rd International Tournament of Young Talents in Active Chess [*XXIII Międzynarodowy Turniej Młodych Talentów w Szachach Aktywnych*]. Another interesting example of stakeholder engagement is the multi-faceted cooperation with the local Municipal Public Library in Chrzanów, the main aim of which is to support the local community and promote the idea of Sustainable Development. In addition to this, the company continues to grow, taking part in training (including *"Energooszczędne Budownictwo oraz Zastosowanie Odnawialnych Źródeł Energii dla MŚP"* [Energy-efficient Construction and Application of Renewable Energy Sources for SMEs]) and conferences (including the 4th edition of the Conference on "Young Researchers' Innovative Ideas: Science Start-Ups in the Industry") to constantly improve the quality of the services it provides, which will be in line with the latest trends and practices, not only in the field of recycling, but also in that of corporate social responsibility and Sustainable Development.

An important element in the company's activity is deliberation with stakeholders, the purpose of which is not only to consult on projects undertaken within the framework of its activity, but also to actively involve stakeholders in processes of change at the company. As part of deliberation with the local community, various types of meetings are held in the form of joint activities for the benefit of the natural environment or the local community combined with educational elements.

During the course of these meetings, the company presents its activities and current plans and educates the local community on environmental issues. It is worth emphasising that the company started to conduct deliberations with its stakeholders based on a sense of intuition by inviting specialists and social experts and demonstrated transparency in its activities by building a forum for the exchange of experiences with its stakeholders. Maintaining relationships and open dialogue as well as encouraging constant evaluation of the activity it conducts are part of what has made the company successful. Thanks to the fact that the UMR maintains a presence with those both in its closer and more distant surroundings, it has earned a great deal of social trust. Deliberation conducted as part of the company's activity with those both in its closer and more distant surroundings has led to the building of pro-ecological attitudes among stakeholders, which in turn has had an impact on the propagation of the CE model.

Given that the Internet has become an important element in the reality of today, making it possible to create media platforms that can serve as a tool for building relations between the company and its stakeholders, the company uses it for deliberation with its stakeholders. UMR's deliberation thus includes both direct meetings with the local community, suppliers, employees, shareholders, local authorities, and the media, as well as a range of activities and campaigns conducted by means of instant messaging that give stakeholders the opportunity to provide feedback. Table 11.1 presents examples of the company's activities for the benefit of its stakeholders, together with details on the form of deliberation. It should be clearly emphasised that deliberation with stakeholders is not only a way of aggregating their preferences, but perhaps most importantly an educational process as well. All activities listed in the table should therefore be included in the deliberation the company conducts with its stakeholders.

## Conclusion

Cooperation and open dialogue with stakeholders are a boundary condition for the CE or, in other words, for an economic model which aims to achieve rational use of resources and limit their negative impact on the environment. The essence of the CE is that materials and raw materials remain in the economy as long as possible, while the waste generated is minimised. For practical reasons, it seems very important to define the concept of company stakeholders and indicate what the process of mapping of the company's stakeholders should look like – so that they are included in processes of implementing the CE, because the opening up of decision-making processes to a broad spectrum of social actors is one of the key principles of the CE.

This chapter has presented the principles of deliberative democracy as a key component in the socialisation of decision-making processes, also pointing to the educational role of processes of public deliberation and to the fact that, as a result of the deliberation process, stakeholders will feel that they are not only the addressees of the newly created law, but also its creators and even co-creators. This is especially valuable in the context of dissemination of the principles of the

*Table 11.1* Examples of activities of Unimetal Recycling in deliberation with stakeholders

| Educational and pro-ecological activities | | Deliberation activities | |
|---|---|---|---|
| *Activities* | *Stakeholders engaged* | *Activities* | *Stakeholders engaged* |
| Regular actions to clean up the local surroundings, such as parks, places for the local community to relax, and cemeteries | Local community/ natural environment/ NGOs/local authorities/ employees | Regular educational meetings with the local community | Local community/ NGOs/local authorities/ employees |
| Charity activities for the benefit of the local community | Local community/ NGOs/local authorities/ employees | Open and transparent process for the valuation of catalytic converters, together with the possibility of presentation of a spectrometer and discussion of how the valuation is carried out | Customers/ suppliers |
| Participation in industry shows/ training. System of training for employee | Employees/suppliers/ shareholders/ customers | Open process of communication with customers/ suppliers/the local community by means of instant messaging | Customers/ suppliers/local community |
| Setting up of an environmental protection department | Employees/suppliers/ shareholders/ customers/ environment/local authorities | Educational and information meetings | Employees/ customers/ suppliers/ shareholders |

CE, the social acceptance of the changes taking place, as well as the implementation of the CE at company level. It was shown that UMR may be considered to be a model example of a company that is building lasting relationships with its stakeholders and becoming a leader among companies which share stakeholders with it. Transparency in the activities it undertakes as well as possessing a fully equipped laboratory and its own software have allowed the company to expand its supply network. In creating an ever broader group of suppliers, UMR is proving that recycling of catalytic converters is something that is concerned not only with the technology of obtaining the monolith from the catalytic converter, but also the company's transparency (Generowicz, Kulczycka, Partyka, & Saługa, 2021). The company's activities allow it to build trust among the company's stakeholders.

Trust is, in turn, an indispensable element of social capital. It gives the company the potential to deliberate with its stakeholders and thereby helps the company to undertake activities in the field of the CE not only on its own premises, but also with its stakeholders. A particularly valuable initiative and best practice is the system of rewards created based on an open system of online assessment and comments allowing reliable, committed, and honest contractors to be selected right across the supply chain.

# References

Dahl, R. A. (2012). *Democracy and its critics*. New Haven: Yale University Press.

Dryzek, J. S. (2001). Legitimacy and economy in deliberative democracy. *Political Theory, 29*(5), 651.

Frith, R. (2008). Cosmopolitan democracy and the EU: The case of gender. *Political Studies, 56*(1).

Gagnon, J. (2011). A potential demarcation between "old" and "new" democratic theory? An attempt at positioning a segment of the extant literature. *Social Alternatives, 30*(3).

Generowicz, N., Kulczycka, J., Partyka, M., & Saługa, K. (2021). Key challenges and opportunities for an effective supply chain system in the catalyst recycling market – a case study of Poland. *Resources, 10*(2), 13. doi:10.3390/resources10020013

Gutmann, A. (1998). Deliberative democracy. *Liberal Education, 84*(1).

Gutman, A., & Thomson, D. (2002). Deliberative democracy beyond process. *The Journal of Philosophy, 10*(2).

Habermas, J. (1999). *Theorie des kommunikativen Handelns*. Frankfurt a.M: Suhrkamp.

Held, D. (2006). *Models of democracy*. Stanford: Stanford University Press.

Janikowska, O., & Słodczyk, J. (2016). *Globalna sprawiedliwość (Global justice)*. Opole: Wydawnictwo Uniwersytetu Opolskiego.

Kapczyńska, K. (2020). *WasteMaster zdobył inwestora*. Retrieved from www.pb.pl/smieci-w-aplikacji-1102201

Kulczycka, J. et al. (Eds.). (2016). *Surowce kluczowe dla polskiej gospodarki (Key raw materials for the Polish economy)*. Kraków: IGSMiE PAN.

Śardecka-Nowak, M. (2008). *Demokracja deliberatywna jako remedium na ponowoczesny kryzys legitymizacji władzy (Deliberative democracy as a remedy for the postmodern crisis of the legitimacy of power)*. Teka Kom. Politol. i Stos. Międzynar. – OL PAN. Retrieved from https://fbc.pionier.net.pl/details/nnlSrZ7

Unimetal Recycling. Retrieved from https://unimetalrecycling.pl/oferta/

Urban, K. (2020). Sprawność katalizatora a jakość powietrza (Catalyst efficiency and air quality). *Dostęp Online*. Retrieved from http://odpowiedzialny-inwestor.pl/2020/12/24/sprawnosc-katalizatora-a-jakosc-powietrza/

# 12 Circular models for sustainable supply chain management

*Magdalena Muradin*

## Introduction

The aim of this chapter is to discuss the possibility of implementing the principles of the Circular Economy in existing sustainable supply chain models. Based on elements of the supply chain which constitute hotspots, for example production, transport, and waste management, the discussion presented in this chapter tries to resolve the dilemma of whether it is possible to fully integrate the Circular Economy and sustainable supply chain management or not (SSCM Supply chains are one of the very important elements of the modern economy that have an influence on achievement of the European Union's goals in implementation of the Green Deal). This is why it is also necessary to take not only the principles of Sustainable Development, but also those of the Circular Economy into account in their design.

This chapter also provides an overview of the most important assumptions concerning the creation of circular business models taking account of different approaches to sustainable supply chain management and discussing factors stimulating and inhibiting SSC development in the context of the CE. It also presents practical examples of business models taken from the field of the bioeconomy, which can provide the industry with support in making the transition from a linear model to a circular model of SSC management. The models characterised in the chapter may be used by managers as an example of the effective implementation of elements of the CE in sustainable supply chains, especially in one of the strategic industries, namely the bioeconomy. The discussion presented in this chapter adopts a critical approach to supply chains in the context of divergence between the principles of Sustainable Development and those of the Circular Economy. The analysis also considers the problem from the point of view of thermodynamics, which is an important and often overlooked perspective in discussion of development of the CE and in the case of implementation of the CE in sustainable supply chains too.

## Integration of the Circular Economy and SSCM

The Circular Economy may be seen as a natural continuation towards achieving a set of previously defined Sustainable Development Goals, or, as in the case

DOI: 10.4324/9781003179788-12

of (Geissdoerfer, Savaget, Bocken, & Hultink, 2017), the concept of the CE is perceived as a complete or partial solution allowing full synergy to be achieved in the economic, environmental, and social spheres. In turn, sustainable supply chains are considered to be key in the context of implementation of the CE, which however requires the involvement of all interested parties in the production and services sector (Muñoz-Torres et al., 2018). They take into account entire industrial ecosystems, allowing for inter-organisational and multi-functional integration of all links in the product life cycle from the design stage, through marketing, logistics, and production to final management.

The foundation of Sustainable Development or green supply chain management is not closing the loop but the "triple bottom line" (TBL) model (Geissdoerfer, Morioka, de Carvalho, & Evans, 2018). It is also for this reason that integration may limit not only the production of waste, but also the creation of self-sustaining production systems in which materials return to the production cycle constituting a closed loop (Genovese, Acquaye, Figueroa, & Koh, 2017). The fundamental question is to determine whether sustainable supply chains (SSCs) can automatically be transformed into circular supply chains (CSCs) or not, or whether significant changes will nevertheless have to be introduced at different levels of decision-making and functionality.

The principles of the CE can be implemented in a Sustainable Development paradigm on the level of three different types of relationships (Geissdoerfer et al., 2017):

1   Conditional;
2   Optional;
3   Potential.

In a conditional relationship, it is assumed that the CE is a condition of Sustainable Development, without which the paradigm cannot be realised. In an optional approach, the CE is only one of many options that can contribute to Sustainable Development but is not a necessary condition for it. The third type of relationship is however based on the assumption that the CE can lead not only to potential benefits, but also to costs related to its implementation within the structure of Sustainable Development.

When including the Circular Economy in supply chain management, the problem should be considered holistically taking different criteria and indicators of efficient management and implementation of the CE into account while maintaining a balance between economic, social, environmental and operative efficiencies (Zeng, Chen, Xiao, & Zhou, 2017).

Figure 12.1 presents the relationships between specific elements of the supply chain in the context of xR or, in other words, the main principles of the Circular Economy – repair – reuse, recycle, redesign, remanufacture, reduce, and recover. Only the integration of sustainable supply chains with the CE allows for all the principles mentioned to be implemented in production systems.

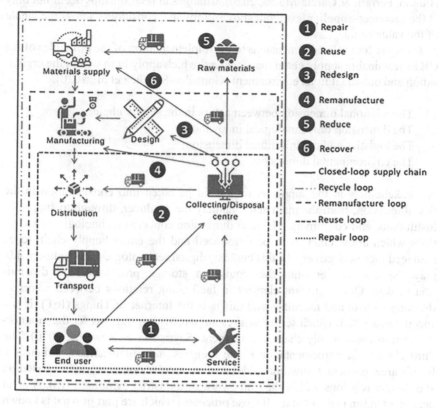

*Figure 12.1* Links between the 6R principles of the CE and elements of the supply chain
Source: Own elaboration based on Zeng et al. (2017).

The Circular Economy, the aim of which is to limit or, as part of a long-term approach, to eliminate negative externalities caused by the consumption of resources, is based on two main pillars (Beu, Ciugudeanu, & Buzdugan, 2018; De Angelis, Howard, & Miemczyk, 2018):

- Extending the period of time for which materials are kept in use;
- Prolonging the durability of those materials by increasing cycles of regeneration, repair, and recycling.

Both pillars require an approach to product design which encourages consumers to use the product for a long time and to repair or regenerate it.

Integration of sustainable supply chains and the CE allows the number of actors in the supply chain to be expanded by taking other sectors of the economy and change in relationships between entities into consideration (González-Sánchez,

Blundo, Ferrari, & García-Muiña, 2020). Changes in relationships occur not only at the producer–supplier level, but above all involve consumers across the whole of the value chain.

There are four related dimensions to the implementation of the principles of the CE in sustainable supply chain management, which apply both within the organisation and outside it in its environment (González-Sánchez et al., 2020):

1   The relational dimension between actors in the supply chain;
2   The dimension of technological innovations;
3   The logistics and organisational dimension;
4   The environmental dimension.

All stakeholders in the supply chain should be taken into consideration in the first dimension, from the main actor, namely the producer, through to suppliers, institutions, and consumers. The next dimension concerns technological innovations which allow changes to be introduced and the entire supply chain to be managed more effectively. In this context, digital technologies and Industry 4.0. play a special role, enabling the acquisition, storage, processing, and distribution of data. One important element in facilitating relations between actors in the supply chain and machines and things is the Internet of Things (IoT). It is a specific ecosystem which serves as a resource providing support for the monitoring of flows and supply chain management (Szymczak & Nowicka, 2020). The third element in implementation is the logistics and organisational dimension. In this area, particular attention is devoted to circular business models based on stakeholder relations within the framework of industrial symbiosis (IS). A great deal of attention is devoted to all those processes which are part of what is known as reverse logistics or, in other words, a model of waste management according to the current hierarchy (from reuse to landfill), maximisation of recovery of materials, and effective methods of recycling. Industrial symbiosis involves creating a network of cooperation between businesses, often from separate branches of industry, in order to exchange different types of resources: materials, energy, water, by-products, and waste. Relationships occurring in IS differ from those in traditional supply chain relationships. Waste management within the framework of IS poses a challenge, because the waste is not produced at the request of the users of the waste, and the quantity and quality of the waste produced depend on the quantity of primary products produced. IS expands the traditional supply chain to include symbiotic suppliers and buyers and companies which exchange resources or information with each other. The issue of industrial symbiosis is also raised in the chapter on the "Interrelationship between Sustainable Manufacturing and Circular Economy". The fourth dimension is that of environmental issues. Here, the priority is to propose appropriate legal regulations and legislative mechanisms. Here, supporting instruments allowing for the effective implementation of the CE have an essential role to play. There are numerous proposed incentives which would appear to be helpful in this area, for example in the form of exempting recycled products from taxes (De Oliveira, Luna, & Campos, 2019) or subsidies to encourage the use of renewable energy. Of course, it is also possible to introduce fees for failure to meet the appropriate levels of emissions, or similar, but this type of activity may have the opposite effect than that expected. Another effective incentive would be the introduction of transparent laws at national and European levels (De Angelis et al., 2018). The key issue in this area then also becomes one of education at all levels of the entire supply chain. Awareness of the need to introduce changes in supply chain management is the first step towards changing the mentality, habits, and behaviours of society as a whole.

Integration of the CE and SSC should take place in all dimensions simultaneously, and it is practically impossible to separate them, without having a negative impact on the final structure of the supply chain. Individual elements in these four areas are interrelated and serve as the foundations for supply chain management in the Circular Economy. Relationships between individual elements have an impact on the effectiveness of implementation of CE. An example of this is the use of IoT in processes of returns or closing the loop at different stages of the supply chain (Zheng, Yang, Yang, & Zhang, 2017) or the development of different types of subsidies supporting technological innovations in the area of the CE.

## Circular business models in sustainable supply chain management

The main difference between conventional business models and circular models lies in their value creation and delivery element (Geissdoerfer et al., 2018). Regardless of which term we choose to use, whether it be sustainable, green, or circular supply chains, primarily in the context of building circular business models (CBMs), we should start considering the circulation of resources in the economy. CBMs are based on closing, slowing, or narrowing flows of materials and energy (Bocken, de Pauw, Bakker, & van der Grinten, 2016). Based on the literature (Bocken et al., 2016; Geissdoerfer et al., 2018; González-Sánchez et al., 2020), it can be assumed that circular business models of sustainable supply chains can be defined as a way of managing all organisational functions of supply chains in order to close the loop of materials and energy and minimise the quantity of resources introduced into the production system and reduce outputs from the system in the form of waste and emissions, while at the same time improving the operative effectiveness and efficiency of the system allowing the creation of additional monetary and non-monetary value and the pro-active management of multiple stakeholders and incorporating a long-term perspective of gaining a competitive advantage in order to be able to achieve optimum results in the area of Sustainable Development.

Table 12.1 presents the most important circular business models, which should be taken into account at individual stages of supply chains. Use of the tools presented at individual stages of the supply chain allows for conditions of integration to be met and goals to be achieved. The involvement of all stakeholders allows for an efficient and effective management in a sustainable way.

*Table 12.1* Elements of circular business models of sustainable supply chains

| Stakeholders | Tools | Processes | Goals | Conditions |
|---|---|---|---|---|
| Investors/ shareholders | Reverse logistics | Marketing | Operative effectiveness and efficiency | Configuration and coordination of organisational functions |
| Employees | Industrial symbiosis | Sales | Competitive advantage | Closing of the loop |
| Customers | Legislative mechanisms/ subsidies | Research and development | Minimisation of material and energy inputs | Slowing of the loop |
| Suppliers | SMART technologies | Production | Reduction of emissions and the generation of waste | Narrowing of the loop |
| Society | Industry 4.0. | Logistics | Social equity | Creation of additional monetary and non-monetary value |
| Environment | Start-ups | IT | Environmental effectiveness | Pro-active multiple stakeholder management |
| Government institutions/ organisations | Educational platforms | Finances | Economic effectiveness | Long-term perspective |
| | | Sales | | |

Source: Based on Geissdoerfer et al. (2018) and González-Sánchez et al. (2020).

Geissdoerfer, Pieroni, Pigosso, and Soufania (2020) proposed four different approaches in circular business models:

1 Cycling strategy – materials and energy are recycled in the system of production through reuse, remanufacturing, refurbishing, or recycling;
2 Extending strategy – implies that the use phase of the product is extended, through long-lasting and timeless design, appropriate marketing, education of consumers, maintenance, and repair;
3 Intensifying strategy – the use phase of the product is intensified through sharing economy solutions – for example car sharing, bicycle hire stations, expansion of public transport;
4 Dematerialising strategy – implemented by substituting material products with service and software solutions.

All four strategies can be combined within one coherent business model of sustainable supply chain management based on the CE. In the case of the cycling strategy, the main role is played by an End-of-Life (EoL) approach to the end of use and management of products or, in other words, reuse, repair, and remanufacturing and reducing or improving the properties of the product (recycling and upcycling). Collaboration in supply chains is thus possible in the context of the CE through effective reverse manufacturing processes, the creation of long-term relationships with customers, the creation of service models, and enhancing value by intangible solutions, such as services and software (Geissdoerfer et al., 2020). In this way, we achieve benefits with an impact on all three areas of Sustainable Development:

- The economy (new pricing mechanisms; new technologies; change in revenue structure; lower costs related to production, resources, and waste);
- The environment (reduced need for producing new products, reduced need for energy, reduced amount of waste generated and of intake of new materials);
- Society (jobs, inter-generational relations).

Sustainable supply chain business models based on the CE may however have a tendency to overlook issues related to the need to maintain the continuity of flows of secondary materials. Closing the loop may thus result in unknown dynamics in the system of supply and demand. This is because there is uncertainty in the volume of flows of raw materials between production–recycling–production systems, which may threaten the operational performance of such supply chains and make them struggle to materialise the said benefits of closed-loop contexts (Ponte, Naim, & Syntetos, 2020). An increase in instability in circular supply chains may result in increased cost at every stage of the chain. Stability of supply of raw materials can however be achieved and depends, among other things, on skills and the possibility of developing forecasting techniques and on transparency in the flow of information between stakeholders. Examples of solutions which may be used here include tools such as industrial symbiosis between companies, as well as digital technologies such as IoT or Big Data.

## Factors stimulating and inhibiting the development of sustainable supply chains

The integration of Sustainable Development with the Circular Economy stimulates the emergence of new factors with an impact on efficient supply chain management. It becomes all the more important to identify drivers and barriers to the extent that developing organisations still do not have a clearly defined direction for the effective implementation and adaptation of Sustainable Development practices. Table 12.2 outlines stimulating factors and barriers to the development of sustainable supply chains in the context of the CE. The factors are divided into groups, namely factors that are either clearly stimulating or inhibiting, and factors, which, depending on the supply chain management strategy and the global approach to the implementation of the CE and Sustainable Development, may act as inhibitors or catalysts of change.

*Table 12.2* Factors stimulating and inhibiting the development of SSCs

| | | |
|---|---|---|
| Stimulating factors | Reduced costs | Long-term goals and a willingness to change in the area of the CE and SD may contribute to a reduction in the costs of production |
| | Ability to manage risk | It is important to correctly assess risk in the area of management of change in SSCs in the context of the next steps to be taken in deployment of the principles of the CE and SD, as this allows the related costs to be kept to a minimum and an appropriate budget to be drawn up |
| | Agile supply chain management | Implementation of innovative management tools and techniques contributes to the more efficient functioning of supply chains |
| | Support from customers, awareness and pressure from society | Increasingly well-informed society demands action in environmental protection, public health, and climate change |
| | Legal provisions in the field of the environment | Legal regulations in the area of environmental protection force decision-makers to care about the quality of the natural environment |
| | Subsidies and economic incentives | Introduction of incentives in the form of lower taxes or subsidies contributes to a significant development of SSCs |
| | Innovative technologies | Broad availability of innovative technologies may be a catalyst for change in supply chain management |
| Inhibiting factors | Limited or low demand | Higher costs of production lead to a rise in prices and have a negative impact on demand |
| | High costs of implementation | Initial costs related to the need to introduce changes may constitute a significant barrier in SSC planning |
| | Lack of strategic thinking in the organisation, short-term goals | Organisation is not focused on development, and short-term thinking obscures the perception of benefits that can be achieved from the implementation of the CE and SD |
| | Too little legislative support/legal chaos | Law is always one step behind innovations and social changes, and so the resulting legal chaos or lack of specific rules may lead to chaos in SSC management |
| | Use of materials for the production of energy | Instead of being put back into the loop, waste materials are used as an alternative energy source. Underdeveloped recycling technologies means there is no alternative to using these materials as raw materials. |

| Factors with a stimulating or inhibiting effect depending on the management strategy | Educated and qualified management personnel | New technologies require qualified personnel, and a lack of such personnel may delay SSC development |
| --- | --- | --- |
| | Commitment of employees and actors in the supply chain | The cooperation and commitment of employees and stakeholders at all levels are necessary for an efficient SSC management |
| | Education of society by the EU and national educational programmes | Education of society about climate change, the depletion of resources, etc., is necessary to bring about a change in mentality and awareness to enable development in this area |
| | Infrastructural resources | Modern infrastructure helps with the implementation of the principles of the CE and SD and increases economic and environmental effectiveness |
| | Supply chain digitisation and virtualisation (Industry 4.0, blockchain technologies) | Increasing computerisation and automation of processes across the entire supply chain will facilitate communication with actors in the supply chain, contribute to more effective management, and generate new jobs but may also have an impact in terms of the professional exclusion of certain social groups. Furthermore, many organisations are not aware of the benefits resulting from the deployment of the concept of Industry 4.0 |
| | Reliability of supply | Where raw materials are obtained from recycling processes, it is possible that there may be disruption in the continuity of supply, so tools are necessary to control the flow of materials between individual elements of the supply chain |

Source: Own elaboration based on Tura et al. (2019), Gupta, Kusi-Sarpong, and Rezaei (2020), Meager, Kumar, Ekren, and Paddeu (2020), and Karmaker et al. (2021).

It is difficult to say unequivocally which of the stimulating factors determines the development of SSC to the greatest degree. All of them are important elements in the chain. However, it would seem that the first priority should be to raise awareness of society about the need for action on climate change or the depletion of resources. Only a change of mentality and the knowledge gained about other elements necessary for the development of SD and the CE will allow for the introduction of structural changes not only in supply chains, but also in all ecosystems and business models.

## Sustainable supply chains in the bioeconomy

The implementation of the Circular Economy is also related to the development of the bioeconomy, a point which is of interest to the European Commission (EC). Only the development of properly balanced supply chains will allow for growth in the significance of the bioeconomy on a global scale, because, as in the case of other industries, here it is also necessary to meet numerous requirements related to environmental, social, and economic problems.

The first definition of the bioeconomy was presented by the European Commission (European Commission, 2012):

> *the production of renewable biological resources and the conversion of these resources and waste streams into value added products, such as food, feed, bio-based products and bioenergy.*

The bioeconomy thus also encompasses the sustainable processing of different types of biomass used among other things for the production of energy, fuels, food, medicines, cosmetics, packaging, and biomaterials and products of other types. In the case of the bioeconomy, we can talk about Sustainable Development of supply chains, since researchers consider the bioeconomy to be sustainable by nature (McCormick & Kautto, 2013) or to be even circular by nature (Sheridan, 2016). However, some claim that closing the loop in this case may have negative effects, for example, in the form of pressure on water systems and natural ecosystems, changes in land use, agricultural intensification and eutrophication, and risks posed by invasive species (Pfau, Hagens, Dankbaar, & Smits, 2014). There are also "hotspots" concerning the use of fossil fuel resources and the related emissions into the environment (Muradin & Kulczycka, 2020). Initial surveys have shown that the sectors considered to be the most promising in the context of development of the CE or having the most potential are related to the production bioplastics, pharmaceuticals and food and feed additives, as well as building materials (Stegmann, Londo, & Junginger, 2020). Bioenergy and biofuels were assessed as having significantly worse potential (Stegmann et al., 2020). It is not possible however to completely overlook the part that these areas play in the development of the CE, because market analyses show that there are projects in the field of bioenergy and biofuels, where supply chains have the potential for Sustainable Development while at the same time fulfilling the requirements of the CE (Stegmann et al., 2020).

### Industrial and agricultural cluster

The first example concerns a model of an industrial and agricultural cluster consisting of a biogas plant with a capacity of 0.5 MW; a spirits distillery with a grain dryer; and a farm which raises livestock, keeps dairy cows, and grows cereal crops (maize and rape) run by a group of agricultural producers. The cluster is located in the voivodeship of Silesia in Poland. By making use of industrial symbiosis, this model helps to significantly limit the use of primary resources, reduce the generation of waste, process by-products, and minimise emissions into the environment. In addition, the group works based on close cooperation between businesses and industrial symbiosis allow the plants involved to be more competitive with regard to other individual entities on the market. Sustainable Development is also to be seen in social issues related to the employment of people from areas at risk of high unemployment. Studies showed the biogas plant analysed to be significantly more effective from an environmental and economic point of view than other installations of that type (Muradin, Joachimiak-Lechman, & Foltynowicz, 2018). The cluster is characterised by a short supply chain. Practically, all operations are performed in a circular loop between neighbouring installations. Some of the raw materials are transported by pipeline, which also limits emissions into the environment (Figure 12.2).

In sustainable supply chain management in the bioeconomy, as in other areas, it is necessary to implement not only the principles of Sustainable Development, but also the principles of the CE related to optimisation of the value of biomass over time. For this reason, it is necessary to have close cooperation across value chains to find trade-offs, for example, between sustainability dimensions or optimising the product for prolonged use or easy repair or recycling (Stegmann et al., 2020), especially given that recycling and recovery do not necessarily result in a reduction in the use of primary resources or $CO_2$ emissions (Daioglou et al., 2014). The introduction of CE principles into sustainable supply chain management allows for more efficient use of waste biomass resources.

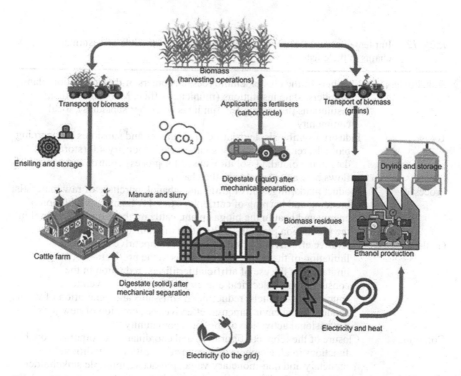

*Figure 12.2* Sustainable supply chain of a biogas cluster operating in a closed loop

In the case of the cluster described, a closing of the natural carbon cycle is observed in nature. Biomass (maize crop) is used in part as silage as animal feed and in part as dried grain that goes to the spirits distillery. The waste biomass in the form of slurry and manure and distiller residues is then sent to the biogas plant, where biogas generated by methane fermentation is transformed into heat and electrical energy used for the needs of operating the production plant and dryer. The remaining electrical energy is sold to the grid. The digestate then undergoes mechanical separation into two fractions: The liquid fraction is used as fertiliser for crops, and the solid fraction is used as litter for dairy cows (Figure 12.2). The relationships described between individual elements of the chain allow the amount of waste generated (slurry and distillery residues) to be minimised, methane emissions to be reduced, and the amount of non-renewable resources used to be limited (production of renewable energy, limited road transport). The supply chain internal to the whole cluster contains elements of a circular business model allowing for sustainable management. A close connection can be observed here between all parts: Stakeholders, tools, processes, goals, and conditions of implementation. Table 12.3 presents elements of circular business models which were directly implemented in the cluster's sustainable supply chain.

*Table 12.3* Implementation of elements of circular business models of sustainable supply chains in the cluster

| | |
|---|---|
| Stakeholders | Members of the cluster, employees, customers of the production plant, suppliers, state institutions (municipal office, Marshal's office, institutions processing EU fundings and other subsidies), local community |
| Tools | Industrial symbiosis, legislative mechanisms and subsidies (concerning biofuels, renewable energy sources, etc.), Industry 4.0 (storage of data in the cloud, IT solutions used for process control and control of flows between elements in the chain) |
| Processes | Product marketing, sale of grain and ethanol, purchase of raw materials, breeding and keeping of cattle, logistics, IT, finance, production of energy and heat in the biogas plant, cattle on the farm and ethanol in the production plant and cultivation of grain and biomass |
| Goals | Operative effectiveness and efficiency, competitive advantage, limitation of the use of raw materials in the production of biogas, limitation of the use of artificial fertilisers, reduction in the consumption of electrical energy and heat from conventional sources – fossil fuels, reduction in emissions and generation of waste, economic and environmental effectiveness, creation of new jobs, professional activation of the local community |
| Conditions | Closure of the loop, configuration and coordination of organisational functions in all elements of the cluster, creation of additional monetary and non-monetary value, pro-active multiple stakeholder management – different shareholders in the cluster |

### BioRen project – processing of waste residues to produce biofuels

Another example of implementation of the CE in the bioeconomy is the BioRen project funded under the Horizon H2020 Programme (Grant no. 818310). It is a planned element of the RenaSci Recycling Center located in Belgium, which consists of recycling municipal waste and utilising each of the fractions as materials which can be reused. The BioRen project concerns the production of biofuels from the residual organic fraction (Refuse-Derived Fuels – RDF) which cannot be recycled. Management of the entire supply chain for the processing of municipal waste is based on two different waste streams: 1) Pre-separated waste fractions suitable for recycling and 2) fractions of mixed waste (RDF) not suitable for recycling. Depending on the type of waste, both fractions are then sent to the appropriate processing/recycling facilities which use a combination of the latest mechanical, thermal, and chemical processes (Figure 12.3). The sorting facility separates plastics into those which are and are not suitable for recycling. Plastics sent for recycling are refined, granulated, and sold in the market. The metal, paper, and cardboard separated out in the process are put back onto the market as valuable raw materials. Here, the RDF fraction, which is currently

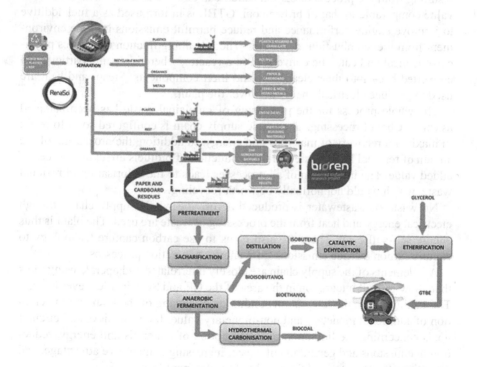

*Figure 12.3* Diagram of the Renasci Recycling Center together with the BioRen section for the processing of organic waste

Source: Own elaboration based on unpublished materials for the BioRen project.

incinerated or put into landfill in most mechanical/biological waste-processing plants, also undergoes further separation into three different waste streams: plastics, an organic fraction, and a residual fraction. Plastics not suitable for recycling from the RDF fraction are chemically converted into hydrocarbons using innovative P2C technology. The fuel produced in the process is used to generate electrical energy for the whole plant. Currently, the organic fraction not suitable for recycling (mainly residues of paper and cardboard) is sent for processing by HydroThermal Carbonisation (HTC) into granulated pellets, also known as biocoal.

The implemented BioRen project also envisages additional processing of the organic fraction into second-generation biofuels. The organic fraction is converted by a two-stage enzymatic process (saccharification and anaerobic fermentation) simultaneously into bioethanol and isobutanol, which are then used to produce glycerol tertiary butyl ether (GTBE) by a process of catalytic synthesis. Only the post-fermentation waste fraction is sent to the HTC facility (Figure 12.3). The aim of this step is to increase the efficiency of the whole waste treatment process and to close the loop for the whole supply chain, while also achieving the high economic and environmental effectiveness of biofuel production. The biocoal resulting from the process of carbonisation can be used as a fuel with a calorific value comparable to that of brown coal. GTBE is in turn used as a fuel additive to improve engine performance and reduce harmful emissions into the environment from fuel combustion in vehicles. The remaining fraction undergoes physical–chemical and catalytic conversion in a reactor, where inorganic particles are converted back into their clean, dry, and inert components. Steam and heat are used to produce electrical energy to power the plant.

The whole process for the processing of municipal waste has been patented as Smart Chain Processing. The whole supply chain is configured so as to result in maximum recovery of materials and energy. In addition, the processing of the stream of residual RDF waste and the production of biofuels are other sources of added value. The integration of all processes leads to the valorisation of residual waste, which would not normally be worth recycling.

No waste or wastewater is produced across the whole supply chain, though electrical energy and heat from the processing of waste are used. The plant is thus energy self-sufficient. There are also plans to use carbon capture technology to capture carbon dioxide emissions from energy-generation processes.

All elements of the supply chain are closely interrelated and operate using tools that allow for full integration in the area of the CE and Sustainable Development. The facility meets the integration conditions of closing of the loop and the creation of additional monetary and non-monetary value. It also achieves all circular goals concerning the limitation of consumption of materials and energy, reduction in emissions and generation of waste, increasing competitive advantage, and operative effectiveness and Sustainable Development Goals.

The presented facility may provide a solution for the management of municipal waste in other countries, as it is capable of being adapted to local conditions and needs concerning the amount and type of waste to be processed. It is thus also a very good example for implementation in both the Polish and Italian systems of waste management.

## Critical analysis of supply chains in the context of the CE

Analysing the possibility of integration, differences between the principles of Sustainable Development and of the Circular Economy should be taken into consideration.

First, it is necessary to consider whether it is technically possible to close the loop completely or not. Processes for the recycling of materials are not free from generating burdens on the environment related, for example, to consumption of resources of different types, for example water, fuels, or emissions of carbon dioxide into the atmosphere. Moreover, practically all operations related to waste management according to the principles of xR require transport processes to be taken into account across the entire supply chain, which may also contribute to the generation of additional burdens.

Muñoz-Torres et al. (2018) even claim that the Circular Economy differs in the approach taken to the sustainability of environmental, social, and economic dimensions, areas which are treated as being equal in Sustainable Development. In support of this hypothesis, the authors suggest that the CE does not take the intergenerational perspective into account, which leaves a gap in the social dimension in the category of generational equity (Muñoz-Torres et al., 2018). In addition to this, the development in Industry 4.0 is having an impact by increasing the computerisation and automation of processes across the entire supply chain. While this creates new jobs, it often also requires a certain level of qualifications, which may have an impact in terms of the professional exclusion of certain social groups.

Furthermore, the main principles of the CE are based on the reduction in the use of raw materials, minimising the generation of waste and limiting wastage and losses across the entire supply chain (Murray, Skene, & Haynes, 2017), which in turn suggest that the main purpose of the CE is to think in environmental terms.

Researchers also suggest that the assumptions of the CE are incompatible with the second law of thermodynamics, which determines the direction of spontaneous thermodynamic changes in nature. One of the most widely known formulations of this is Clausius' Second Law: "There is no thermodynamic process whose sole effect is to extract a quantity of heat from a colder reservoir and deliver it to a hotter reservoir."

Entropy is however a thermodynamic function and may be defined as a measure of "disorder" of a system. This means the higher the entropy, the higher the disorder. It follows that the efficiency of energy conversion must be less than 1, and every process is a source of "waste heat". This is also why, for every process, there is a certain basic, maximally achievable efficiency of energy conversion. This is also why to reduce the disorder of a given system or, in other words, to decrease the entropy of the system – that is to make a product – it is necessary to do some work or, in other words, provide energy from an external source. Considering the assumptions of the Circular Economy in this context, it is not possible to attain a completely isolated system without a huge amount of work energy. Of course, it is possible to reverse the dissipation of matter by recovering all raw materials, for example rare earth metals, but these processes also require enormous amounts of energy. Thus, the more we succeed in closing the loop, the more energy we are going to need for it to function. Of course, at this stage, these are purely theoretical considerations, and taking things further in this direction, one could safely assume that in order to maintain all processes within the loop for the collection and recovery of materials and recycling processes, we will be able to use infinite solar energy (Ayres, 1999).

The aforementioned considerations do not however change the fact that the introduction of principles of the CE concerning the use of renewable energy resources and limiting the consumption of raw materials makes it possible to achieve a more sustainable model of supply chains. It is however worth drawing attention to the fact that closing of the loop is not unequivocally beneficial and does not guarantee a sustainable result per se, and every project should be carefully analysed for its impact on the Sustainable Development of the economy as a whole (Korhonen, Honkasalo, & Seppälä, 2018).

## Conclusion

Taking into consideration the assumptions regarding implementation of the Circular Economy in accordance with the regulations of the European Commission, the main models of sustainable supply chain management must integrate elements of both approaches. Despite the fact that the Circular Economy has certain shortcomings, it is the best solution proposed to date to limit the consumption of resources and to reduce the production of waste and emissions into the environment. The CE also meets one of the EC's main objectives to be climate-neutral with net-zero $CO_2$ emissions by 2050.

Of course, the vision of creating a completely Circular Economy is a utopian one, and, in the light of research into thermodynamics of the CE, it even seems to be one that is impossible to achieve. However, it remains an option for making supply chains more sustainable and for reducing the use of primary resources.

The examples of sustainable supply chain models in the bioeconomy presented here do not achieve all of the goals concerning the integration of the CE and Sustainable Development, and, from a thermodynamic point of view, they confirm the thesis that it is always necessary to supply energy from an outside source to a closed system with lower entropy, thus meaning that the whole system in fact proves not to be self-sufficient and not to be a closed loop. However, the main benefit achieved from the creation of this type of model is what is most important for the protection of the planet as a place for current and future generations to live, and that is to limit its degradation while maintaining continuous economic growth.

## References

Ayres, R. U. (1999). The second law, the fourth law, recycling and limits to growth. *Ecological Economics*, *29*(3), 473–483. Retrieved from https://econpapers.repec.org/RePEc:eee:ecolec:v:29:y:1999:i:3:p:473-483

Beu, D., Ciugudeanu, C., & Buzdugan, M. (2018). Circular economy aspects regarding LED lighting retrofit – from case studies to vision. *Sustainability.* doi:10.3390/su10103674

Bocken, N. M. P., de Pauw, I., Bakker, C. A., & van der Grinten, B. (2016). Product design and business model strategies for a circular economy. *Journal of Industrial and Production Engineering, 33*(5), 308–320. doi:10.1080/21681015.2016.1172124

Daioglou, V., Faaij, A. P. C., Saygin, D., Patel, M. K., Wickea, B., & van Vuurenab, D. P. (2014). Energy demand and emissions of the non-energy sector. *Energy & Environmental Science, 7*(2), 482–498. doi:10.1039/C3EE42667J

De Angelis, R., Howard, M., & Miemczyk, J. (2018). Supply chain management and the circular economy: Towards the circular supply chain. *Production Planning & Control, 29*(6), 425–437. doi:10.1080/09537287.2018.1449244

De Oliveira, C. T., Luna, M. M. M., & Campos, L. M. S. (2019). Understanding the Brazilian expanded polystyrene supply chain and its reverse logistics towards circular economy. *Journal of Cleaner Production, 235*, 562–573. doi:10.1016/j.jclepro.2019.06.319

European Commission. (2012). *Innovating for sustainable growth: A bioeconomy for Europe.* Brussels: Author.

Geissdoerfer, M., Morioka, S., de Carvalho, M., & Evans, S. (2018). Business models and supply chains for the circular economy. *Journal of Cleaner Production, 190*, 712–721. doi:10.1016/j.jclepro.2018.04.159

Geissdoerfer, M., Pieroni, M. P., Pigosso, D. C. A., & Soufania, K. (2020). Circular business models: A review. *Journal of Cleaner Production, 277*, 123741. doi:10.1016/j.jclepro.2020.123741

Geissdoerfer, M., Savaget, P., Bocken, N. M. P., & Hultink, E. J. (2017). The circular economy – a new sustainability paradigm? *Journal of Cleaner Production.* doi:10.1016/j.jclepro.2016.12.048

Genovese, A., Acquaye, A. A., Figueroa, A., & Koh, S. C. L. (2017). Sustainable supply chain management and the transition towards a circular economy: Evidence and some applications. *Omega.* doi:10.1016/j.omega.2015.05.015

González-Sánchez, R., Blundo, S., Ferrari, A. M., & García-Muiña, F. E. (2020). Main dimensions in the building of the circular supply chain: A literature review. *Sustainability.* doi:10.3390/su12062459

Gupta, H., Kusi-Sarpong, S., & Rezaei, J. (2020). Barriers and overcoming strategies to supply chain sustainability innovation. *Resources, Conservation and Recycling, 161*, 104819. doi:10.1016/j.resconrec.2020.104819

Karmaker, C. L., Ahmed, T., Ahmed, S., Mithun Ali, S., Moktadir, A., & Kabire, G. (2021). Improving supply chain sustainability in the context of COVID-19 pandemic in an emerging economy: Exploring drivers using an integrated model. *Sustainable Production and Consumption, 26*, 411–427. doi:10.1016/j.spc.2020.09.019

Korhonen, J., Honkasalo, A., & Seppälä, J. (2018). Circular economy: The concept and its limitations. *Ecological Economics, 143*, 37–46. doi:10.1016/j.ecolecon.2017.06.041

McCormick, K., & Kautto, N. (2013). The bioeconomy in Europe: An overview. *Sustainability.* doi:10.3390/su5062589

Meager, S., Kumar, V., Ekren, B., & Paddeu, D. (2020). Exploring the Drivers And Barriers To Green Supply Chain Management Implementation: A study of independent UK restaurants. *Procedia Manufacturing, 51*, 1642–1649. doi:10.1016/j.promfg.2020.10.229

Muñoz-Torres, M. J., Fernández-Izquierdo, M. Á., Rivera-Lirio, J. M., Ferrero-Ferrero, I., Escrig-Olmedo, E., Gisbert-Navarro, J. V., & Marullo, M. C. (2018). An assessment tool to integrate sustainability principles into the global supply chain. *Sustainability.* doi:10.3390/su10020535

Muradin, M., Joachimiak-Lechman, K., &Foltynowicz, Z. (2018). Evaluation of eco-efficiency of two alternative agricultural biogas plants. *Applied Sciences*, *8*(11), 2083, Multidisciplinary Digital Publishing Institute. doi:10.3390/app8112083

Muradin, M., & Kulczycka, J. (2020). The identification of hotspots in the bioenergy production chain. *Energies*. doi:10.3390/en13215757

Murray, A., Skene, K., & Haynes, K. (2017). The circular economy: An interdisciplinary exploration of the concept and application in a global context. *Journal of Business Ethics*, *140*(3), 369–380. doi:10.1007/s10551-015-2693-2

Pfau, S. F., Hagens, J. E., Dankbaar, B., & Smits, A. J. M. (2014). Visions of sustainability in bioeconomy research. *Sustainability*. doi:10.3390/su6031222

Ponte, B., Naim, M., & Syntetos, A. (2020). The effect of returns volume uncertainty on the dynamic performance of closed-loop supply chains. *Journal of Remanufacturing*, *10*. doi:10.1007/s13243-019-00070-x

Sheridan, K. (2016). Making the bioeconomy circular: The biobased industries' next goal? *Industrial Biotechnology*, *12*(6), 339–340, Mary Ann Liebert, Inc. Publishers. doi:10.1089/ind.2016.29057.ksh

Stegmann, P., Londo, M., & Junginger, M. (2020). The circular bioeconomy: Its elements and role in European bioeconomy clusters. *Resources, Conservation & Recycling: X*, *6*, 100029. doi:10.1016/j.rcrx.2019.100029

Szymczak, M., & Nowicka, K. (2020). Logistyka i łańcuchy dostaw w obliczu czwartej rewolucji przemysłowej. *Studia BAS*, *63*, 61–84. doi:10.31268/StudiaBAS.2020.22

Tura, N., Hanski, J., Ahola, T., Ståhlec, M., Piiparinen, S., & Valkokari, P. (2019). Unlocking circular business: A framework of barriers and drivers. *Journal of Cleaner Production*, *212*, 90–98. doi:10.1016/j.jclepro.2018.11.202

Zeng, H., Chen, X., Xiao, X., & Zhou, Z. (2017). Institutional pressures, sustainable supply chain management, and circular economy capability: Empirical evidence from Chinese eco-industrial park firms. *Journal of Cleaner Production*, *155*, 54–65. doi:10.1016/j.jclepro.2016.10.093

Zheng, B., Yang, C., Yang, J., & Zhang, M. (2017). Pricing, collecting and contract design in a reverse supply chain with incomplete information. *Computers & Industrial Engineering*, *111*, 109–122. doi:10.1016/j.cie.2017.07.004

# 13 Determinants of consumer behaviour – towards sustainable consumption

*Kamila Pilch and Małgorzata Miśniakiewicz*

## Introduction

Sustainable consumption and sustainable production, as pillars of Sustainable Development, are currently one of the most important developing trends on the consumer goods market (Haller, Lee, & Cheung, 2020; Mintel, 2021). Consumers are taking an increasing interest in ethical, social, and environmental issues and looking for products and brands that fit with their value systems. On a global scale, over the period from 2014 to 2019, there has been 68% growth in sustainable and environmentally friendly investment, and the awareness of global environmental problems has led to systematic change in consumer habits. In 2019, 6 out of 10 consumers were willing to change their shopping habits to reduce negative impact on the natural environment, and 8 out of 10 said that Sustainable Development is important to them. In this latter category, over 70% consumers were ready to pay on average 35% more for products, which are sustainable and environmentally friendly (Haller et al., 2020) (cf. Chapter 5).

In the time of the COVID-19 epidemic, the grounds for consumer choices have changed somewhat – with the safety and hygienic quality of products now becoming the most important, though ever more consumers are identifying with what is known as the LOHAS (Lifestyle of Health and Sustainability) movement or, in other words, persons who live in accordance with nature and the principles of Sustainable Development. This is confirmed by the results of a Gallup poll, according to which as many as 94% of consumers surveyed declared that they prefer to use products from companies which care about the natural environment (De Neve & Sachs, 2020).

On the one hand, consumer awareness and knowledge related to Sustainable Development, manifested amongst other things in limited purchasing of products, limited use of natural resources, or limiting food waste, are having an impact on consumption (Geiger, Fischer, & Schrader, 2018). On the other hand, consumer psychology and consumerism are also resulting in higher consumption, which can be seen across the globe after the outbreak of the COVID-19 pandemic. In response, the United Nations is calling for social change in order to "build back better" (UN, 2020). In this context, sustainable production and sustainable consumption, as well as ethical issues are becoming increasingly important and

DOI: 10.4324/9781003179788-13

require special attention (Fernández-Rovira, Álvarez Valdés, Molleví, & Nicolas-Sans, 2021).

The aim of this chapter is to present the essence of sustainable consumption, the determinants of its development and implementation and the attitudes that consumers adopt towards it, paying particular attention to the factors that contribute to the shaping of sustainable and environmentally friendly consumer choices. An analysis of the results of our own research conducted with regard to how the concept of sustainable consumption is perceived by young Polish consumers made it possible to determine the factors contributing to their pro-ecological market behaviours. The aim of the analyses carried out was to formulate substantive recommendations for practices to break down barriers to sustainable consumption and the promotion of activities conducive to its implementation in practice.

## Determinants of sustainable consumption

Analysing the determinants of the development of sustainable consumption, it should be considered from multiple dimensions, including in the context of changes in the needs, attitudes, motivation, hierarchy of values, and lifestyles of contemporary consumers. Socio-cultural, demographic, and economic changes occurring in society, as well as education, including education about the environment and the state of the natural environment and consumer awareness of the impact of its pollution on health are also very important. Elements conducive to the development of sustainable consumption also include macroeconomic factors, state policy, and corporate strategies. In this area, marketing activities and CSR initiatives undertaken by representatives of business practice or government are important (Maciejewski, 2020; Mazurek-Łopacińska & Sobocińska, 2014).

The essence of sustainable consumption lies in the optimum, conscious, and responsible use of available natural resources, products, and services at the level of individual consumers, households, local communities and society, local governments, business, governments of individual countries, or international structures (Tunn, Bocken, van den Hende, & Schoormans, 2019). The purpose of this is to meet the needs of current and future generations and improve quality of life both locally and globally, while at the same time respecting human and labour rights and preserving and restoring natural capital. Sustainable consumption is an attitude which takes the limitation of wastage and of the production of waste and pollution into consideration, along with the choice of goods and services, which, to the maximum extent possible, fulfil specific ethical, social, and environmental criteria (Cooper, 2013; Mont & Plepys, 2008; Akenji et al., 2015).

Choosing ecological products and promoting sustainable consumption may stimulate firms and society in the pursuit of sustainability, balancing the demand and supply side of the business cycle with addressing the ecological and social benefits (Jaiswal & Singh, 2018). Determinants of the development of sustainable consumption should be examined while paying attention to the following considerations (Neale, 2015; Patrzałek, 2016):

• Rationalisation of consumer behaviours – limitation of consumption in an uncertain economic situation or in order to rationalise it, limitation of the quantity of products purchased in favour of their higher quality, or limitation of consumption in the materials sphere in favour of that in the immaterial sphere;
• The greening of consumer behaviours – for example economical use of consumer products, the purchasing and consumption of goods, which generate a low quantity of post-consumption waste or ecological products;
• Growth in the ecological awareness of consumers – taking action to protect the natural environment by, for example avoiding buying too much food, the segregation of waste, and the use of multiple-use bags when shopping.

The behaviours identified earlier are to a large extent promoted by sustainable development-oriented marketing, which should take account of the ecological and social conditions at each of the individual stages of management of value for the customer and thereby create a specific market message which helps to bring about a change in customer preferences and develop sustainable consumption. Socially and environmentally responsible practices may lead to the company being perceived more positively by consumers and also to an increase in its profitability (Luo & Bhattacharya, 2006; Olsen, Slotegraaf, & Chandukala, 2014). This stimulates innovation and helps to identify opportunities for the development of products and markets, the use of new technologies, increasing organisational efficiency, and motivating and building employee loyalty (Berns et al., 2009).

The contradiction between marketing activities (understood as the creation of value for the customer by encouraging purchases) and sustainable consumption is illusory. In practice, companies undertaking marketing activities enhance the benefits resulting from sustainable consumption (White, Habib, & Hardisty, 2019). These activities have a positive impact on brand image and, as a result, in an increase in profits (Jung, Kim, & Kim, 2020).

According to K. White, R. Habib, and D. J. Hardisty, the factors, which contribute to the shaping of sustainable and environmentally friendly consumer choices, include Social Influence, Habit Formation, Individual Self, Feelings and Cognition, and Tangibility (White et al., 2019). They were identified by authors based on a review and critical analysis of 320 scientific articles from major international journals in sustainable consumption, behavioural marketing, and consumer behaviour. The most important observations are presented in Table 13.1.

In order to achieve the desired effects, it is often advisable to use several strategies at once in one action, to take actions on a small scale, and then extend their scope if they prove to be successful. The consumer should have at least minimal knowledge of the problem at hand of the social norm, of threats resulting from a given behaviour, or of the benefits that may result from a change in behaviour.

*Table 13.1* Factors and examples of activities influencing sustainable consumption

| Factor | Description | Possibilities of activities supporting socially responsible consumption |
| --- | --- | --- |
| Social influence | In dealing with ethical issues, consumers are often guided by the activities of others. They choose sustainable options to make a positive impression on those around them and often act in a socially desirable manner in public contexts in which other people can observe and evaluate their actions (Green & Peloza, 2014). | Introduction of a "brand ambassador" – a person takes a specific action first and then finds followers in his/her immediate vicinity. Use of the peer effect – developing new social norms where the consumption of sustainable products is well perceived, without having to be prescriptive or guilt inducing (cf. Lucas, Salladarré, & Brécard, 2018). |
| Habit formation | Habits, which are said to form a second nature, have to change in period of overconsumption, and in order to form new, sustainable habits, it is necessary to get rid of any previous bad habits. Pro-environmental actions should be easy to implement and carry out, be economically attractive, and should provide specific information about the non-economic benefits that the consumer is able to gain, for example the reduction of $CO_2$ emissions. | Use of important moments in life, e.g. wedding, moving house, or starting a new job, as an opportunity to change existing habits (Verplanken & Roy, 2016; Walker, Thomas, & Verplanken, 2015). Imposition of penalties, for example failure to care about the environment, or the segregation of waste, though there is a concern that people will return to old habits if the penalty is lifted, and the new habit has not yet been developed. The most effective way of encouraging people to change habits seems to be a system of incentives and rewards. Provision of feedback about specific actions in real time, for example when information on energy consumption was made available to householders directly in real time, electrical consumption fell by 5 to 15% (Darby, 2006). |
| Individual self | Sustainable development becomes more attractive when the personal benefits for the consumer are emphasised, for example the impact of decisions taken on the health of loved ones. It is important to be able to maintain internal consistency. People make choices, which fit with their perception of who they want to be. It has been | Informing consumers that their actions have significance and that they are effective at what they do. This makes them more likely to make more ecological choices (White et al., 2019). Consumers want to be consistent in what they say and what they do and be able to see and value that consistency in others. Taking one pro-environmental decision often brings a series of subsequent actions along with it, which are |

| Factor | Description | Possibilities of activities supporting socially responsible consumption |
| --- | --- | --- |
| | shown, among other things, that being environmentally friendly is often perceived as being feminine, which may discourage some men who are in favour of division into traditional gender roles (Brough, Wilkie, Ma, Isaac, & Gal, 2016). | a consequence of it (Lanzini & Thøgersen, 2014). They expect companies and state government entities to be similarly consistent and thorough in their pursuit of sustainable development. Presentation of ecology in a more neutral way, as a form of protection and preservation of the natural environment, is attractive to both men and women, thereby eliminating the gender gap often seen in sustainable development (Brough et al., 2016). |
| Feelings and cognition | In the context of sustainable development, emotional and pragmatic arguments have to be weighed up against each other. Consumers look for positive emotions, such as happiness, pride, and joy resulting from good deeds – caring for the natural environment. | Making the option of sustainable consumption attractive – then people will naturally want to choose it. Conversely, negative emotions, such as fear, a feeling of guilt, or shame can be effective when they are used in a balanced, subtle way. At the same time, an over-emotional message triggering a feeling of guilt is exclusionary or will be actively ignored or will result in behaviour that is the opposite of that planned (White et al., 2019). Consumers must be provided with specific information allowing them to identify real benefits and at the same time feel good about themselves. Well-thought out eco-labelling is one way of providing consumers with information about the possibility of achieving sustainable development (cf. Chapter 5). Consumers care about future losses than future gains (Hardisty, Johnson, & Weber, 2010), so labels on energy-efficient appliances should provide information about energy costs rather than savings (Min, Azevedo, Michalek, & de Bruin, 2014). |
| Tangibility | Consumers want tangible solutions. Abstract, future consequences of overconsumption do not make that much of an | Effects of engaging in sustainable consumption should be real, measurable, and verifiable here and now – e.g. concrete financial savings, reduction of environmental footprint. |

*(Continued)*

*Table 13.1* (Continued)

| Factor | Description | Possibilities of activities supporting socially responsible consumption |
|---|---|---|
| | impression on them, because they do not feel that they may be directly concerned. | Information should be provided about the local effects of pro-environmental actions taken, for example by how much it has been possible to reduce $CO_2$ emissions locally, how many trees have been planted, and what has been gained as a result. Actions taken should be framed with specific examples, showing individual cases, which illustrate the scale of changes. |

Source: Own elaboration based on White et al. (2019), Lucas et al. (2018), Green and Peloza (2014), Verplanken and Roy (2016), Walker et al. (2015), Darby (2006), Lanzini and Thøgersen (2014), Brough et al. (2016), Hardisty et al. (2010), and Min et al. (2014).

This creates a real opportunity to engage in a specific form of sustainable consumption (Gifford & Nilsson, 2014).

What distinguishes sustainable consumption from typical consumer behaviours is a certain trade-off, which is necessary in consumer actions, by accepting a long time horizon and the requirement of collective action. Only actions taken on a suitably large scale will make it possible to achieve the desired outcome. There is also a need to take a conscious approach to making everyday purchase decisions and to replace certain automatic, impulse actions with controlled and well-thought-out ones (Johnstone & Tan, 2015). There is no one best solution leading to sustainable consumption. It is necessary to understand the specific behaviour that has to be changed, the context in which the given behaviour occurs, the intended target, and the associated barriers and benefits. Combining different strategies often allows better effects to be achieved than sticking to a single strategy (White et al., 2019).

As shown in the first part of the study, the literature on the subject emphasises the importance of various factors influencing pro-ecological behaviour. In this regard, it is interesting to examine sustainable consumption from the customer's perspective. Of particular importance in this respect may be the view of young consumers, who are indicated as being those who are the most open to new experiences in sustainable consumption (Morris & Venkatesh, 2000; Roberts, Walton, & Viechtbauer, 2006; Wiernik, Dilchert, & Ones, 2016), have the potential to influence the behaviour of others in the field of environmental protection (Lee, 2008; Lukman, Lozano, Vamberger, & Krajnc, 2013; Waas, Verbruggen, & Wright, 2010), and moreover form a group, which is willing to pay more for products made in a sustainable way (Nielsen, 2015). In the next part of the study, the analysed issue will be presented based on the example of the behaviours of young Polish consumers.

## Research design

The aim of the surveys conducted was to understand how young Polish consumers perceive the concept of sustainable consumption and to identify the factors contributing to their pro-ecological behaviours. It was possible to achieve the defined aim by answering the following research questions:

- How are consumption patterns shaped among young consumers (including what is most important for them, what do they pay attention to when making purchases)?
- How is the concept of the CE understood by young consumers?
- How is socially responsible consumption understood?
- What factors constitute barriers for the respondents when purchasing products made in accordance with the principles of the Circular Economy?
- What benefits do the respondents see when purchasing products made in accordance with the principles of the Circular Economy?

The research conducted was of an explorative nature. Due to the type of questions asked, a qualitative methodology was used, employing a technique based on focus group interviews (conducting of online focus group interviews – FGIOs). The applied methodology allowed data to be gathered in a social context and to understand the problem analysed from the point of view of participants in the study. Six focus group interviews were conducted with representatives of young consumers, who were selected to take part in the study in a purposeful way. Six to nine respondents took part in each of the individual groups. The study was conducted in the second quarter of 2020. Each of the interviews lasted 60 minutes. The data collected was transcribed and encoded using a computer program for qualitative data analysis. The most important conclusions from the analyses conducted with reference to the research questions asked are presented in three main thematic blocks:

- The consumption patterns of young consumers;
- How the concepts of the "Circular Economy" and socially responsible consumption are understood; and
- Identified barriers and benefits when purchasing products made in accordance with the principles of the Circular Economy.

The conclusions are illustrated with quotes taken from the interviews.

## Results

### *Consumption patterns*

In solutions related to the transition to the Circular Economy, emphasis has been placed on the need for change not only in the way that economies function (at a national and supranational level) and in how companies function, but also in the way that consumers function too (Gifford & Nilsson, 2014). In the first part of this study, factors influencing consumer behaviours were distinguished. In the focus group studies conducted, they were compared with declarations of the respondents on the topic of the determinants of their purchasing choices.

In their statements, the respondents drew attention to several basic factors determining the purchase decisions made. The results of discussion are summarised in a diagram form in Figure 13.1.

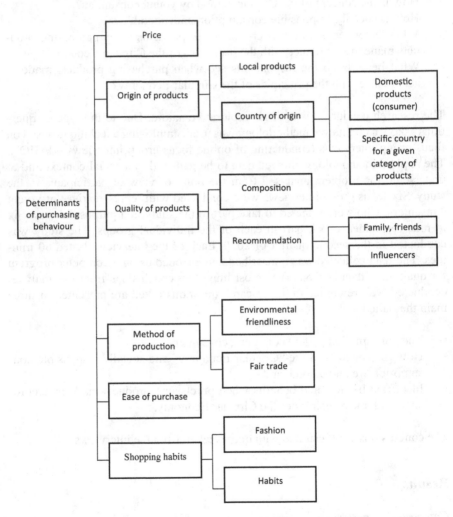

*Figure 13.1* Determinants of purchasing behaviour of young consumers – conclusions from studies

The most frequently mentioned determinant of purchasing behaviour of young consumers was price, which is best illustrated by the answer provided by one of the respondents:

> I mainly pay attention to the price, because that is what is most important to me. If something is too expensive then I don't buy it.
>
> (R1, M, 21 years of age)

At the same time, it was underlined that there are situations or product categories in which the lowest price is not a motivator of the decisions made. The respondents nevertheless emphasised that price should be appropriate to the quality of the purchased goods. In their opinion, this quality is confirmed in the composition of the products, while recommendations of friends, family, or influencers are often considered to be a guarantee of quality.

Another factor influencing the purchase decisions taken was habit formation related to the positive experience of purchasing specific brands of product.

Importantly, this experience may result from some sort of purchasing routine or from the influence and experience of people making recommendations (such as recommendations by influencers, for example). The respondents emphasised that, in the case of many products, especially those for daily use, their availability, understood as ease of purchase, is important. They included food products in this category, while emphasising that the country of origin is important when choosing them. Domestic and regional products are preferred (considered to be more ecological). One exception is in situations, where the purchased food is used as an ingredient in specific dishes from the cuisine of a specific country or region of the world. Determinants of purchasing behaviour also included environmental issues most often associated with the production method of products. Respect by producers for the rules of fair trade and the materials from which a given product is made (especially packaging) are also important to the respondents (see Chapter 2). It is however worth emphasising that in the surveyed groups, there were many opinions voiced pointing to the secondary importance of environmental issues in making purchasing choices.

The identification of factors shaping the purchasing behaviour of young consumers was an introduction to discussion on the topic of the Circular Economy and the perception of socially responsible consumption.

## *Understanding of the concept of the "Circular Economy" and "socially responsible consumption"*

The respondents had not come across the idea of the "Circular Economy". Due to the fact that, in all the focus groups, there was only one person who had heard of the term and was able to explain what it was about, it was not possible to gain any insight into spontaneous, free associations or examples, which would illustrate the phenomenon discussed. For this reason, the respondents were presented with one of the existing definitions of the Circular Economy, and, on that basis, they tried to give examples of actions/behaviours, which they knew or which they associated the term with. Most frequently, the respondents associated the Circular Economy with recycling (above all packaging):

> More and more producers of mineral water . . . somehow they are very strongly emphasising that those bottles have been made from recycled materials, or at least some percentage of them are made from recycled materials.
>
> (R2, M, 20 years of age)

At the same time, attention was paid not only to the material from which products are made, but also to the possibility of purchasing them without unnecessary packaging (refills). Another way of identifying products manufactured in accordance with the concept of the CE is environmental labelling (cf. Chapter 5). However, most often, it was emphasised that they are of marginal importance when making purchasing choices, because in most cases the respondents were not familiar with them or did not believe in their reliability.

The respondents did not have any problem defining the distinguishing features of socially responsible consumption. Emphasis was above all placed on seeing it as a way of purchasing products, which are compatible with the concept of "zero waste". A socially responsible consumer was identified to be an eco-consumer, who can be characterised by this answer provided by one of the respondents:

> An eco-consumer is a more conscious person, who is aware of the impact some products, and some materials from which they are made, have on the environment. Just like with the production even of just a simple pair of trousers or a blouse, it's not just like it that we can just produce them and everything will be OK, but in fact litres of water have to be used to do that and, for example, those people don't have any problem with buying used clothes to give them a second lease of life.
>
> (R11, F, 27 years of age)

In studies of socially responsible consumption, it is often analysed seen through the prism of the purchasing of eco-innovative products (Jansson, 2011; Noppers, Keizer, Bolderdijk, & Steg, 2014) or sustainable products (Kumar, Manrai, & Manrai, 2017). For the respondents, socially responsible consumption is associated with the conscious purchasing of products, paying attention to how the companies that produced them operate and to the materials from which they were made (both composition and packaging). In their opinion, a socially responsible consumer is more likely a young person, who, especially due to the influence of social media, is familiar with the following trends.

It was necessary to characterise socially responsible consumption and to attempt to identify products or initiatives related to the CE in order to know the opinion of the respondents relating to the perceived benefits of or barriers to purchasing them.

## *Identified barriers to or benefits of purchasing sustainable products*

Determinants of purchases of sustainable products are an issue, which has been raised in many scientific works in the area analysed (Alexander, 2012; Gleim, Smith, Andrews, & Cronin, 2013; Malodia & Bhatt, 2019). As highlighted before, for those surveyed, one of the main determinants in making purchases is price. Products manufactured in accordance with the concept of the CE are often associated by the respondents with ecological or BIO products, and these are often considered to be clearly more expensive than conventional products. This is why the most frequently cited barrier to purchasing products manufactured in accordance with the concept of the CE is their higher price:

> Bio products, all those ones which are packed not in plastic packaging but in paper, are usually more expensive. . . . it's not always a good alternative to buy something which is more healthy, but more expensive. So we choose the cheaper alternative, which may perhaps be full of chemicals and are themselves packed in plastic packaging. So I think that is the main factor which basically discourages us from caring about the environment, because it is pretty weird that caring should be the more expensive alternative.
>
> (R13, F, 23 years of age)

Price was often linked with a lack of consumer awareness and habits.

To bring about a change in consumption patterns, in the opinion of those surveyed, there is a need for information and knowledge about the method of production and the values of the company, which manufactures a given product. As the respondents, this can be expressed in the form of certificates. However, during the discussion, many people said they had doubts about their reliability.

One of the respondents pointed out that *being eco requires more effort*. On the one hand, there is the lack of convenience related to products manufactured in accordance with the concept of the CE, while, on the other, there is a lack of information, also in the form of social campaigns, which would emphasise the importance of pro-environmental consumer behaviours (cf. Environmental Locus of Control; Cleveland, Kalamas, & Laroche, 2012).

The benefits of sustainable consumption were clearly identified by the respondents. Attention was drawn to the positive environmental consequences and the positive impact on the individual's health. In the opinion of the respondents, there is a currently a trend or a fashion to be eco-friendly. Activities and patterns of consumption, which contribute to the implementation of the idea of the Circular Economy, may therefore have an impact on how an individual is perceived and received socially:

> [Eco-consumers] are perceived to be so nice . . . that they are great, so cool, that they are doing something and taking action.
>
> (R17, F, 24 years of age)

One of the determinants of purchasing behaviour distinguished in the study was that of habit formation. In this respect, the respondents drew attention to the fact that the promotion of products manufactured in accordance with the concept of the CE may lead to a change in purchasing habits and thus help to develop loyalty to certain brands:

> I think I could find a benefit for a company, because if such a company would act in a way that takes such aspects into account . . . then I think customers will be more willing to purchase their products, precisely for that reason. And they will also be that bit more loyal . . . and they may also become more attached to a specific brand and, when they go back into the store, they will reach out for it again automatically.
>
> (R18, M, 22 years of age)

Respondents note the changes taking place, but they emphasise the lack of activity by the state in Poland to support such changes. This is accompanied by the belief that the reluctance and slow pace of changes in consumption patterns are also associated with insufficient marketing activity on the part of companies and the lack of facilitating measures. It is precisely such facilitating measures that, in their opinion, could lead to greater consumer engagement in caring for the environment and developing patterns of socially responsible consumption.

The analyses conducted made it possible to identify the way in which socially responsible consumption is perceived and how the concept of the CE is understood. Based on the identified barriers, it was possible to draw up recommendations for practices in the promotion of socially responsible consumption and promotion of the CE. The described barriers perceived by the consumers surveyed can be grouped into three main categories: social barriers, those related to habits, and those relating to a lack of knowledge/informed beliefs. The development of social norms supporting sustainable consumption can be considered to be a desirable effect of the promotional activities undertaken (Figure 13.2). The nature of the goal formulated in this way means it is necessary for action to be taken on the part of producers, distributors, and the state. Activities that could allow the goal to be achieved should be of an ongoing nature, as the development of or change to existing norms is a long-term process. At an operational level, it above all seems to be necessary to conduct information campaigns, which would point to the need for a change in consumption habits and the impact of the behaviours of individuals functioning in different socio-economic contexts on the planet. In this respect, activities aimed at people living in smaller towns and with lower incomes may prove to be especially important. As one of the respondents underlined

> I, for example, come from a small town . . . looking at me, at my family around me here, I think that no-one, absolutely no-one pays any attention to whether something is bio, or not bio, or to what type of packaging something is in. Absolutely no one.
>
> (R2, M, 20 years of age)

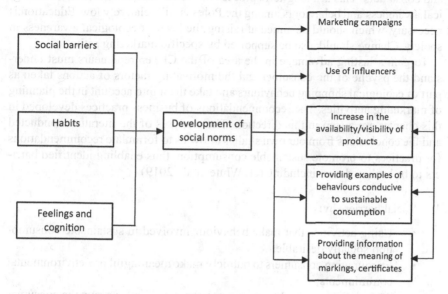

*Figure 13.2* Elements and activities supporting the development of socially responsible consumption

Especially with reference to young consumers, who are active users of social media, it seems to be justified to use influencers, who can be ambassadors in spreading everyday behaviours that are conducive to sustainable consumption. At the same time, it is worth emphasising what the consequences of specific behaviours are. Those surveyed emphasised that a better knowledge of environmental labels would allow them to assess the quality of products and their impact on the environment. This is why, as part of the activities undertaken to promote and inform, it seems to be necessary to make knowledge in this area more widely available. Furthermore, something else that seems to be a factor, which could be of key importance in developing social norm supporting the CE, is the introduction of facilitating measures allowing the consumer to purchase products easily and the application of a system of penalties and rewards for socially responsible behaviours.

## Conclusion

Currently, in an era of competitiveness and Sustainable Development of companies and whole countries, the phenomenon of Sustainable Development is something more than just being environmentally friendly. It is about balancing demand and supply in the cycle of production and consumption. In many countries, as in Poland that is the subject of the analysis in this work, Sustainable Development and consumption require changes in the attitudes and beliefs of both producers and consumers. This is a long-term process due to the fact that the level of ecological awareness and behaviours among the Poles is still relatively low. Education is necessary, which should be aimed at raising the level of ecological awareness in society. Change should also be supported by specific marketing activities.

To gain a lasting advantage in the area of the CE, entrepreneurs must understand the psyche of the consumer and the motivating factors of actions taken as part of ecological shopping behaviours and take them into account in the planning of marketing activities. The recommendations of business practices developed in this chapter are a step in this direction. The analyses of the literature conducted and the conclusions from our own studies allow us to formulate recommendations for practices to promote sustainable consumption, thus enabling identified barriers to be broken down, including (cf. White et al., 2019):

1   Social barriers by:

- Using activities that make behaviour involved in sustainable consumption socially desirable;
- Encouraging consumers to publicly make meaningful pro-environmental commitments;
- Associating sustainable consumer behaviours with a group the consumer identifies with or which they aspire to belong to;

2   Barriers in the area of habit formation by:

- Using special occasions and events, including in private life, to eliminate or limit habits which place a burden on the environment and to encourage specific actions to be taken in the area of sustainable consumption;
- Establishing the sustainable choice as the default option when shopping, choosing services, etc.;
- Generating and making individual and comparative feedback available emphasising the favourable impact of decisions taken on the environment;
- Providing information about sustainable actions and the results of them in a way which reflects local and real effects, so that everyone feels he/she has an influence on the current state of affairs, and that his/her actions are helping to improve the state of the environment;
- Directing the attention of consumers towards the future, so that they pursue the goal of Sustainable Development, including through sustainable consumption;

3   Mental barriers by:

- Communicating in a way which triggers negative emotions but, at the same time, helps to build a feeling of self-agency and self-efficacy and, thereby, through action, allows them to be limited or even eliminated; and
- Generating a sense of pride as a result of engagement in sustainable consumer behaviours.

It is however worth emphasising that the research project conducted has its limitations. The use of qualitative techniques to collect research material means that the results obtained cannot be generalised and treated as being statistically significant. Moreover, due to the purposeful selection of the sample, it should be noted that the recommendations formulated in the study are based solely on an analysis of the consumption patterns of the surveyed group (young Polish consumers). It would be of cognitive value to carry out quantitative studies, which would allow the identified dependencies to be statistically verified. Nevertheless, the identified barriers to the purchasing of products manufactured in accordance with the concept of the CE point to the need for support for socially responsible consumer behaviours from both companies and from the authorities at a national level. This should be achieved through activities that support the development of new social norms, which would emphasise the importance of pro-environmental consumer behaviours.

# References

Akenji, L., Bengtsson, B., Briggs, E., Chiu, A., Daconto, G., Fadeeva, Z., . . . Tabucanon, M. (2015). *Sustainable consumption and production: A handbook for policymakers*. Geneva: UNEP.

Alexander, S. (2012). Degrowth implies voluntary simplicity: Overcoming barriers to sustainable consumption. *SSRN Electronic Journal*. doi:10.2139/ssrn.2009698

Berns, M., Townend, A., Khayat, Z., Balagopal, B., Reeves, M., Hopkins, H. S., & Kruschwitz, N. (2009). The business of sustainability: What it means to managers now. *MIT Sloan Management Review*, *51*, 20–26.

Brough, A. R., Wilkie, J. E. B., Ma, J., Isaac, M. S., & Gal, D. (2016). Is eco-friendly unmanly? The green-feminine stereotype and its effect on sustainable consumption. *Journal of Consumer Research*, *43*(4), 567–582. doi:10.1093/jcr/ucw044

Cleveland, M., Kalamas, M., & Laroche, M. (2012). "It's not easy being green": Exploring green creeds, green deeds, and internal environmental locus of control. *Psychology & Marketing*, *29*(5), 293–305. doi:10.1002/mar.20522

Cooper, T. (2013). Sustainability, consumption and the throwaway culture. In *The handbook of design for sustainability* (pp. 137–155). London: Bloomsbury.

Darby, S. (2006). *The effectiveness of feedback on energy consumption: A review for defra of the literature on metering, billing and direct displays*. Oxford: Environmental Change Institute, University of Oxford.

De Neve, J. E., & Sachs, J. D. (2020). *Sustainable development and human well-being* (pp. 113–128). World Happiness Report 2020. Retrieved from https://happiness-report.s3.amazonaws.com/2020/WHR20.pdf

Fernández-Rovira, C., Álvarez Valdés, J., Molleví, G., & Nicolas-Sans, R. (2021). The digital transformation of business: Towards the datafication of the relationship with customers. *Technological Forecasting and Social Change*, *162*, 120339. doi:10.1016/j.techfore.2020.120339

Geiger, S. M., Fischer, D., & Schrader, U. (2018). Measuring what matters in sustainable consumption: An integrative framework for the selection of relevant behaviors. *Sustainable Development*, *26*(1), 18–33. doi:10.1002/sd.1688

Gifford, R., & Nilsson, A. (2014). Personal and social factors that influence pro-environmental concern and behaviour: A review. *International Journal of Psychology*. doi:10.1002/ijop.12034

Gleim, M. R., Smith, J. S., Andrews, D., & Cronin, J. J. (2013). Against the green: A multi-method examination of the barriers to green consumption. *Journal of Retailing, 89*(1), 44–61. doi:10.1016/j.jretai.2012.10.001

Green, T., & Peloza, J. (2014). Finding the right shade of green: The effect of advertising appeal type on environmentally friendly consumption. *Journal of Advertising, 43*(2), 128–141. doi:10.1080/00913367.2013.834805

Haller, K., Lee, J., & Cheung, J. (2020). *Meet the 2020 consumers driving change: Why brands must deliver on omnipresence, agility, and sustainability*. IBM Institute for Business Value. Retrieved from www.ibm.com/downloads/cas/EXK4XKX8

Hardisty, D. J., Johnson, E. J., & Weber, E. U. (2010). A dirty word or a dirty world?: Attribute framing, political affiliation, and query theory. *Psychological Science, 21*(1), 86–92. doi:10.1177/0956797609355572

Jaiswal, D., & Singh, B. (2018). Toward sustainable consumption: Investigating the determinants of green buying behaviour of Indian consumers. *Business Strategy & Development, 1*(1), 64–73. doi:10.1002/bsd2.12

Jansson, J. (2011). Consumer eco-innovation adoption: Assessing attitudinal factors and perceived product characteristics. *Business Strategy and the Environment, 20*(3), 192–210. doi:10.1002/bse.690

Johnstone, M. L., & Tan, L. P. (2015). Exploring the gap between consumers' green rhetoric and purchasing behaviour. *Journal of Business Ethics, 132*(2), 311–328. doi:10.1007/s10551-014-2316-3

Jung, J., Kim, S. J., & Kim, K. H. (2020). Sustainable marketing activities of traditional fashion market and brand loyalty. *Journal of Business Research, 120*, 294–301. doi:10.1016/j.jbusres.2020.04.019

Kumar, B., Manrai, A. K., & Manrai, L. A. (2017). Purchasing behaviour for environmentally sustainable products: A conceptual framework and empirical study. *Journal of Retailing and Consumer Services, 34*, 1–9. doi:10.1016/j.jretconser.2016.09.004

Lanzini, P., & Thøgersen, J. (2014). Behavioural spillover in the environmental domain: An intervention study. *Journal of Environmental Psychology, 40*, 381–390. doi:10.1016/j.jenvp.2014.09.006

Lee, K. (2008). Opportunities for green marketing: Young consumers. *Marketing Intelligence & Planning, 26*(6), 573–586. doi:10.1108/02634500810902839

Lucas, S., Salladarré, F., & Brécard, D. (2018). Green consumption and peer effects: Does it work for seafood products? *Food Policy, 76*, 44–55. doi:10.1016/j.foodpol.2018.02.017

Lukman, R., Lozano, R., Vamberger, T., & Krajnc, M. (2013). Addressing the attitudinal gap towards improving the environment: A case study from a primary school in Slovenia. *Journal of Cleaner Production, 48*, 93–100. doi:10.1016/j.jclepro.2011.08.005

Luo, X., & Bhattacharya, C. B. (2006). Corporate social responsibility, customer satisfaction, and market value. *Journal of Marketing, 70*(4), 1–18. doi:10.1509/jmkg.70.4.001

Maciejewski, G. (2020). Consumers towards sustainable food consumption. *Marketing of Scientific and Research Organizations, 36*(2), 19–30. doi:10.2478/minib-2020-0014

Malodia, S., & Bhatt, A. S. (2019). Why should I switch off: Understanding the Barriers to sustainable consumption? *Vision: The Journal of Business Perspective, 23*(2), 134–143. doi:10.1177/0972262919840197

Mazurek-Łopacińska, K., & Sobocińska, M. (2014). Determinanty rozwoju zrównoważonej konsumpcji w Polsce – Wybrane zagadnienia. *Zeszyty Naukowe Uniwersytetu Szczecińskiego: Problemy Zarządzania, Finansów i Marketingu, 35*(824), 169–179.

Min, J., Azevedo, I. L., Michalek, J., & de Bruin, W. B. (2014). Labeling energy cost on light bulbs lowers implicit discount rates. *Ecological Economics, 97*, 42–50. doi:10.1016/j.ecolecon.2013.10.015

Mintel. (2021). *Global consumer trends: The now, next, and future global consumer*. Mintel. Retrieved from https://www.mintel.com/press-centre/social-and-lifestyle/mintel-ann ounces-global-consumer-trends-for-2021

Mont, O., & Plepys, A. (2008). Sustainable consumption progress: Should we be proud or alarmed? *Journal of Cleaner Production, 16*(4), 531–537. doi:10.1016/j.jclepro. 2007.01.009

Morris, M. G., & Venkatesh, V. (2000). Age differences in technology adoption decisions: Implications for a changing work force. *Personnel Psychology, 53*(2), 375–403. doi:10.1111/j.1744-6570.2000.tb00206.x

Neale, A. (2015). Sustainable consumption: Sources of concept and implementation. *Prace Geograficzne, 141*, 141–158. doi:10.4467/20833113PG.15.014.4066

Nielsen. (2015). *The sustainability imperative*. Nielsen Report. Retrieved from https:// engageforgood.com/2015-nielsen-global-sustainability-report/

Noppers, E. H., Keizer, K., Bolderdijk, J. W., & Steg, L. (2014). The adoption of sustainable innovations: Driven by symbolic and environmental motives. *Global Environmental Change, 25*, 52–62. doi:10.1016/j.gloenvcha.2014.01.012

Olsen, M. C., Slotegraaf, R. J., & Chandukala, S. R. (2014). Green claims and message frames: How green new products change brand attitude. *Journal of Marketing, 78*(5), 119–137. doi:10.1509/jm.13.0387

Patrzałek, W. (2016). Pro-environmental behaviour of households. *Marketing i Zarządzanie, 3*(44), 157–166.

Roberts, B. W., Walton, K. E., & Viechtbauer, W. (2006). Patterns of mean-level change in personality traits across the life course: A meta-analysis of longitudinal studies. *Psychological Bulletin, 132*(1), 1–25. doi:10.1037/0033-2909.132.1.1

Tunn, V. S. C., Bocken, N. M. P., van den Hende, E. A., & Schoormans, J. P. L. (2019). Business models for sustainable consumption in the circular economy: An expert study. *Journal of Cleaner Production, 212*, 324–333. doi:10.1016/j.jclepro.2018.11.290

UN. (2020). *United Nations comprehensive response to COVID-19: Saving lives, protecting societies, recovering better*. Retrieved from www.un.org/pga/75/wp-content/uploads/sites/100/2020/10/un_comprehensive_response_to_covid.pdf

Verplanken, B., & Roy, D. (2016). Empowering interventions to promote sustainable lifestyles: Testing the habit discontinuity hypothesis in a field experiment. *Journal of Environmental Psychology, 45*, 127–134. doi:10.1016/j.jenvp.2015.11.008

Waas, T., Verbruggen, A., & Wright, T. (2010). University research for sustainable development: Definition and characteristics explored. *Journal of Cleaner Production, 18*(7), 629–636. doi:10.1016/j.jclepro.2009.09.017

Walker, I., Thomas, G. O., & Verplanken, B. (2015). Old habits die hard: Travel habit formation and decay during an office relocation. *Environment and Behaviour, 47*(10), 1089–1106. doi:10.1177/0013916514549619

White, K., Habib, R., & Hardisty, D. J. (2019). How to SHIFT consumer behaviours to be more sustainable: A literature review and guiding framework. *Journal of Marketing, 83*(3), 22–49. doi:10.1177/0022242919825649

Wiernik, B. M., Dilchert, S., & Ones, D. S. (2016). Age and employee green behaviours: A meta-analysis. *Frontiers in Psychology, 7*. doi:10.3389/fpsyg.2016.00194

# 14 Characteristics of sustainable consumption from an economic perspective

*Jakub Głowacki, Piotr Kopyciński,*
*Mateusz Malinowski, and Łukasz Mamica*

## Introduction

As shown in the previous chapter, the implementation of the principles of sustainable consumption (SC) means consumers taking responsibility for shopping in an ethical way, in order to reduce the negative impact on the environment (Evans, 2011), which as a result should guarantee a decent life for all, within the boundaries of the earth's resources (Lorek & Spangenberg, 2014). The implementation of sustainable consumption in the area of the CE is often considered from two dimensions: the social and the ecological, which are mutually related, but which constitute separate decision-making factors (Bruska, 2016). The social dimension of the implementation of sustainable consumption above all consists of the fulfilment of needs as a result of the conscious purchasing of products. Environmentally friendly consumption is however associated with the saving of energy (including by using systems which employ renewable sources of energy) and water, the reduction of food waste, the prevention and minimisation of waste produced, including guiding consumers in the direction of recycling and the sharing of goods, including on redistribution markets (Balderjahn, Peyer, & Paulssen, 2013; Bruska, 2016; Gullstrand Edbring, Lehner, & Mont, 2016). The implementation of sustainable consumption in the CE should bring measurable economic benefits (the third dimension), including the creation of new jobs (Głowacki, Kopyciński, Mamica, & Malinowski, 2019). This results directly from the character of the actions that the consumer/user can take to close the cycle and put resources back into the system. These actions however require not only the appropriate knowledge, but also awareness and active engagement, which translates into responsible consumer decisions and behaviours (Jastrzębska, 2019). Measurement of these actions/activity will provide an answer as to the direction of development of the CE in the area of SC and also the pace and dynamics of this development. This is knowledge that can be used both by local government authorities, local leaders, and also company managers and directors.

## Monitoring of implementation of the CE in the area of sustainable consumption

Monitoring of the transition of current linear economic systems towards the Circular Economy (CE) is made difficult due the lack of any one commonly recognised

DOI: 10.4324/9781003179788-14

indicator for measuring progress in the implementation of the CE at the level of companies, sectors, cities, regions, or states. In recent years, individual countries, scientists, and institutions have presented a range of guidelines (principles) for the construction of indicators and patterns (models) providing the basis for assessment of selected activities in the implementation of the CE at different levels of the system (Głowacki et al., 2019). As a result, a wide range of indicators have been developed. These are often controversial, being based on indicators of achievement of Sustainable Development (Camana, Manzardo, Toniolo, Gallo, & Scipioni, 2021) and of the limitation of damage to the planet (Hofmann, 2019; Desing et al., 2020; Calisto Friant, Vermeulen, & Salomone, 2021), but no consensus has yet been found in this area (De Oliveira, Eduardo, Dantas, & Soares, 2021). In this context, the European Circular Economy Action Plan clearly emphasises the importance of identifying instruments for monitoring the consumption of materials and their environmental impacts in order to assess the "interlinkages between circularity, climate neutrality and the zero pollution ambition" (EU, 2020).

The majority of indicators known to date concern the analysis of trends in industrial activity (above all in the processing of raw materials) and the creation of new business models (Saidani, Yannou, Leroy, Cluzel, & Kendall, 2019); however, they do not represent the systemic and multi-disciplinary nature of the CE (Saidani, Yannou, Leroy, & Cluzel, 2017). Moreover, in the opinion of Pauliuk (2018), the number of indicators is growing fast, resulting in too many indicators for measuring the effectiveness of implementation of the CE. The European Academies' Science Advisory Council (EASAC) provides a list with more than 300 indicators that could potentially be used to measure progress in the CE (EASAC, 2016). Iacovidou et al. (2017) found more than 60 environmental, economic, social, and technical metrics that can be used to assess waste management and resource recovery systems alone.

It is however important that, at the centre of the CE stands the consumer who uses different types of products and makes use of services available on the market and whose choices (decisions) may be more or less compatible with the concept of the CE. The CE should thus been seen as a means of building a responsible society (Smol, Avdiushchenko, Kulczycka, & Nowaczek, 2018), which requires a thorough reorganisation of how we handle available resources. This is why progress in the development of social awareness in the area of the CE must be monitored and systematically assessed, so that barriers to its implementation can be identified on an ongoing basis (Guo et al., 2017). Practically, every global organisation currently dealing with issues of the CE has proposed its own solutions for the monitoring of sustainable consumption. These organisations include the OECD, the World Bank, EUROSTAT, Yale and Columbia Universities, the Ellen MacArthur Foundation, and EURES (EASAC, 2016). The main indicators used for assessment of the CE include (EASAC, 2016):

- Domestic extraction (DE). Material extracted within the territories of the EU;
- Direct Material Input (DMI) comprises all materials with economic value, which are directly used in production and consumption activities. DMI equals the sum of domestic extraction and direct imports;

- Raw Material Input (RMI) adds the Raw Material Equivalents (RME) of imports to DMI;
- Total Material Requirement (TMR) comprises all types of input flows;
- Domestic Material Consumption (DMC) measures the total quantity of materials used within an economic system, excluding indirect flows;
- Raw Material Consumption (RMC) deducts from RMI the export of materials plus the RME of exports; and
- Total Material Consumption (TMC) adds the unused extraction related to RMEs of both imports and exports to RMC.

The *Ellen MacArthur Foundation* (EMF, 2015) proposes the application of the following circularity indicators at an initial stage:

- Resource productivity (GDP/1 kg of DMI);
- Indicators of recycling rate and eco-innovation indexes;
- Waste generation – the two indicators proposed are weight of waste generated in relation to GDP (excluding major mineral waste) and the weight of municipal waste generated per capita; and
- Energy consumption (electricity and heat) and greenhouse gas emissions in the form of indicators of consumption of energy from renewable sources and greenhouse gas emissions per GDP output.

The Italian proposal for the measuring implementation of the CE in the area of sustainable consumption identifies the following indicators (Leoni et al., 2019): Domestic material consumption for Italy, final energy consumption by households, share of renewable energy consumed for domestic purposes, collection of used clothes, and the number of companies which perform repairs on electronic goods, as well as on other personal possessions (clothing, footwear, watches, jewellery, furniture). In addition to this, in Italy, the CE is monitored by taking into account the weight of municipal solid waste in cities, the percentage of recycling of all waste and urban waste, employment in sectors of the Circular Economy, and eco-innovation activity (Ghisellini & Ulgiati, 2019).

The Dutch strategy for the CE focuses on five elements of the CE, namely biomass and food, the construction sector, plastics, the manufacturing industry, and consumer goods. Ultimately, the decision was taken to measure quantity of consumption of raw materials and materials, use of secondary materials, security of supply of raw materials, environmental effects (water consumption and greenhouse gas emissions), and the level of employment in relation to priority themes for the Netherlands (Nowaczek et al., 2019).

In Austria, transition to the CE in the area of sustainable consumption is measured *inter alia* by the following indicators: reduction in the use of fossil fuels, share of renewable energy in the energy mix, and recycling of waste suitable for such processing (Nowaczek et al., 2019).

In Spain, as many as 187 indicators have been determined for assessing the country's transition to the CE, including consumption of energy produced from renewable sources as a share of gross energy consumption (%), consumption of diesel oil in the transport sector (KTep), number of students trained in the field of the CE, and the number of training courses on renewable energy sources.

In France, the CE is monitored in the area of responsible consumption by the popularity of car-sharing (eco-mobility), the limitation of food waste, household spending on repair and maintenance, quantities of waste sent to landfill, and the level of employment in the CE (Ministry of the Environment, Energy and Marine Affairs, 2017).

Slovenia has drawn up a roadmap, which divides the economy into four sectors: Food, value chains, the manufacturing industry and mobility. Indicators have been identified for each of these sectors. These include the amount of waste food per capita, the share of renewable energy in final energy consumption, the share of $CO_2$-neutral fuel (from wood), the use of renewable energy sources, the number of vehicles (passenger vehicles) per 1,000 inhabitants, the number of passengers on public transport, the use of car sharing, and the use of bicycles (Nowaczek et al., 2019).

## CE assessment indicators in key areas of sustainable consumption

Based on an analysis of the literature, it is possible to say that the construction of well-chosen indicators for assessment of the implementation of the CE in the area of sustainable consumption must be preceded by a discussion between all interested economic entities and the world of science. Under the conditions of the Polish economy, Głowacki et al. (2019) and Malinowski, Głowacki, Kopyciński, and Mamica (2019) propose that the country's transition in the area of sustainable consumption to be monitored should be based on five pillars: the prevention of municipal waste, responsible food consumption (limitation of food waste), education in the area of the CE, the sharing economy in the area of mobility, and energy based on renewable energy sources (RESs). The aforementioned indicators were selected for the assessment and monitoring of the country's transition to the CE. The aforementioned proposal was preceded by expert studies in which local government representatives, scientists, and representatives of business took part. On the basis of the assessments, indicators which were surplus to requirements or did not adequately cover the thematic area of sustainable consumption to be assessed were eliminated.

### *Municipal waste management – waste prevention*

In accordance with the Framework directive on waste, which is a key act of EU law in the area of waste management, the EU's ambition is to create a "recycling society", whose goal will be to avoid waste generation and the use of waste as a resource (Directive, 2018). At several reprises, the Directive emphasises the importance of waste prevention as the overarching goal of municipal waste management programs. Moreover, EU Member States were required to establish waste prevention programmes (KPZPO, 2014). The main goal set out by the EU in this area is the breaking of the link between economic growth and waste generation (or the separation thereof). This goal is being achieved with differing degrees of effectiveness in EU countries. Progress in the area of waste prevention is dependent on many social factors, such as economic growth, income, place of residence, or age. Patterns of consumption and ecological awareness also have an impact on the amount of waste generated (KPZPO, 2014). This area is very important in the assessment of the implementation of the CE in Poland in the area of sustainable consumption and has been written into the roadmap for the country's transition to the CE (MPiT, 2019).

The first group of proposed indicators are those most frequently used in EU countries for the measurement of implementation of the CE in the area of sustainable consumption, namely the weight of municipal waste generated, as well as the weight of municipal waste in relation to domestic consumption of goods, efficiency of segregation, and the index of waste accumulation by weight. The next group of indicators is related to municipal waste management. Here, how much of such waste that is recycled, reused, or recovered is determined not only by how well the country is equipped in terms of waste treatment facilities, but, above all, by the model of consumption and lifestyle of its inhabitants. The remaining indicators proposed are of a postulative nature (their assessment will be possible only as a result of systematic consumer research), and these include (Malinowski, Głowacki, Kopyciński, & Mamica, 2021):

- Share of financial spending on the repair and maintenance of household appliances as a part of total household expenditure;
- Share of household expenditure on used goods;
- Willingness to purchase products produced from or containing recycled materials;
- Share of waste sent to landfill; and
- Number of entities providing repair services or the revenues of such companies.

### *Conscious food consumption*

Food is wasted at every stage of the food chain "from farm to fork", starting from primary agricultural production, followed by storage, production, and food processing, and ending with the distribution and consumption of food. In households, food waste is produced before, during, or after the preparation of meals. This includes the following substances: vegetable peelings, spoiled meat, excess ingredients, etc. (KPZPO, 2014). In addition, this particular group also includes waste that is generated on commercial sites and in catering establishments (e.g. restaurants) and in food-processing plants. As Kopeć, Gondek, and Hersztek (2018) point out, growth in social awareness of the rules for and possible ways of processing of food waste is important, because it allows this process to be made more efficient. Food waste in general does not pose a problem from an environmental point of view. Such waste is well-known for its value as fertiliser. Food waste is organic matter, which should be put back into cultivated soils, closing the cycle of elements.

Factors that determine food waste in households to the greatest extent include increase in wealth measured for example by the share of expenditure on food and non-alcoholic drinks in household consumption expenditure and demographic changes consisting of a decrease in the number of persons per household (Bilska, Wrzosek, Krajewski, & Kolozyn-Krajewska, 2015). Taking the development of Polish society into consideration, it should be assumed that the concept of sustainable consumption and production is key to limiting food waste (Stępień & Dobrowolski, 2017). Its implementation must include educational activities on a much wider scale while disseminating knowledge among consumers on combating food waste and implementing basic principles, such as the 4Ps, that is planning purchases in advance [*planuj*], processing food to extend its shelf life [*przetwarzaj*], storing products in appropriate conditions [*przechowuj*], and sharing surplus food with those in need [*podziel się*] (MPiT, 2019). It is difficult to assign specific indicators for the assessment of this area in terms of the implementation of the CE due to the lack of reliable data on the weight of such waste. The indicators proposed here are therefore of a postulative nature (Malinowski et al., 2021):

• Share by weight of waste discarded in households as a part of waste generated;
• Weight of discarded food in relation to the weight (or value) of food purchased per household;
• Share of households composting food waste (into fertiliser) or managing it for their own needs (e.g. animal feed);
• Share of food waste collected (generated in the household) sent for recovery; and
• Index of food sharing or, in other words, the share of food waste suitable for consumption by those in need, for example food banks.

### Education in the CE

Education plays a very important role in the process of the transition of society towards the Circular Economy. Its importance was already noted many years ago, at the turn of the 1980s and 1990s when the concept of Sustainable Development was being elaborated. The importance of education was emphasised in the first publications on the topic (Brundtland, 1987) and at events where the concept of Sustainable Development was defined and promoted (including the 1992 Earth Summit in Rio de Janeiro). Measurement of the effects of educational activities is an important element in the assessment of the transition of countries towards the CE in the area of sustainable consumption.

Above all, education fulfils the role of building knowledge and raising awareness of the Circular Economy. Its task is to educate individuals to behave in a certain way, as a result not only of the knowledge they possess, but also based on their beliefs and attitudes. At the same time, education can also be treated as a sector of the economy which uses specific resources and generates waste and pollution. This second aspect is one that should not be forgotten, because, as a result of analyses conducted in Glasgow, Scotland (EESC, 2019), this sector was deemed to be one of the three economically most important areas and was found to affect materials, emissions, and wastewater to an extent similar to another key sector, healthcare. As indicators for the assessment of a country's transition to the CE, the following measures are proposed with regard to educational activities (Malinowski et al., 2021):

- Educational activities in the CE – This indicator is calculated as the number of educational projects completed in the field of the CE, financed from both public and private funds;
- Funds allocated to promoting and providing information about the CE – This indicator is calculated as the total amount of funding allocated to activity directly promoting and providing information about the principles of the CE. Reporting in this and the previous area can be conducted by expanding the system for the transfer of data to the Polish Central Statistical Office (GUS) by local government authorities and businesses;
- Scientific publications concerning the CE – This indicator is calculated based on the number of scientific publications directly concerning the CE (Circular Economy or similar should be included in the keywords of the article) published in High Impact Factor scientific journals;
- Scientific activity in the CE – This indicator is calculated as the percentage share of scientific employees who have had a scientific article published in a prestigious journal with the concept of Circular Economy (or similar) in the keywords;
- Share of research projects financed from public funds concerning the CE as a percentage of all projects – This indicator is calculated as the percentage share of research projects in the area of the CE which have received funding from government institutions; and
- Innovation in the CE – This indicator is calculated as the number of registered patents concerning the CE (e.g. concerning the improvement of the recycling process or eco-design).

### The sharing economy in the area of mobility

Sharing covers different areas, including consumption, production, and financing. The sharing economy covers three categories of participants (Poniatowska-Jaksch & Sobiecki, 2016):

1  Service providers, who share assets, resources, time, or skills (they may be private individuals offering services on an occasional basis or professional service providers operating within the scope of their own professional activity);
2  Users; and
3  Intermediaries which connect service providers with users by operating online internet platforms and handling transactions between them.

Studies concerning the CE do not examine mobility itself but indicators of it (e.g. in the area of sharing and pollution). The importance of sharing is however emphasised in national strategies for the CE. For example in the document published by the Danish Government on this subject, it is possible to reduce transport costs and thus increase the economic profit from recycling. In addition, there is also the possibility of better utilisation of excess capacity through sharing of, for example production equipment between enterprises, which may contribute to increased productivity and savings for each enterprise (Ministry of Environment, Food and Ministry of Industry, 2018). French experts in turn identify car-sharing as one of the 10 key indicators for monitoring the CE (Ministry of the Environment, Energy and Marine Affairs, 2017), as it helps to reduce the consumption of fossil fuels (petrol and diesel) and materials (vehicle manufacturing). Both the use of public transport and the shared use of vehicles subscribe to the concept of use over possession. The growing interest in this form of transport results on the one hand from the establishment of suitable electronic platforms for the rental of vehicles and on the other hand from the growth in digital services connecting drivers with potential passengers (Ministry of the Environment, Energy and Marine Affairs, 2017). Other factors which are of not inconsiderable importance in this respect are difficulties with finding available parking spaces and parking fees or fees for entering specific zones in one's own vehicle, which are often dependent on meeting specific emissions standards (preference for low- and zero-emission vehicles). Ultimately, people who rarely use the car prefer to pay a relatively small fee for renting a vehicle from time to time, rather than pay a larger one-off amount for the purchase of a car. Moreover, the expansion of companies representing a specific sector, for example the outsourcing/offshoring industries (cf. Kopyciński, 2021) which brings a specific corporate culture and lifestyle along with it, may be conducive to the development of rental systems.

The negative effects of the growing popularity of car sharing should also not be forgotten. There is a risk that this is a form of mobility used by people who to date have preferred another model (e.g. public transport or bicycle), as a result of which, vehicle traffic and occupancy of parking spaces may increase or decrease less than expected. The potential rebound effect must also be remembered. Ottelin, Heinonen, and Junnila (2017) point out that a reduction in the number of cars owned may be a contributory factor leading to an increase in the number of journeys by air. In turn, Amatuni, Ottelin, Steubing, and Mogillon (2020, p. 8) speculate that if the annual costs of car sharing and the substituting modes of transport are lower than the average costs of car ownership and use, the indirect rebound effect would be positive (undesirable), meaning that there are actually additional emissions due to an increased consumption in other consumption categories. At the end, the long-term effects of the COVID-19 pandemic on mobility, and so also on the development of car sharing, are not known.

As it is possible to conclude from the points raised earlier, the issue of mobility is an important component of the CE. This area is not however limited just to cars, but also includes other means of transport such as the bicycle or personal light electric vehicles (PLEVs), including electric scooters. The indexes for the assessment of a country's transition to the CE in the area of sharing of means of transport include the following postulative indicators (Malinowski et al., 2021):

- Cars available in the rental-by-the-minute system per capita;
- Average market price for rental of a car for 1 hour in the rental-by-the-minute system;
- Rate of usage (as a percentage) of one car in the rental-by-the-minute system;
- Number of bicycles available in the rental-by-the-minute system per capita;
- Number of bike stations available in the rental-by-the-minute system per capita;
- Rate of usage (as a percentage) of one bicycle in the rental-by-the-minute system;
- Number of scooters and other personal light electric vehicles (PLEVs) available in the rental-by-the-minute system per capita;
- Rate of usage (as a percentage) of one PLEV in the rental-by-the-minute system;
- Percentage of inhabitants using public transport;
- Number of cars registered to private users per capita (destimulant); and
- Occupancy rate (as a percentage) of cars on longer journeys.

### Renewable energy sources in households

The issue of sustainable energy consumption in households can be analysed in terms of attitudes limiting its consumption as well as the willingness to pay more for energy were it to come from renewable sources. The need for decisive action in this area is underlined by the projected increase in energy demand, which according to the U.S. Energy Information Administration is estimated to be almost 50% over the period from 2018 to 2050 (EIA, 2019). If it does not prove to be possible to limit energy consumption or use renewable sources, this will result in a deepening of the phenomenon of irreversible climate changes (Hertzberg, Siddons, & Schreuder, 2017).

The main determinants of energy-saving behaviour are socio-demographic factors, including income levels and the size of households (Abrahamse, Steg, Vlek, & Rothengatter, 2005). These behaviours are also determined by the energy commodity price, the location of the household, or other parameters relating to it (Quaglione, Cassetta, Crociata, & Sarra, 2017). Persons with a higher level of knowledge about the natural environment more frequently presented pro-environmental attitudes and are more motivated to save energy (Pothitou, Hanna, & Chalvatzis, 2016). It is also pointed out that individuals with empathic and moral predispositions to knowledge about environmental protection are capable of considering the long-term good of humanity as a whole over the short-term good of those closest to them (Sharma & Christopoulos, 2021).

Desired energy-saving attitudes may also be stimulated by appropriate psychological stimuli in the form of "nudges" (Beshears & Kosowsky, 2020). These often consist of information presented in an interesting graphic form to prompt specific behaviours, and their purpose is more to propose a choice than to impose bans (DellaValle & Sareen, 2020). The use of nudges used is based to a certain degree on the phenomenon of "learning by looking" (Kendel, Lazaric, & Maréchal, 2017). Like nudges, the effectiveness of the use of this type of tool decreases the occurrence of habits to perform specific actions which are contradictory to desired behaviours (Venema, Kroese, Verplanken, & Ridder, 2020). The use of nudges has an impact in terms of an increase in knowledge on recipients, though it does not always lead to changes in behaviours (Abrahamse et al., 2005).

## The impact of the CE on economic growth and socio-economic development

One of the key issues in the process of construction of indexes of the impact of the CE on socio-economic development is the distinction between the concept of economic growth and that of socio-economic development (Balderjahn et al., 2013; Gullstrand Edbring et al., 2016; Pellet, 2018; Ghisellini & Ulgiati, 2019). In economic theory, economic growth is a category which serves the purpose of describing quantitative changes. It is a process that consists of an increase in national wealth over time and relates only to the measurable sphere of the economy. Most frequently, the measure of economic growth is growth in real gross domestic product (GDP). It is the sum of the value of all final goods and services produced in the country over a specified period of time (usually a year). "Indirect" goods and services, used in the process of further production, are not included in GDP (Łapacz, 2015).

While remaining aware of the numerous shortcomings of metrics based on the material sphere of the economy, people started to search for categories which reflect the qualitative aspects of wealth. One way of responding to this need is the concept of socio-economic development, which describes the changes taking place both in the economy and society. Its essence lies in positive transformations in quantitative and qualitative terms, which contribute, *inter alia*, to an improvement in living conditions, as well the conditions for conducting business and cultural activities (Pisarski, 2014). As a matter of principle, it is more difficult to quantify than economic growth. As part of certain approaches, socio-economic development is understood as a process of transformation of low-income economies into modern economies (Kubiczek, 2014). However, even modern economies and societies are continuously developing, thus setting ever higher standards of development, which means that for highly developed nations it is also worth making the assumption of progressive economic development. To achieve satisfactory results in this area, it is necessary to have a well-thought-out industrial policy that takes existing institutional frameworks into account (Dolfsma & Mamica, 2020).

### *Research conducted to date into the impact of the CE on GDP*

In the scientific literature, there are several studies which conduct an analysis of the impact of the Circular Economy on economic growth (in the macroeconomic sense). In particular, it is worth citing the studies conducted by the European Commission, University College London, and the Ellen MacArthur Foundation. It should be noted that the cited studies use different definitions of what exactly the Circular Economy is. They include an assessment of the efficiency of use of resources and energy, recycling and regeneration, and new business models of the Circular Economy. A summary of the three main studies is presented later in the chapter.

The first example is that of macroeconomic modelling conducted for countries of the European Union, which showed that it is possible to improve resource productivity understood as GDP per unit of raw material consumption (RMC), excluding costs of energy, labour, and capital, by around 2–2.5% pa with a positive net impact on the GDP of the European Union (Cambridge Econometrics, 2014), whereby it is worth drawing attention to the fact that growth in GDP is not so much related to growth in resource productivity, but to the mechanism of implementation of appropriate public policy, and also with initial investment in technology which makes efficient use of resources. In the report, it is however noted that beyond a rate of 2.5%, further improvements in resource productivity are associated with net costs to GDP which outstrip savings.

Another example of the estimation of the impact of the CE on GDP is the report prepared within the framework of the "Policy Options for a Resource-Efficient Economy" (POLFREE) project from University College London. It uses a model similar to that described before in the studies for the European Commission. Different scenarios are analysed, ranging from ordinary business through to global corporations and post-consumerist society. The reports assume a scenario of global cooperation, and it is estimated that efficient means of resource management may potentially result in an increase in the GDP of the European Union by 8% in 2050 compared to activities to date (Meyer, Distelkamp, & Beringer, 2015). However, in the Ellen MacArthur Foundation report (Schulze, 2016), it was found that the implementation of efficient resource management in buildings, food waste, and transport may result in an increase in European GDP of 11% by 2030 and of 27% by 2050 compared with 4% and 15% based on the traditional model. This quite spectacular result was also due to the effect of assuming technical progress leading to a reduction in the costs of reuse of resources.

It is worth drawing attention to the confirmation in empirical studies of the positive relationship occurring in highly developed countries between the Circular Economy and economic growth. This conclusion is important in the context of the three areas identified in the concept of the TBL.

## The impact of the CE on socio-economic development

A tool that is helpful in determining the impact of the CE on socio-economic development is the concept of the triple bottom line (TBL). It is a type of accounting structure, which takes three parts into account for analysis: social, environmental (ecological), and financial. The concept is considered to have been coined by John Elkington, who first wrote about it in 1994.

*Table 14.1* Indicators of the impact of the CE on socio-economic development

| | | |
|---|---|---|
| The economic dimension | Household expenditure on the repair and maintenance of products | Share of financial spending of households on the repair and maintenance of products as a part of overall expenditure |
| | Eco-mobility | Percentage of zero-emissions or low-emissions travel/journeys, broken down into public transport, bicycle/pedestrian transport, personal light electric vehicles (PLEVs), and electric cars. |
| | Indicator of availability of servicing | Number of entities offering repair and maintenance services |
| | Indicator of willingness to pay for renewable energy | Share of people who are willing to pay 10/25/75/100% more for energy from renewable energy sources |
| | Willingness to invest in renewable energy sources | Value of investment in renewable energy sources by prosumers, housing cooperatives, and associations of residential property owners |
| The social dimension | Consumer awareness relating to the CE | Indicator of the level of consumer awareness or, in other words, among children, young people, and other social and professional groups in the area of the CE |
| | Indicator of food waste in households | Percentage of waste discarded/suitable and not suitable for consumption/ in the morphological structure of municipal waste generated |
| | Indicator of sharing | Number of people making shared use of cars or other sharing services at least once a month |
| | Education in the CE | Expenditure on promoting and providing information about the CE |
| | Indicator of energy saving | Consumption of energy per person in the household |
| | Used goods | Household expenditure on used goods |
| The environmental dimension | Composting of waste | Share of households declaring that they compost food waste |
| | Level of recycling | The quotient of the weight of all municipal waste that is recycled /including that which is reused/and the total weight of municipal waste collected |
| | Share of waste sent to landfill | The weight of all municipal waste sent to landfill/including after processing/ in relation to the total weight of municipal waste collected |
| | Wastewater treatment by consumers | Quantity of wastewater treated per capita |
| | Carbon dioxide emissions | Total annual amount of carbon dioxide emissions per capita |
| | Indicator of share of energy from renewable energy sources | Share of renewable energy sources in the mix of consumption of electrical energy, thermal energy, and in transport |

At the turn of the twentieth and the twenty-first centuries, an accounting concept based on the concept of the "triple bottom line" became popular in NGO-related circles. The concept of TBL assumes that the ultimate success of a company can and should be measured not only on the basis of traditional financial results, but also on its social, ethical, and environmental effects. Based on this model, it is assumed that the impact of the CE on socio-economic development should be considered from these three perspectives: Economic, social and environmental. A set of 17 indicators is presented in Table 14.1. They attempt to represent the impact of the CE on socio-economic development understood in this sense.

## Conclusion

The indicators presented in this chapter are a key element in the measurement of a country's transition towards the CE in the area of sustainable consumption. Their use by local government authorities, local leaders, or managers of companies should provide them with information about the areas in which activities to implement the principles of the CE should be intensified at consumer level. The cited indicators may be used as components in the construction of one or more indexes, which present in summary form what progress is being made in the area of sustainable consumption. In this chapter, an analysis was also conducted, the purpose of which was to attempt to determine the impact of the CE on socio-economic development. In the opinion of the authors, the effect of implementation of the concept of the CE should be considered not only from a purely economic perspective, but also with regard to their social and economic dimensions. These three perspectives are inextricably linked and provide the foundations that form the basis for an improvement in the quality of life of society. The development of a country based on the principles of sustainable consumption should thus contribute to socio-economic development understood in this sense.

## References

Abrahamse, W., Steg, L., Vlek, C., & Rothengatter, T. (2005). A review of intervention studies aimed at household energy conservation. *Journal of Environmental Psychology*, *25*(3), 273–291. doi:10.1016/j.jenvp.2005.08.002

Amatuni, L., Ottelin, J., Steubing, B., & Mogillon, J. M. (2020). Does car sharing reduce greenhouse gas emissions? Assessing the modal shift and lifetime shift rebound effects from a life cycle perspective. *Journal of Cleaner Production, 266*.

Balderjahn, I., Peyer, M., & Paulssen, M. (2013). Consciousness for fair consumption: Conceptualization, scale development and empirical validation. *International Journal of Consumer Studies, 37*, 546–555.

Beshears, J., & Kosowsky, H. (2020). Nudging: Progress to date and future directions. *Organizational Behavior and Human Decision Processes, 161*, 3–19. doi:10.1016/j.obhdp.2020.09.001

Bilska, B., Wrzosek, M., Krajewski, K., & Kolozyn-Krajewska, D. (2015). Zrównoważony rozwój sektora żywnościowego a ograniczenie strat i marnotrawstwa żywności [Sustainable development of food sector and limitations of food losses and its waste]. *Journal of Agribusiness and Rural Development, 2*(36), 171–179 [in Polish].

Brundtland, G. H. (1987). Our common future – Call for action. *Environmental Conservation, 14*(4), 291–294.

Bruska, A. (2016, April). Zrównoważona konsumpcja: Istota – formy – nabywcy [Sustainable consumption: Facts – forms – buyers]. *Logistyka odzysku, 21*, 27–31 [in Polish].

Calisto Friant, M., Vermeulen, W. J. V., & Salomone, R. (2021). Analyzing European Union circular economy policies: Words versus actions. *Sustainable Production and Consumption, 27*, 337–353.

Camana, D., Manzardo, A., Toniolo, S., Gallo, F., & Scipioni, A. (2021). Assessing environmental sustainability of local waste management policies in Italy from a circular economy perspective: An overview of existing tools. *Sustainable Production and Consumption, 27*, 613–629.

Cambridge Econometrics. (2014). *Study on modelling of the economic and environmental impacts of raw material consumption.* European Commission Technical Report 2014–2478. Cambridge: Author.

DellaValle, N., & Sareen, S. (2020). Nudging and boosting for equity? Towards a behavioural economics of energy justice. *Energy Research & Social Science, 68*, 101589. doi:10.1016/j.erss.2020.101589

De Oliveira, C. T., Eduardo, T., Dantas, T., & Soares, S. R. (2021). Nano and micro level circular economy indicators: Assisting decision-makers in circularity assessments. *Sustainable Production and Consumption, 26*, 455–468.

Desing, H., Brunner, D., Takacs, F., Nahrath, S., Frankenberger, K., & Hischier, R. (2020). A circular economy within the planetary boundaries: Towards a resource-based, systemic approach. *Resources Conservation and Recycling, 155*, Article 104673.

Dolfsma, W., & Mamica, Ł. (2020). Industrial policy – an institutional economic framework for assessment. *Journal of Economic Issues, 54*(2), 349–355.

Directive. (2018). *Directive (EU) 2018/851 of the European Parliament and of the Council of 30 May 2018 amending Directive 2008/98/EC on waste.* Retrieved from https://eur-lex.europa.eu/legal-content/EN/TXT/HTML/?uri=CELEX:32018L0851&from=PL

EASAC. (2016). *Indicators for a circular economy.* European Academies' Science Advisory Council. Halle (Saale), Germany: Author.

EESC. (2019). *Circular economy strategies and roadmaps in Europe: Identifying synergies and the potential for cooperation and alliance building.* European Economic and Social Committee. Retrieved from https://www.eesc.europa.eu/sites/default/files/files/qe-01-19-425-en-n.pdf.

EIA. (2019). *International energy outlook 2019.* Washington, DC: Author.

EMF (Ellen MacArthur Foundation). (2015). *Delivering the circular economy: A tool-kit for policymakers.* Danish: Author.

EU. (2020). *European Union: Circular economy action plan for a cleaner and a more competitive Europe (2020).* Retrieved January 12, 2020, from https://ec.europa.eu/environment/circular-economy/index_en.htm

Evans, D. (2011). Thrifty, green or frugal: Reflections on sustainable consumption in a changing economic climate. *Geoforum, 42*, 550–557.

Ghisellini, P., & Ulgiati, S. (2019). Circular economy transition in Italy: Achievements, perspectives and constraints. *Journal of Cleaner Production, 243*, 118360.

Głowacki, J., Kopyciński, P., Mamica, Ł., & Malinowski, M. (2019). Identyfikacja i delimitacja obszarów gospodarki w obiegu zamkniętym w ramach "zrównoważonej konsumpcji". In J. Kulczycka (Ed.), *Gospodarka o obiegu zamkniętym w polityce i badaniach naukowych* (pp. 167–179). Kraków: Wydawnictwo IGSMiE PAN [in Polish].

Gullstrand Edbring, E., Lehner, M., & Mont, O. (2016). Exploring consumer attitudes to alternative models of consumption: Motivations and barriers, *Journal of Cleaner Production, 123*, 5–15.

Guo, B., Geng, Y., Jingzheng, R., Zhu, L., Liu, Y., & Sterr, T. (2017). Comparative assessment of circular economy development in China's four megacities: The case of Beijing, Chongqing, Shanghai and Urumqi. *Journal of Cleaner Production, 162*(20), 234–246.

Hertzberg, M., Siddons, A., & Schreuder, H. (2017). Role of greenhouse gases in climate change. *Energy & Environment, 28*(4), 530–539.

Hofmann, F. (2019). Circular business models: Business approach as driver or obstructer of sustainability transitions? *Journal of Cleaner Production, 224*, 361–374.

Iacovidou, E., Velis, C. A., Purnell, P., Zwirner, O., Brown, A., Hahladakis, J., . . . Williams, P. T. (2017). Metrics for optimising the multi-dimensional value of resources recovered from waste in a circular economy: A critical review. *Journal of Cleaner Production, 166*, 910–938.

Jastrzębska, E. (2019). Konsument w gospodarce o obiegu zamkniętym [The consumer in a circular economy]. *Studia i prace Kolegium Zarządzania i Finansów, 172*, 53–69 [in Polish].

Kendel, A., Lazaric, N., & Maréchal, K. (2017). What do people "learn by looking" at direct feedback on their energy consumption? Results of a field study in Southern France. *Energy Policy, 108*, 593–605. doi:10.1016/j.enpol.2017.06.020

Kopeć, M., Gondek, K., & Hersztek, M. (2018). Gospodarka o obiegu zamkniętym w kontekście strat i marnowania żywności [Circular economy in the context of food losses and wastage]. *Polish Journal for Sustainable Development, 22*(2), 51–58.

Kopyciński, P. (2021). The embeddedness of firms and employees in Central Europe: Krakow as an offshoring and outsourcing centre. In Ł. Mamica (Eds.), *Outsourcing in European emerging economies: Territorial embeddedness and global business services* (pp. 155–165). London and New York: Routledge.

KPZPO. (2014). *Krajowy Plan zapobiegania powstawaniu odpadów*. Warszawa: Author [in Polish].

Kubiczek, A. (2014). Jak mierzyć dziś rozwój społeczno-gospodarczy krajów? *Nierówności Społeczne a Wzrost Gospodarczy, 38*, 40–56.

Łapacz, D. (2015). Udział małych i średnich przedsiębiorstw w wytwarzaniu PKB – Polska na tle Unii Europejskiej. *Współczesna Gospodarka, 6*(1), 43–51.

Leoni, S., Ronchi, E., Aneris, C., Bienati, M., Pettinao, E., & Vigni, F. (2019). *Report on circular economy in Italy – 2019*. ENEA [in Italian]. Retrieved from https://circu lareconomynetwork.it/wp-content/uploads/2019/04/Proposals-and-Research-Summary-Report-on-circular-economy-in-Italy-2019.pdf

Lorek, S., & Spangenberg, J. H. (2014). Sustainable consumption within a sustainable economy – beyond green growth and green economies. *Journal of Cleaner Production, 63*, 33–44.

Malinowski, M., Głowacki, J., Kopyciński, P., & Mamica, Ł. (2019). Wskaźniki oceny wdrażania gospodarki o obiegu zamkniętym w obszarze zrównoważonej konsumpcji. In J. Kulczycka Joanna (Ed.), *Gospodarka o obiegu zamkniętym w polityce i badaniach naukowych* (pp. 181–192). Kraków: Wydawnictwo IGSMiE PAN.

Malinowski, M., Głowacki, J., Kopyciński, P., & Mamica, Ł. (2021). Ocena transforamcji kraju w kierunku gospodarki o obiegu zamkniętym – propoycja wskaźników pomiaru dla obszaru "zrównoważona konsumpcja". In A. Krakowiak-Bal, M. Malinowski, & J. Sikora (Eds.), *Infrastruktura I środowisko w gospodarce o obiegu zamkniętym*. Kraków: Published by PAN – in press.

Meyer, B., Distelkamp, M., & Beringer, T. (2015). *Report about integrated scenario interpretation: GINFORS*. Osnabrück: LPJmL Results, Gesesllschaft für wirtschaftliche Strukurforschung (GWS).

Ministry of the Environment, Energy and Marine Affairs. (2017). *10 key indicators for monitoring the circular economy*. Retrieved from https://inis.iaea.org/search/search.aspx?orig_q=RN:50078470

Ministry of Environment, Food and Ministry of Industry, Business and Financial Affairs. (2018). *The Danish government: Strategy for circular economy*. Copenhagen: Author.

MPiT. (2019). *Mapa drogowa transformacji w kierunku gospodarki o obiegu zamkniętym*. Warszawa. Retrieved February 2021, from www.gov.pl/web/przedsiebiorczosc-technologia/rada-ministrow-przyjela-projekt-mapy-drogowej-goz

Nowaczek, A., Kulczycka, J., & Pędziewiatr, E. (2019). Przegląd wskaźników gospodarki o obiegu zamkniętym w dokumentach strategicznych wybranych krajów UE. In J. Kulczycka (Ed.), *Gospodarka o obiegu zamkniętym w polityce i badaniach naukowych* (pp. 21–33). Kraków: Wydawnictwo IGSMiE PAN.

Ottelin, J., Heinonen, J., & Junnila, S. (2017). Rebound effects for reduced car ownership and driving. *Nordic Experiences of Sustainable Planning: Policy and Practice*, 263–283.

Pauliuk, S. (2018). Critical appraisal of the circular economy standard BS 8001:2017 and a dashboard of quantitative system indicators for its implementation in organizations. *Resources, Conservation and Recycling, 129*, 81–92.

Pellet, P. F. (2018). *Economic growth vs. Socioeconomic development: South Florida case study*. Retrieved from https://nsuworks.nova.edu/hcbe_facpres/1356/

Pisarski, M. (2014). Wzrost gospodarczy a rozwój społeczno-gospodarczy w Chinach. *Społeczeństwo i Ekonomia, 1*, 173–182.

Poniatowska-Jaksch, M., & Sobiecki, R. (2016). Przedsiębiorczość w sharing economy. In M. Poniatowska-Jaksch & R. Sobiecki (Eds.), *Sharing economy (gospodarka współdzielenia)* (pp. 11–26). Warszawa: Oficyna Wydawnicza SGH.

Pothitou, M., Hanna, R. F., & Chalvatzis, K. J. (2016). Environmental knowledge, proenvironmental behaviour and energy savings in households: An empirical study. *Applied Energy, 184*, 1217–1229. doi:10.1016/j.apenergy.2016.06.017

Quaglione, D., Cassetta, E., Crociata, A., & Sarra, A. (2017). Exploring additional determinants of energy-saving behaviour: The influence of individuals' participation in cultural activities. *Energy Policy, 108*, 503–511.

Saidani, M., Yannou, B., Leroy, Y., & Cluzel, F. (2017). How to assess product performance in the circular economy? Proposed requirements for the design of a circularity measurement framework. *Recycling, 2*.

Saidani, M., Yannou, B., Leroy, Y., Cluzel, F., & Kendall, A. (2019). A taxonomy of circular economy indicators. *Journal of Cleaner Production, 207*, 542–559.

Schulze, G. (2016). Growth within: A circular economy vision for a competitive Europe. *Ellen MacArthur Foundation and the McKinsey Center for Business and Environment*, 1–22.

Sharma, S., & Christopoulos, G. (2021). Caring for you vs. caring for the planet: Empathic concern and emotions associated with energy-saving preferences in Singapore. *Energy Research & Social Science, 72*, 101879. doi:10.1016/j.erss.2020.101879

Smol, M., Avdiushchenko, A., Kulczycka, J., & Nowaczek, A. (2018). Public awareness of circular economy in southern Poland: Case of the Malopolska region. *Journal of Cleaner Production, 197*, 1035–1045.

Stępień, S., & Dobrowolski, D. (2017). Straty i marnotrawstwo w łańcuchu dostaw żywności – propedeutyka problemu. *Progress in Economic Sciences, 4*, 305–316.

Venema, T. A. G., Kroese, F. M., Verplanken, B., & Ridder, D. T. D. de (2020). The (bitter) sweet taste of nudge effectiveness: The role of habits in a portion size nudge, a proof of concept study. *Appetite, 151*, 104699. doi:10.1016/j.appet.2020.104699

# 15 The role of universities in development of the Circular Economy

*Barbara Campisi, Patrizia De Luca, Janina Filek, Gianluigi Gallenti, and Magdalena Wojnarowska*

## Introduction

Universities have always contributed, in a direct and indirect way, through the opinions of their representatives, their research, and education activities, to public debate on core values and the emerging problems of societies.

Nowadays, university action is increasingly formalised and publicly recognised in terms of education, research and technology transfer, and public engagement, which shape the role that universities are expected to play in modern society.

The contribution of universities to Sustainable Development (SD) and the Circular Economy (CE) can be found in this system of relations between public functions and higher education (HE).

Formulating a new mission, that is not related to scientific research and education alone, has strengthened the role of universities in shaping the vision of social development (Shek & Hollister, 2017).

Moreover, with the globalisation, this new role of university systems called University Social Responsibility (USR), previously perceived as being local or national in character, has increasingly begun to take on a global relevance, particularly for environmental issues.

As longstanding institutions, universities command the appropriate authority, expertise, and experience to serve in this capacity. Furthermore, SD is the paradigm which reinforces and profiles the role of universities in the challenge of transition from the traditional unsustainable economic growth to the CE.

The aim of the analysis is to compare the experience of these challenges for universities in Poland and Italy. Both countries share the characteristics of having important historical university institutions, and, after following two different paths of economic, social, and political developments after World War II, they now share the same EU educational and research framework.

## The concept of the Circular Economy in the context of the three missions of a university

### *The educational mission*

As indicated in target 4.7 of the UN 2030 Agenda, the conditions shall be established for all learners to acquire the necessary knowledge and skills needed to

DOI: 10.4324/9781003179788-15

promote SD, including, among others, through education for SD and sustainable lifestyles, human rights, gender equality, promotion of a culture of peace and non-violence, global citizenship and appreciation of cultural diversity, and of culture's contribution to SD.

Universities prepare and train new cadres of employees, and thus they are a perfect vector for spreading new ideas through the development of human capital and the skills required for tackling challenges and capitalising on the opportunities related to SD. Through a number of initiatives, universities exert significant influence over the formation of new civil attitudes, particularly in the field of SD and the CE. Furthermore, universities have a significant influence over the lower tiers of national education systems (teaching training universities or colleges are where future teachers are educated). Therefore, universities have the potential to influence children and young people to form proper attitudes and, by the same token, to indirectly influence the attitudes of future consumers with respect to responsible and sustainable consumption.

## The scientific research mission

Implementation of SD and the CE is mainly dependent on the development of scientific research and the resulting technological innovations. The role of universities in these fields is pivotal. Within the scientific research mission, higher education institutions (HEIs) are purveyors and authors of new theories, as well as practical solutions; cooperation between universities and economic entities is thus growing in importance (Andersson & Hellerstedt, 2009). In cooperation between science and business, transfer of knowledge depends on the proximity or geographical density of businesses, research institutions, and related branches of industry (Hermannsson, Scandurra, & Graziano, 2019). The level of university research expenditure (Whalley & Hicks, 2014) and cooperation between businesses (Wasiluk, 2017) is of the utmost importance. Although there are studies of academic knowledge and its transfer dating back to the 1980s, it is only recently that we have been able to observe growth in the number of debates and discussions in academic, political, and business spheres concerning this type of cooperation (Galán-Muros & Plewa, 2016; Hernández-Trasobares & Murillo-Luna, 2020; Orazbayeva, Davey, Plewa, & Galán-Muros, 2020). When companies and universities cooperate to exchange knowledge, they not only develop and reinforce their competitive advantage, but also become a major driver for innovation and economic growth.

Universities thus play a crucial role in the process of implementation of SD and the CE "because they establish a connection between generating knowledge and transfer of knowledge to society, by both educating future policymakers and informational activities and social services" (Adomßent & Michelsen, 2006, pp. 87–88).

Thus, quoting Zampetakis and Moustakis (2006), we may state that universities support the development of both creativity and entrepreneurship. Moreover, already in 1973, the European Industrial Research Management Association (EIRMA) emphasised the importance now assigned to the transfer of knowledge and cooperation between enterprises and universities due to the level of their contribution to regional development (EIRMA, 1973). Universities therefore contribute to supporting productivity and innovation, the key factors for boosting the level of regional development and competitiveness, as confirmed by the studies conducted, for example, by Fernandes and Ferreira (2013) and Mueller (2006).

### The social mission

In relation to the third mission – defined as "a university being open to the social-economic context through valorisation and transfer of knowledge" – universities may support the development of a society and communities by applying their professional knowledge, authority, and external leadership to influence policy and practices defined by various stakeholders in accordance with the 17 Sustainable Development Goals (SDGs). Universities are becoming change leaders by incorporating SD-oriented practices into their management processes and monitoring progress and the results of these practices, as well as informing the public about their obligations with regard to SDGs. Furthermore, universities have a significant influence on the development of regions due to their role in supplying the labour market with appropriately prepared graduates. They thus generate demand for a diverse range of services among local businesses (Harloe & Perry, 2004). The fundamental reorganisation and changes in the area of new concepts such as the CE require not only a far-reaching and fundamental shift in thinking but also necessitate changes in the manner in which society operates as a whole. Thus, it is expected that actions taken by a university to promote SD and the CE will lead to improvements in both the social awareness and the skills required for independent participation in shaping future development by making an innovative contribution to all spheres of economic, social and environmental cultural affairs.

## The role of higher education in the implementation and promotion of the concept of SD

### Higher education in Poland

In Poland, there are both public HEIs (established and operated by public entities such as the state or local self-governments and maintained through public funding) and non-public (private) HEIs (established by a natural or a legal person) in operation. In the 2019/2020 academic year, there were 370 HEIs operating in Poland, including 133 public institutions where 71.6% of all students, a clear majority, studied. Poland is at the forefront of European Union member states with regard to the number of HEIs per 1 million residents (11.3). The educational offering of Polish HEIs has been steadily growing from 5,892 subject courses in 2012 to 6,794 in 2018 (Państwowy Instytut Badawczy, 2019).

*University Social Responsibility*

Due to the popularity of the concept of Corporate Social Responsibility (CSR) in Poland and with the rapid growth in student numbers (over the period from 2001 to 2010) and over-frequent reforms of the HE system, an interesting debate has emerged over the last decade concerning the subject of USR, which laid the foundations for the emergence of the third university mission – the social mission. Given that this is an issue which has only emerged relatively recently, a singular binding interpretation of the definition of USR has not yet been developed. Each of the universities tackling this issue has attempted to provide its own definition. For instance, in one of the first reports concerning USR, USR was taken to consist primarily of the social obligations exceeding the obvious responsibilities resulting from legal regulations, as well as the act of becoming increasingly open to notions concerning the natural environment in order to develop a civic attitude among students and increase the influence of a university over the form and character of social development.

As this definition indicates, the new approach to the USR is beneficial for the wider propagation of issues of environmental protection and thus for the promotion of numerous SD goals, and, by the same token, it facilitates the introduction of issues related to the CE into education. Furthermore, one of the more valuable aspects of USR is the move away from preparing students solely for meeting the demands of the labour market and redirecting efforts towards preparing graduates for thoughtful work with the aim of contributing to the future social and economic development of the country and the world, in order to secure and preserve opportunities for future generations independently of the subject of HE.

What appears to be beneficial for more complete development of the concept of the CE in Poland is that the position of the concept of CSR has been consolidated in both theory and practice. In recent years, this idea has been further reinforced by the academic initiative of preparing a USR Declaration which can be signed by any of Poland's universities. When this Declaration was first made in 2017, it was signed by only 23 universities (6.2%), but in 2019 it was signed by a further 58 universities, a total number which altogether represents 21.8% of all Polish HEIs, with even more universities planning to sign the declaration in the future. Furthermore, as part of a cooperation between the Ministry of Development Funds and Regional Policy, the USR Workgroup – representing 90% of scientific researchers from Polish HEIs – and the Commission for Communication and Social Responsibility operating under the Conference of Rectors of Academic Schools in Poland,[1] 35 questions were defined based on the 12 rules of the Declaration to allow each university to prepare a self-assessment report regarding these issues (MNiSW, 2020). Currently, work is in progress on preparing an appropriate application and a website where the reports will be submitted independently of being published on universities' websites. Within the framework of this initiative, USR is understood to be as the voluntary participation of an HEI in promoting the idea of SD and SR in its education programmes, scientific studies, and the managerial and organisational solutions employed by a university (Cracow University of Economics, 2020).

## Higher education in Italy

Higher Education (HE) in Italy encompasses two main sectors: the University System and the system of Higher Education for Fine Arts, Music and Dance (AFAM) (MUR, 2020). In particular, the Italian University system consists of:

- Ninety-seven university institutions, of which 67 are public or state universities;
- Nineteen private (legally recognised) universities;
- Eleven private (legally recognised) online universities since 2003.

These university institutions include some of the world's oldest universities, most notably the University of Bologna, which was established in the last decades of the eleventh century. In the academic year 2020/21, there were 1.75 million students enrolled on a study programme at Italian universities, where women make up 56% of the total student population. In the same year, the total educational offering of Italian Universities included 5,277 study programmes (87% at public universities, 10% at private universities, 20% in a foreign language, and 3% at online universities).

## SD and the CE in the Italian university system

The diffusion of the ideas of SD and CE in the Italian university system has followed different paths.

Unlike other EU's countries, Italy has not developed a national strategy for implementation of the European CE action plan, though, in 2017, it did adopt its Bioeconomy Strategy that includes the CE approach. This promoted technological transfer within some technological clusters. Also, the National Sustainable Development Strategy (SNSvS), adopted in 2017 in order to implement the UN 2030 Agenda, is coherent with the CE approach. At the end, the "National Recovery and Resilience Plan" was adopted in spring 2021 to implement the measures of the European Green Deal (EGD) include the CE as one of its specific strategies. These processes are creating a growing demand for knowledge transfer and professional assistance from innovative firms and public authorities towards university system.

The Bioeconomy and CE Strategies and the EGD operate within the existing sectoral and cross-sectoral policies of the EU, which have already gradually taken the principles of SD into account. Thus, Italian policies financed from EU funds have contributed to the diffusion of SD concepts within the Italian public sector and society, in particular, within university research focused on EU policies (e.g. Horizon 2020 programme).

Moreover, Italian legislation has intervened over the years on environmental, economic, and social issues by including ethical aspects, which today form part of the SDGs (inclusion, gender equity, work, health, etc.). Therefore, it has regulated, consistent with the SDGs, the life of public bodies and private relationships in society, including the activity of academic communities.

At the end, the civil society has increased along with the 2030 Agenda approval process its role that ensures government accountability based on responsibility, answerability, and enforceability (Sénit, 2020). In Italy, the traditional role of NGOs with environmental interests (WWF, Legambiente, Italia Nostra, and others) has broadened to include other stakeholders with wider interests in the global issues of SD, such as the "Foundation for sustainable development" and the "Alliance for sustainable development" (ASviS), and this has aroused the interest of universities, which also have been directly involved (as in the case of ASviS) through their public engagement activities. Moreover, Italian universities have set up RUS – the Italian University Network for Sustainable Development – which acts as a model to encourage collaboration between universities and society to spread social innovation at local level and to provide cultural awareness at national level, furthering the recognisability and the value of the experience of Italian Universities at an international level. This is increasingly bearing fruit thanks also to the participation of a growing number of Italian universities in international rankings related to SD issues such as GreenMetric (UI GreenMetric, 2021) and the Times Higher Education Impact Rankings (THE, 2021).

## Results

### Data and methods

In this study, we aim to explore the implementation of sustainability at universities in a comprehensive way, considering its different dimensions, to integrate knowledge not yet developed in adequate depth on this topic (Fissi, Romolini, Gori, & Contri, 2021).

Here, we describe some cases of Polish and Italian Universities exploring the sustainability approach in HEIs. In order to capture some differences and similarities among various situations and journeys towards sustainability in the sphere of HEIs, we focus on different cases ranging from the legal, public or private, generalist, or specialist nature of the educational offering to the size of the institution, based on the number of students enrolled, academic staff, and employees.

We chose to consider three Polish and four Italian universities in the timeframe January–May 2021. The universities are assigned codes as follows:

- University $A_P$ – a private economics university in Warsaw, employing 251 members of academic staff and 181 administrative and technical employees, educating approximately 6,400 students in 2020.
- University $B_P$ – a public state economics university in Warsaw, employing 815 members of academic staff and 540 administrative and technical employees, educating 10,198 students in 2020.

- University $C_P$ – a public state economics university in Cracow, employing 730 members of academic staff and 431 administration and technical employees, educating 12,158 students in 2020.
- University $A_I$ – a large, public state comprehensive university in Bologna, educating 81,459 regular students in 2020, employing 3,987 members of academic staff, of which 2,386 are with a permanent contract (58.84%), 2,894 are administrative and technical employees, of which 2,811 are with a permanent contract (97.13%). It is Italy's top university and ranked tenth in the UI GreenMetric World University Ranking 2020 (Universitas Indonesia, 2020).
- University $B_I$ – a medium-sized, public state generalist university in Trieste, educating over 15,864 regular students in 2020, employing 644 members of academic staff, of which 561 are with a permanent contract (87.11%), and 600 are administrative and technical employees, of which 594 are with a permanent contract (99%).
- University $C_I$ – a large (among non-public universities), private specialist (economics, law, and social sciences) university in Milan, educating 13,606 regular students, employing 668 members of academic staff, of which 234 are with a permanent contract (35%), and 523 are administrative and technical employees, of which 501 are with a permanent contract (95.79%). It came top in the Italian Census 2020–21 ranking of large-sized private Italian universities (Census, 2020).
- University $D_I$ – a medium-sized (among private universities), private specialist (economics, law, and social sciences) university in Rome, educating 9,504 regular students, employing 1,592 members of academic staff, of which only 90 are with a permanent contract (5.65%), and 237 are administrative and technical employees, of which 222 are with a permanent contract (93.67%). It is at the top of the Italian Census 2020–21 ranking of medium-sized private Italian universities (Census, 2020).

We adopted a qualitative method for the reading, organisation, and reporting of qualitative and quantitative data according to other studies (Bonaventura Forleo & Palmieri, 2017; De Marco, Gonano, & Pranovi, 2017; Sonetti, Barioglio, & Campobenedetto, 2020).

For data collection and analysis, a set of 81 qualitative and quantitative indicators was used to assess the level of integration of sustainability issues, with different emphasis on SD/SDGs, SR, and the CE in the university's governance, strategies, policies, education and research activities, third mission, and public engagement. For Polish universities, the framework was provided to the selected universities which directly collaborated in the data collection process. For Italian universities, secondary sources, such as the official university website, the ministry database (MUR, 2021), and, if published, the university's sustainability or social report, were used.

For the research activities of all these universities, data collection was carried out using the search engine of the Scopus database to search for the terms "SD, SGDs, SR and CE" in titles, abstracts, or keywords of scientific publications listed there, from January 2018 to December 2020.

## Results of the studies conducted in Poland

The studies carried out indicate that among the universities surveyed, only University $A_P$ refers to the concepts of SD/SDGs and the CE in its strategic papers. Currently, University $B_P$ does not refer to SD, SCR, or the CE in its strategy documents but intends to change this status as part of its work to develop a new strategy. Furthermore, University $C_P$ is only now planning to refer to CSR in its strategic documents; however, it does not plan to refer to SD or CE issues.

All the universities surveyed gave an affirmative answer to the question "Has Sustainable Development/Have Sustainable Development Goals or social responsibility been taken into account and incorporated into the values, regulations, standards, and behavioural norms of the university?"

The analysed universities also confirmed the filter question concerning the existence of mechanisms (e.g. regulations) ensuring compliance with values, standards, and behavioural norms, as well as the functioning of a website dedicated to ethics and integrity issues.

The subsequent questions concerned the issue of the university's management of SD/SDGs. At Universities $A_P$ and $C_P$, one person has been appointed to deal with SD issues. In addition, an SD committee/commission operates within Universities $A_P$ and $B_P$, while an SD unit/office operates within University $C_P$ only.

In the case of questions concerning universities engaging or being involved in cooperation with external stakeholders (e.g., enterprises, businesses, public bodies, communities, and other universities) also with regard to SD, SDGs, and CE, all the universities surveyed indicated that they engage in or are involved in such forms of activity.

As concerns reporting and the university ranking regarding SD/SDGs and CE, regrettably, only University $C_P$ prepares and possesses a Social Responsibility Report, but the contents of this report do not refer in any meaningful manner to the CE. In turn, only $A_P$ participates and is listed in the rankings concerning SD/SDGs other than GreenMetric or THE Impact Rankings.

The educational programmes of bachelor's, master's, and PhD degree courses were also examined with a focus on the course contents related to SD/SDGs, CSR, and CE issues. The detailed results are presented in Table 15.1.

Concerning the scientific articles published, in the case of all the universities surveyed, the studies conducted display a similar, comparatively low percentage of publications concerning the issue of the CE – less than 1% of the total number

*Table 15.1* Educational programmes related to issues of SD/SDGs, SR, and the CE

| Educational programmes | $A_P$ | $B_P$ | $C_P$ |
| --- | --- | --- | --- |
| Total number of bachelor's/master's/PhD programmes offered | 1,395 | 5,108 | 2,373 |
| Total number of bachelor's/master's/PhD programmes on SD/SDGs | 5 | 0 | 47 |
| Total number of bachelor's/master's/PhD programmes on SR | 9 | 12 | 17 |
| Total number of bachelor's/master's/PhD programmes on the CE | 2 | 0 | 16 |

of publications. In the case of Universities $B_P$ and $C_P$, we observed a greater interest in the subject of SD in scientific publications indexed in the Scopus database – 4.6% and 6.3%, respectively – than in the subject of CSR. In the case of both these universities, the subject of CSR accounted for 1.7% of all scientific publications. In the case of University $A_P$, we noted a marginally higher number of publications on CSR (2.6%) compared to the number of publications tackling SD (2.4%).

In the context of the studies conducted with the selected Polish universities, it is appropriate to explain that initially, the source of Polish interest in and understanding of CSR was based on theoretical considerations with practical sources emerging only significantly later. Thus, two various groups of educators took part in developing this idea and in the process of providing education on this subject. The first group consisted of the academic and teaching staff of Polish universities who were initially engaged in theoretical reflections on the ethical aspects of economic activity. Most frequently, such people were specialists in the field of economic ethics working on determining the ethical and social obligations of economic entities, and the process of education consisted of making future managers aware of their individual responsibility and the responsibility of a company for the positive and adverse influence the company can have on society. The second group of CSR protagonists and educators consisted of employees of major corporations which transplanted specific CSR practices to their Polish branches. In the initial period, Polish entrepreneurs were focused primarily on the necessity of accumulating capital and gaining competitive advantage and took a very reluctant stance towards the notion of CSR. Acting as antagonists opposing the notion of CSR, they most frequently deemed CSR to be either excessively utopian or inhibiting free-market competition and thus too costly.

In turn, the idea of SD was (in terms of overall recognition) taken note of in Poland in 2015 when the 2030 Agenda containing 17 SDGs was announced as the strategy for world development. However, a regular rise in the number of related scientific publications, which increased by tenfold over the period from 2001 to 2011, was seen as early as in 2001. Initially, supporters of this idea related the notion of SD to the need for limiting the adverse effect of the economy on the natural environment and frequently referred to SD as eco-development. Only with time did the idea of SD take more concrete shape. Apart from the goals related to protection of the environment, social goals were also taken note of. Currently, the notion of SD is emerging with increasing frequency as the main topic of discussions concerning social and economic development and is becoming a horizontal principle that is reflected in all of the country's developmental policies (Ministerstwo Rozwoju, Pracy i Technologii, 2018). Its first active protagonists were representatives of the pro-environmental organisations on one hand and economists who are proponents of a heterodox approach (in opposition to a neoliberal approach) on the other. The individual components of this notion are, to an increasing extent, being incorporated into the subjects covered and taught at Polish universities. Its rapid introduction into social and economic life was made possible primarily due to state institutions popularising the notion of SD and owing to the involvement of representatives of business and industry who, after implementing the principles of CSR in their business strategies, did not have to change anything apart from preparing better profiling.

In turn, the concept of the CE reached Poland last, approximately in 2016 when the Polish government adopted the EU premises regarding this issue (Ministerstwo Rozwoju, 2016). Its emergence in social discourse and its subsequent acceptance required an even shorter amount of time due to the implementation of both of the aforementioned concepts, that is CSR and SD, which laid the appropriate foundations for acceptance of the concept of the CE. Furthermore, the rapid adoption of the concept of the CE in Poland was the result of significant support on the part of the government and, second, the major interest that entrepreneurs themselves had in reducing the costs of their operations. This time, entrepreneurs initiated and continued cooperation with university representatives, primarily with those involved in production engineering.

In summary, due to the political transformation initiated in 1989 and based on the radical shift from the socialist economy towards the free-market economy and, owing to the efforts of numerous academic researchers, primarily from economic universities, the notion of CSR was, after the period of reluctance and misunderstanding, adopted and ultimately became firmly entrenched in Polish social awareness and has become a specific form of legitimisation of companies' operations. Since the idea of SD reached Poland late, once the economic transformation was already concluded and the circumstances of stable social conditions allowed, it was adopted significantly more rapidly. This was primarily because, for numerous companies, it, in practice, overlapped in various areas with CSR and because it was more strongly supported by government institutions and non-governmental organisations (primarily those involved in tackling environmental problems).

However, recently, the number of scientific publications dedicated to the topic of SD has been increasing exponentially. That means that the changing economy requires a new approach and new practical solutions. Therefore, pragmatic changes, instead of ethical changes, which played a significant role in the adoption of CSR, are becoming the main driving force for change.

This is confirmed by the rapid establishment of the concept of the CE, fuelled by the threat of resource scarcity, which is a real issue in Poland. In the case of this particular concept, the agents of change have mainly been public institutions, representatives of companies involved in production based on resources that are being depleted at an ever-increasing rate, and representatives of universities, most frequently technological but also economic universities.

### Results of the studies conducted in Italy

Towards SD, the integration of sustainability in strategic plans or other documents is considered to be an important starting point for business organisations and public administration, including HEIs. Among the four Italian universities considered here, it emerged that both the state universities ($A_I$ and $B_I$) and the private universities ($C_I$ and $D_I$) have included a clear reference to sustainability and social responsibility in their strategic plans. It is notable that social and environmental sustainability issues are also considered in other governance documents, such as Performance Plans (Universities $A_I$ and $B_I$) or Green Procurement Policy (established only by University $D_I$). Instead, the CE is mentioned only in the Strategic Plan of University $A_I$.

Regarding the inclusion of issues relating to SD, SDGs, or SR in university values, regulations, standards and behavioural norms, all universities considered have a public statute, where specific statements about social commitments are traceable. Moreover, all the universities here studied have one or more web pages dedicated to initiatives and/or committees appointed for ethics and integrity. For example web pages were published for Ethics Committees (e.g. for Bioethics and Animal Welfare) and for the Ethical Code of Conduct (University $A_I$); for Ethics in Research (University $B_I$); and for presenting the reference documents published (Universities $C_I$ and $D_I$).

Concerning governance-level oversight of sustainability, social responsibility, or SDGs, both the public universities studied have identified a Rector's Delegate: a Delegate for Construction and Environmental Sustainability (University $A_I$) and a Delegate for Sustainability (University $B_I$). Moreover, University $A_I$ has also established a working group involved in the UN's SDG-reporting initiative, including the Rector, Director-General, and other Rector's delegates. University $A_I$ has a Green Office, a sustainability hub based on the model of the Maastricht University Green Office (GO), to foster active student engagement in spreading a culture of sustainability and implementing a sustainability programme, together with university staff. University $C_I$ also has a Sustainability Committee made up of faculty, staff, and students to promote and coordinate sustainability projects and initiatives, especially for reducing environmental impacts. HEIs can also decide to dedicate administrative personnel to a sustainability-related office, as for University $D_I$ with an Office of Ethics, Responsibility, and Sustainability in operation.

Concerning external relationships, namely engagement, cooperation, and partnership with external parties, all the universities considered describe and list many initiatives, among which specific SD, SDGs, and CE activities can be identified. All four universities have the participation of external members on their Boards of Governors – in charge of strategic planning, as well as financial and staff planning, and the implementation of consultation processes with external stakeholders, especially in the design and periodic review phases of their study programmes.

It is noteworthy that, at University $A_I$, the stakeholder consultation process is even more structured with the introduction of a Sponsor's Committee at a central level, which promotes and develops scientific and learning activities and knowledge transfer. Moreover, all four universities have an internship and job placement service to provide students with support in acquiring work experience based on specific agreements with national and international companies.

Regarding public engagement, it is now recognised as an essential element of Universities' Third Mission. Italian Universities are, therefore, encouraged to develop such activities, and all four universities have included a specific objective for public engagement in their strategic plans.

As for innovation, the universities considered have either a knowledge transfer office or a dedicated team for promoting and supporting innovative projects and a web page to communicate better not just about the initiatives carried out, but also about the innovation outcomes, such as the number of registered patents or spin-offs. For Universities $A_I$ and $B_I$, data about patents and spin-offs was found, and also the percentages of green patents were about 25 per cent for University $A_I$ and nearly 10 per cent for University $B_I$. More difficulties were encountered however with the identification of green spin-off initiatives.

All universities also have an international relations office to promote agreements and to collaborate with international partners on teaching and research projects. The main agreements are listed on dedicated web pages that are also available in English.

As for external reporting activities, University $A_I$ is deeply engaged in monitoring and disclosing social and environmental performance via various annual reports such as a Social Responsibility Report (eighth edition in 2020), a Report on UN SDGs (fourth edition in 2020), and a Gender Report (fifth edition in 2020). The two private universities also report on their commitments to the sustainability agenda and the results achieved, namely with the publication of a GRI (Global Reporting Initiative) standard-based report – where the CE is also included (University $C_I$) and an Impact Report (University $D_I$) to assess the impact and value generated by the institution for people, citizens, professionals, and the wider community in general is also included. University $B_I$ is working on the next issue of its first Sustainability Report, after having developed a social reporting initiative in 2008 and 2009. Regarding the inclusion in the leading sustainability rankings, all these universities are included in at least one sustainability ranking, specifically the UI Green Metric (all four universities). Universities $A_I$, $C_I$, and $D_I$ are also featured in the Times Higher Education Impact Rankings, where the results of one of the universities considered placed it sixth in the world and first in Europe in 2020.

As regards the educational programmes for bachelor's, master's, and PhD degree courses, this study focused on the explicit contents of the course related to issues of SD/SDGs, CSR, and the CE. The results are presented in Table 15.2.

The Italian universities considered in this study have a limited number of courses dedicated to sustainability. Concerning CSR, the offer is currently non-existent. The CE is present in a very low number of courses only at University $A_I$.

The final stage of the research process was conducted aiming at determining the share of scientific publications on the topic of SD, CSR, and the CE on the

*Table 15.2* Educational programmes related to SD/SDGs, SR, and the CE

| Educational programmes offered | $A_I$ | $B_I$ | $C_I$ | $D_I$ |
| --- | --- | --- | --- | --- |
| Total number of bachelor's/master's/PhD degrees | – | 396 | 127 | 42 | 69 |
| Total number of bachelor's/master's/PhD degrees concerning SD/SDGs | 22 | 5 | 2 | 3 |
| Total number of bachelor's/master's/PhD degrees concerning SR | 0 | 0 | 0 | 0 |
| Total number of bachelor's/master's/PhD degrees concerning CE | 5 | 0 | 0 | 0 |

total number of scientific articles published by the considered universities. Data were collected by performing a search in the Scopus database, looking for the terms included in the study contained in the titles, abstracts, or keywords of scientific publications over the period from January 2018 to December 2020.

In terms of publications, the results for the universities considered showed the situation to be non-uniform.

Overall, the two generalist universities have a higher number of publications than the two specialist universities: $A_I$ – 22,518; $B_I$ – 6,237; $C_I$ – 1,573; $D_I$ – 703. The situation within the two groups is also very different: the number of publications of $B_I$ is about 30% of the total number of publications of $A_I$; the number of publications of $D_I$ is about 50% of the publications of $C_I$. Regarding the topics studied, publications on sustainability are the most widespread in the four universities considered but with different weights. The highest share of these was for University $B_I$ (7.30%), followed by $D_I$ (5.69%), $C_I$ (3.88%), and $A_I$ (2.53%). The number of publications on CSR is significantly lower than the number of those on sustainability. Again, $B_I$ has the highest share of these (2.66%), followed by $C_I$ (1.27%), $D_I$ (1%), and $A_I$ (0.17%). Publications on the CE are still extremely limited at all the universities considered, compared to their total number of university publications. The figures for this are as follows: $A_I$ – 0.24%; $B_I$ – 0.18%; $C_I$ – 0%; $D_I$ – 0.28%. It can be seen that, for University $A_I$, the percentage of publications on the CE exceeds the percentage of publications on SR, albeit only slightly.

## Conclusion

With the introduction of the EGD in late 2019, the EU's ambition to being the first climate-neutral continent by 2050 has resulted in a concrete target to be pursued. In this context, a "clean and circular economy" has been recognised as a key element of the EU strategy for SD.

Although the contribution of this chapter has been to outline the role of universities in the development of the CE, it was necessary to frame the CE in the context of SD and all the related concepts. In particular, the state of the art in Poland and Italy was outlined, also by analysing some cases of universities selected from among the public and private HEIs of both countries. The positive role of HEIs in the context of sustainability and, as a result, in the transition to the CE has been widely acknowledged. Universities are important players in providing and transferring knowledge to society and businesses and by educating not only decision-makers, leaders, and managers, but also everyday citizens.

Scientific literature on the CE is still limited. The total number of publications in Scopus focusing on the CE for the period from 2018 to 2020 is 6,557, with Italy in the first place with 975 publications and Poland also featuring in the top ten countries with 258 publications. All this confirms the involvement of Polish and Italian universities in the transition to the CE.

With regard to these findings, it is necessary to highlight that the CE, despite having scientific origins, has acquired importance thanks to a series of similar political and institutional initiatives such as the EC action plans and the Ellen

MacArthur Foundation network. Moreover, it is also necessary to consider that these initiatives are based on topics (i.e. plastic pollution, waste recycling, biomass use, sustainability of buildings, eco-innovation) that have long been the subject of important university research, even though they do not refer directly to the concept of the CE. Therefore, there are a high number of relevant studies that are not explicitly labelled as CE research, and which it is thus difficult to identify by means of a meta-analysis, but which can legitimately be included as dealing with the concept of the CE. These "latent" studies on the CE will emerge thanks to the growing diffusion of the concept of SD and the accountability and reporting initiatives of universities in line with the principles and policies of USR.

With their implementation at national level, the EGD and other EU policies will also increase the demand for technological know-how, and knowledge transfer will stimulate growth in CE research and, consequently, growth in educational programmes focused on the CE.

As far as the educational offering is concerned, the number of study programmes focusing on sustainability issues has been on the increase in recent years, especially postgraduate courses, even if this is the case in more limited measure for the CE. Given the transdisciplinary nature of the topics to be addressed, what is needed is a closer integration between teaching and research activities and the implementation of innovative educational and training projects. That means investing in networking and partnership activities to favour the definition of a renewed model of interaction between universities and its external partners driving sustainable and circular innovation. Unfortunately, organisational and structural factors, especially in public HEIs, can be a significant barrier to engagement in third mission activities and to the development of a partnership-based approach. A policy-driven impetus is without doubt a key element that can make the difference engagement of the university systems.

## Note

1  The Conference of Rectors of Academic Schools in Poland is the most important organisation representing academic rectors in Poland (a member of European University Association – EUA).

## References

Adomßent, M., & Michelsen, G. (2006). German academia heading for sustainability? Reflections on policy and practice in teaching, research and institutional innovations. *Environmental Education Research*, *12*(1), 85–99. doi:10.1080/13504620500527758

Andersson, M., & Hellerstedt, K. (2009). Location attributes and start-ups in knowledge-intensive business services. *Industry & Innovation*, *16*(1), 103–121. doi:10.1080/136 62710902728126

Bonaventura Forleo, M., & Palmieri, N. (2017). University value for sustainability: What do stakeholders perceive? An Italian case study. *Rivista di Studi sulla Sostenibilità*, *2*, 104–118.

Censis. (2020). *La classifica Censis delle Università italiane (Edizione 2020/2021)*. Roma: Fondazione Censis. Retrieved from www.censis.it/sites/default/files/downloads/classi fica_universit%C3%A0_2020_2021.pdf

Cracow University of Economics. (2020). *W poszukiwaniu społecznej doskonałości: Raport społecznej odpowiedzialności uniwersytetu.* Krakow: Cracow University of Economics.

De Marco, F., Gonano, M., & Pranovi, F. (2017). La sostenibilità nell'Università: Il caso di Ca' Foscari. In M. Fasan & S. Bianchi (Eds.), *L'azienda sostenibile: Trend, strumenti e case study* (pp. 159–181). Venezia: I libri di Ca' Foscari. doi:10.14277/6969-188-1/ LCF-4-8

EIRMA News. (1973). *R&D Management, 4*(1), 61. doi:10.1111/j.1467-9310.1973. tb01034.x

Fernandes, C. I., & Ferreira, J. J. M. (2013). Knowledge spillovers: Cooperation between universities and KIBS. *R&D Management, 43*(5), 461–472. doi:10.1111/radm.12024

Fissi, S., Romolini, A., Gori, E., & Contri, M. (2021). The path toward a sustainable green university: The case of the University of Florence. *Journal of Cleaner Production, 279*, 1–9.

Galán-Muros, V., & Plewa, C. (2016). What drives and inhibits university-business cooperation in Europe? A comprehensive assessment. *R&D Management, 46*(2), 369–382. doi:10.1111/radm.12180

Harloe, M., & Perry, B. (2004). Universities, localities and regional development: The emergence of the "mode 2" university? *International Journal of Urban and Regional Research, 28*(1), 212–223. doi:10.1111/j.0309-1317.2004.00512.x

Hermannsson, K., Scandurra, R., & Graziano, M. (2019). Will the regional concentration of tertiary education persist? The case of Europe in a period of rising participation. *Regional Studies, Regional Science, 6*(1), 539–556. doi:10.1080/21681376.2019. 1680313

Hernández-Trasobares, A., & Murillo-Luna, J. L. (2020). The effect of triple helix cooperation on business innovation: The case of Spain. *Technological Forecasting and Social Change, 161*, 120296. doi:10.1016/j.techfore.2020.120296

Ministerstwo Rozwoju. (2016). *Komunikat Komisji Do Parlamentu Europejskiego, Rady, Europejskiego Komitetu Ekonomiczno-Społecznego I Komitetu Regionów, Zamknięcie obiegu – plan działania UE dotyczący gospodarki o obiegu zamkniętym.* Retrieved from http://cima.ibs.pw.edu.pl/?page_id=158&lang=pl

Ministerstwo Rozwoju, Pracy i Technologii. (2018). *Zrównoważony rozwój.* Retrieved from www.gov.pl/web/rozwoj-praca-technologia/zrownowazony-rozwoj

MNiSW. (2020). *Deklaracja Społecznej Odpowiedzialności Uczelni.* Retrieved from http://wsinf.edu.pl/assets/img/podstorny/Deklaracja%20Spolecznej%20Odpowiedz ialno%C5%9Bci%20Uczelni.pdf

Mueller, P. (2006). Exploring the knowledge filter: How entrepreneurship and university – industry relationships drive economic growth. *Research Policy, 35*(10), 1499–1508. doi:10.1016/j.respol.2006.09.023

MUR. (2020). *Istituzioni universitarie accreditate.* Retrieved from www.miur.gov.it/web/ guest/istituzioni-universitarie-accreditate

MUR. (2021). *Open data: Portale dei dati dell'istruzione superior.* Ministero dell'Istruzione dell'Università e della Ricerca. Retrieved from http://ustat.miur.it/opendata/

Orazbayeva, B., Davey, T., Plewa, C., & Galán-Muros, V. (2020). Engagement of academics in education-driven university-business cooperation: A motivation-based perspective. *Studies in Higher Education, 45*(8), 1723–1736. doi:10.1080/03075079.2019.15 82013

Państwowy Instytut Badawczy, O. P. I. (2019). *Szkolnictwo wyższe w Polsce w latach 2012– 2018.* Retrieved from https://radon.nauka.gov.pl/analizy/szkolnictwo-wyzsze-w-Polsce

Sénit, C. A. (2020, June). Leaving no one behind? The influence of civil society participation on the sustainable development goals. *Environment and Planning C, 38*(4), 693–712. doi:10.1177/2399654419884330

Shek, D. T. L., & Hollister, R. (2017). University social responsibility and promotion of the quality of life. In *University students: Promotion of holistic development in Hong Kong.* New York: Nova Science Publishers.

Sonetti, G., Barioglio, C., & Campobenedetto, D. (2020). Education for sustainability in practice: A review of current strategies within Italian universities. *Sustainability, 12,* 1–23.

THE. (2021). The times higher education impact rankings. *Times Higher Education.* Retrieved from www.timeshighereducation.com/impactrankings#!/page/0/length/25/sort_by/rank/sort_order/asc/cols/undefined

UI GreenMetric. (2021). *UI greenmetric world university rankings.* Retrieved from https://greenmetric.ui.ac.id/

Universitas Indonesia. (2020). *UI green metric world ranking 2020.* Retrieved from http://greenmetric.ui.ac.id/wp-content/uploads/2015/07/press-release-UI-GreenMetric-World-University-Rankings-2020.pdf

Wasiluk, A. (2017). Pro-innovative prerequisites for establishing the cooperation between companies (in the perspective of creation and development of clusters). *Procedia Engineering, 182,* 755–762. doi:10.1016/j.proeng.2017.03.195

Whalley, A., & Hicks, J. (2014). Spending wisely? How resources affect knowledge production in universities. *Economic Inquiry, 52*(1), 35–55. doi:10.1111/ecin.12011

Zampetakis, L. A., & Moustakis, V. (2006). Linking creativity with entrepreneurial intentions: A structural approach. *The International Entrepreneurship and Management Journal, 2*(3), 413–428. doi:10.1007/s11365-006-0006-z

# 16 Resilience of the Circular Economy

*Marek Ćwiklicki*

## Introduction

The purpose of the chapter is to discuss the resilience of the Circular Economy to the occurrence of different risks, many resulting from the nature of the very essence of the Circular Economy itself. The issue shall be presented from the perspective of the organisation, of the company engaging in the manufacturing of sustainable products, as well as industry more widely and the economy as a whole. The topic of the resilience of new business models is important because of their durability and viability. There can be no doubt about the benefits of the Circular Economy for society (Wijkman & Skånberg, 2015), yet the issue of its resilience to change has not been raised comprehensively in the literature. Some potential weaknesses of circular business models have been noted (Linder & Williander, 2017). It is also for this reason that resilience to potential risks which may arise is of key importance to managers who are trying to transform their business into a circular one. The premise for the preparation of this chapter is to seek to answer the following question: How resistant are circular business models to potential internal risks related to the models themselves, and to external risks concerning socio-economic changes, which could pose a threat to the functioning of such models?

In order to achieve this goal, recognised Circular Economy business models to be found in the literature and developed by various researchers will be used, such as the 9R model (Lüdeke-Freund, Gold, & Bocken, 2019; Potting, Hekkert, Worrell, & Hanemaaijer, 2017). These models are the embodiment of implementation of the principles of the Circular Economy in general. These models will be used as the first element of a theoretical canvas. The second element will be the risks most often accompanying the implementation of these models and arising during the course of their functioning, the occurrence of, and changes to which may have an influence on the status quo and pose a threat to the durability of the business model. The risks were determined on the basis of a literature review which indicated the existence of the following types of risks: internal process, technical, market, regulatory, and financial risks (cf. Bianchini, Rossi, & Pellegrini, 2019).

The structure of this chapter is made up of the following points. First, the concept of resilience in relation of the Circular Economy is discussed. Next, the types

DOI: 10.4324/9781003179788-16

of factors posing a threat to the functioning of circular business models will be characterised. For illustrative purposes, examples described in the literature will be used. The next point presents an assessment of the importance of a given risk for a given element of the Circular Economy model. Last, recommendations are formulated for managers and political decision-makers, drawing attention to the key issues resulting from the analysis conducted.

## Resilience and the Circular Economy

Resilience is understood as being characteristic of a system responsible for that system's reaction to change, enabling it to continue to remain active in its existing or a modified form. On this interpretation, it is emphasised that it is the ability to adjust to both anticipated events and to unexpected events and shocks (Labaka, Hernantes, Laugé, & Sarriegi, 2012). Depending on the research perspective adopted, resilience can be perceived as dynamic of the ecosystem of the flow of knowledge, factors, and robustness related to those processes. The ecosystem perspective results from initial consideration on the subject of resilience in the environmental sciences (Folke, 2006), which has since come to be used in other disciplines. This is why it may also concern the recovery of the system, its robustness, as well as adaptive capacity and learning, and transformability (ibid.). Table 16.1 presents some selected definitions of system resilience.

Thus, in the light of the aforementioned terminological considerations, resilience in the context of the Circular Economy, and at the level of the organisation – of circular business models, will involve determining the areas susceptible to potential damage or loss of its original properties, which may pose a threat to the functioning of the economic ecosystem as a whole.

*Table 16.1* Selected definitions of system resilience

| Author, year | Definitions |
| --- | --- |
| Allenby & Fink, 2000 | "Capability of a system to maintain its functions and structure in the face of internal and external change and to degrade gracefully when it must." |
| Haimes, 2009 | "Ability of system to with stand a major disruption within acceptable degradation parameters and to recover with a suitable time and reasonable costs and risks." |
| Vugrin et al., 2010 | "Given the occurrence of a particular disruptive event (or set of events), the resilience of a system to that event (or events) is that system's ability to reduce efficiently both the magnitude and duration of deviation from targeted system performance levels." |
| Pregenzer, 2011 | "Measure of a system's ability to absorb continuous and unpredictable change and still maintain its vital functions." |

Source: Own elaboration based on Hosseini, Barker, and Ramirez-Marquez (2016).

To determine the degree of resilience, it is useful to identify vulnerabilities in the circular business model. These may be susceptible to lose their properties due to change exerted by the external business environment, as well as due to the impact of an element of the system with which they are related. For example problems with the timeliness of production will result from the availability of raw material for recycling, and this in turn will result in a lack of performance in logistics activities, which are not necessarily related to irregularities in the supply process itself. This is why the first task in determining the resilience of the Circular Economy is to distinguish the elements which can be subjected to an assessment of risk and the occurrence of change.

## Characteristics of the Circular Economy business model

There are several Circular Economy business models described in the literature on the subject. The following section provides a descriptive summary allowing the common characteristics of that model to be determined. They were identified based on queries on databases of articles in Google Scholar, Scopus, and Web of Science, without any time restrictions on a search for the keyword "Circular Economy business model". This chapter treats circular business models as an operationalisation, as an expression of implementation of the CE.

The criteria for the breakdown of circular business models are related to the customer value proposition and the value network (Urbinati, Chiaroni, & Chiesa, 2017). In this way, we obtain a basic breakdown into a customer-oriented area, expressed by the introduction of circularity to the product and service offer, such as a sustainable product, for example, and an area external to the organisation and its participation in the supply chain, for example sustainable production, or sustainable supply chains.

A more structured description of the Circular Economy resulting in the expression of a larger number of components of the model is obtained in canvas-based models (Lewandowski, 2016). According to Lewandowski, they can also be identified based on the ReSOLVE framework proposed by the Ellen MacArthur Foundation. In the report published by that organisation (Ellen MacArthur Foundation, 2015), the main action areas for business and countries are Regenerate, Share, Optimise, Loop, Virtualise, and Exchange.

Publications aiming to summarise the knowledge of models have led to more advanced classifications, including the 9R model (Reike, Vermeulen, & Witjes, 2018). For this reason, in what follows in this text, the analysis will be based on this model, and its individual blocks will be discussed in the context of the risks to the functioning of the Circular Economy.

## Threats for the Circular Economy

Potential weaknesses of the Circular Economy were identified based on an analysis of the literature focussed on the risks, threats, and barriers to the implementation of circular business models. I will start by presenting a review of those publications and then give a summary, which will be used in the remaining chapters as part of a theoretical framework.

The first weak element of circular business models is said to be the necessity to find an appropriate group of *consumers*. After all, not every consumer will be ready to approve of a change in the type or nature of the product. Given that the issue of the importance of the customer in the Circular Economy is discussed in Chapters 13 and 14, I will only address that issue in brief here.

A list of consumer factors relating to the Circular Economy was compiled by L. Chamberlin and C. Boks (2018) based on an analysis of the literature. It is as follows based on the number of sources: tangible value (e.g. cost) (10), contamination, newness, disgust (9); quality (8), brand image (7), availability/convenience (6), ownership (5), environmental impact (5), customer service and support (5), warranty (3), and the opinion of other customers (3). These may be divided into internal and external factors.

Another summary of a review of barriers to the acceptance of the circular business model points to (in order of the number of publications mentioning the given barrier): hygiene issues (4), beliefs (4), worse performance of services (3), limited value for the customer (3), norms (2), lack of ownership (2), lack of awareness (1), and risk aversion (1) (Singh & Giacosa, 2019). As a result, the authors identified the main factors limiting consumer acceptance of change to a circular business model as:

- Individual factors, including person–product relationship, psychological ownership, habits and behaviours, and negative attitude;
- Social factors: Norms, lifestyle (consumerism), psychological essentialism expressed as valuing exclusivity; and
- Cultural factors: Power distance belief.

In accordance with the Push–Pull–Mooring model, change in consumer behaviour is related to macroeconomic factors influencing a propensity to change, for example price. Mooring factors of behaviour are attitudes. Pull factors for change are government incentives and environmental benefits (Hazen, Mollenkopf, & Wang, 2017). The studies showed that the key factor influencing consumers in making the switch from traditional to circular products is dependent on attitudes. This micro-level factor includes personal, social, and cultural values, which ties into the typology presented before.

To sum up on the topic of the consumer, it should be said that it is a factor which is directly under the control of the organisation. In a way similar to consumer behaviours in relation to any other product, here it is also possible to take demographic, economic, psycho-social, and cultural factors into consideration, along with socio-material conditions. Despite the fact that it is an important element in the circular business model, being at the boundaries of the organisation, it remains to a large extent susceptible to variations resulting from external influences. Broadening the approach to this barrier to the human factor in the organisation allows it to be related to attitudes understood more widely.

This risk comes along with other risks related to *fashion* (Linder & Williander, 2017). Fashion is something which is beyond the company's control. This is something which concerns industries where the aesthetics of the product is important. Attributes conducive to long-term satisfaction include the following aspects: quality, aesthetic, functional and emotional aspects (Niinimäki, 2017). These are aspects which concern the consumer, a subject which has already been addressed before.

The third factor is the requirement to possess greater *technological knowledge*. This is related to having an understanding not only of how the original product works, but also how its modified version, for example one that has undergone remanufacturing, can be successful (Linder & Williander, 2017). The risk related with this factor may be associated with a loss of knowledge and know-how in manufacturing or a lack of readiness in terms of competence to manage change to a circular business model. In general, this may be a question of circular literacy (Zwiers, Jaeger-Erben, & Hofmann, 2020). This concept includes system knowledge, required to provide an understanding of complexity; target knowledge, including how to set up the Circular Economy; and transformation knowledge, in other words know-how that is necessary to bring about change. Another element is learning from previous experiences. In this area, it may be equally appropriate to break down competencies by type with regard to the design of a Circular Economy into systems thinking (e.g. holistic thinking), as well as anticipatory (e.g. taking the use of the product in future cycles into consideration), normative (e.g. assessment of the environmental impact), and strategic (considering circular logistics, connecting processes) competencies (Sumter, de Koning, Bakker, & Balkenende, 2020).

This factor remains under the control of the organisation, and identification of one's own potential to make the transition to the Circular Economy should not pose any problems. A more important question is to point to the given type of knowledge and also to determine the probability of its occurrence. Studies indicate that a lack of specialist knowledge (expertise) in implementation of the Circular Economy is related to holding back change to a circular business model in small- and medium-sized firms in Europe (García-Quevedo, Jové-Llopis, & Martínez-Ros, 2020).

The next area vulnerable to threats is the *flow of raw materials* and thus the occurrence of interruptions in the supply of materials (secondary raw materials) for manufacturing. This factor is external to the organisation and is dependent on other entities on the market. This issue relating to the product life cycle shows that it is important to forecast trends in the ability to obtain raw materials (Östlin, Sundin, & Björkman, 2009). For example in the UK, Xerox encountered this problem when implementing a remanufacturing programme to use parts from old photocopiers to make new products (King, Miemczyk, & Dufton, 2006). It turned out that changes in supplies of used products due to changes in business demographics were hard to predict. This led to problems in maintaining an appropriate supply of labour.

Another example of differences in the flow of raw materials relates to changes in the level of recycling of municipal waste. As shown in Figure 16.1, the rate of

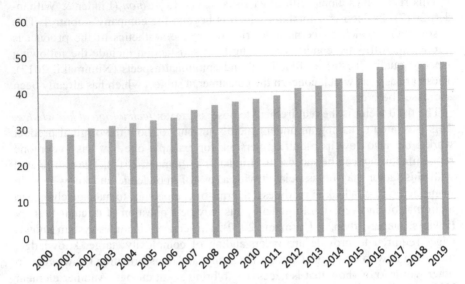

*Figure 16.1* Recycling rate of municipal waste in Europe (27 countries) for the years 2000 to 2019

Source: Eurostat (2021).

recycling is growing, which allows one to conclude that materials are available for further processing. However, the situation is different for each secondary raw material. For example the rate of return of aluminium cans in Poland in 2018 was 80% (RECAL, n.d.).

A study conducted in Suzhou, China, showed differences in the methods and scale of recycling of raw materials (Mo, Wen, & Chen, 2009). This gave grounds to suggest the implementation of an integrated IT system allowing for the flexible coordination of recycling, the introduction of tax incentives, and the encouragement to do research and development work on the recycling of raw materials with a low rate of recovery, such as textile waste or e-waste, for example. This factor shows the relationship with the consumer, because the attitude of the customer towards the product of recycling determines whether the purchase is made or not. Studies show that the image of product together with safety is of key importance to the consumer and not perceived quality or environmental benefits (Calvo-Porral & Lévy-Mangin, 2020).

The next factor which is critical from the perspective of organisation is the *specificity of products* offered. Different products are suitable to varying degrees to pro-sustainable activities. In this area, it is possible to distinguish between circular products with a traditional sales model (e.g. office furniture, desktop computers) and traditional products offered via circular business models (e.g. clothing, exterior building cleaning) (van Loon, Diener, & Harris, 2021). The potential factors identified in the studies contributing to product circularity are the possibility of extending product life, an increase in efficiency and a reduction in environmental impact during use of the product, and determining the period of product obsolescence (ibid.). This is related to the factor associated with the design for a Circular Economy approach which is the subject of Chapter 2. It thus concerns the susceptibility of the product to slowing resource loops, including the design of long-life goods and product-life extension, and closing resource loops (Bocken, de Pauw, Bakker, & van der Grinten, 2016). Other activities related to the introduction of circular products include the usage of renewable sources of energy, an increase in resource efficiency, and exploitation of waste as a resource (Urbinati, Chiaroni, & Toletti, 2019).

The aforementioned factor can be considered from the point of view of its specificity and susceptibility to change and the related circumstances necessary for its production. There are internal risks related to the company's own system of production and external risks resulting from the relationship between the company and its business partners. This is why, in the transition to a Circular Economy, it is recommended to have knowledge not just about the product itself, but also about its basis or, in other words, business models and the social context, the government environment, the social one, and the one accompanying the business model which sets out the framework for the product life cycle (European Environment Agency, 2017).

The next risk resulting from internal characteristics of the Circular Economy is referred to as *cannibalisation* (Linder & Williander, 2017). This refers to the potential decrease in sales as a result of continuing to use products which last longer. This is thus related to planned product obsolescence causing the consumer need to make a new purchase (product replacement) (Guiltinan, 2009). This also leads to the concept of creative destruction, which the author invokes and is based on Schumpeter's theory. Problems related to longer-lasting products arise on saturated markets. Product durability contributes to the emergence of new problems for the producer, because it stimulates competition between old and new versions of the product, with an impact on their price and ultimately on revenue.

The next risk is associated with *capital tied up* as a result of taking another approach to the product offering in the form of rental and leasing, and not sales. This issue is related to change in how the product is processed and – judging by the references to publications made by Linder and Williander – concerns product-service systems. A review of the literature concerning the strategies used showed that, in such systems, the following occurs most often: operational support for the product, for example through dedicated monitoring; product maintenance, for example through upgrades and servicing; product sharing; responsibility for reuse of product at end of life; and optimisation of result of use of product through offering additional functionality (Kjaer, Pigosso, Niero, Bech, & McAloone, 2019). The aforementioned activities require additional investment and thus have an impact on price. This once again raises the issue of the consumer and his/her willingness to finance the use of the product in a different way.

The next risk is that of a *lack of regulatory support*. For example in Poland, the classification of a material as waste means that it has reached the end of its life cycle. This results in legal barriers relating to it being put back onto the market as a secondary raw material. The main legal barriers include limited circular pro-curement, obstructing laws (including those which have an impact on keeping the price of natural raw materials low as a result of government subsidies), and a lack of global consensus (e.g. the measures to facilitate international trade and cross-border business cooperation on circular products) (Kirchherr et al., 2018).

Another weakness of the Circular Economy mentioned by Linder and Willian-der (2017) is that of *partner restrictions* in cooperation within the framework of a circular business model. This applies to the situation where there is cooperation between suppliers and sellers, whose own models should therefore be harmonised and adjusted. This applies in particular to cooperation within the supply chain. Barriers to the implementation of the Circular Economy identified based on a systemic review of the literature conducted by Govindan and Hasanagic (2018) showed that, in the case of the supply chain, critical factors include the technolog-ical capability needed to have a circular product manufactured by actors involved in the supply chain, employees possessing the appropriate skills, product quality, and a lack of standards concerning the manufacturing of such products (ibid.).

The restrictions identified before may be identified with the main categories of barriers to implementation of the Circular Economy: cultural, market, technologi-cal, and regulatory (Kirchherr, Hekkert, et al., 2017). However, what differenti-ates them is the source of their existence which lies in the characteristics of the Circular Economy.

The last weak point of the Circular Economy is the increase in *operating risk* related to the activity described before or, in other words, the extension of prod-uct monitoring, and thus the product-service system, but which may also occur independently. Classifications of operational risk for the CE are concerned with (Ethirajan, Arasu, Kandasamy, Kek, Nadeem, & Kumar, 2021):

- Delays in the supply of material resulting from the need for cooperation between different independent entities, for example recycler-producer;
- Coordination of workers resulting from different performances of the organi-sation within the supply chain;
- Worsening in the quality of materials related to a lack of a guarantee of conti-nuity in the procurement of raw materials contributing to a decrease in manu-facturing performance;
- Violation of safety measures adversely affecting the performance of the whole production process;
- Problems with the availability of specialised machinery resulting from the incompatibility of production systems between business partners participat-ing in the same supply chain.

## Assessment of resilience of the Circular Economy using the 9R model

One of the most advanced models for interpretation of the Circular Economy is considered to be the 9R model, which illustrates the transition from the linear economy to the Circular Economy. It is made up of three component parts, which can be assigned to specific strategies to be followed (Kirchherr, Reike, & Hekkert, 2017; Potting et al., 2017). This model was discussed in Chapter 1, but for the purposes of maintaining an orderly narrative flow, I will present the main categories here in tabular form along with a short explanation (Table 16.2).

The elements in Table 16.2 also characterise the circular business model with regard to the activities (functions) performed within it. In the part of the text which follows, these functions are listed together with the areas prone to weakness, resulting in the system being thrown out of balance. It is worth adding that areas that pose a risk of disrupting the activity of the system are of a static nature and concern concrete characteristics.

Table 16.3 presents the results of the correspondence drawn up by the author based on an assessment of the importance of a given risk for a given element of the model.

Areas vulnerable to threat do not have a major impact on the performance of activities in the form of R9. To this dimension of the Circular Economy, it is possible to assign technological knowledge, flow of raw materials, product specificity, and a lack of regulation. This is due to the relationships of these areas with energy recovery from incineration. In the case of R8, changes in a greater number of areas may have an impact on its activity. In particular, this applies to the flow of raw materials, the specificity of products, partner restrictions, and a lack of regulation. Such an assessment justifies the aspect most

*Table 16.2* Main elements of the 9R model of the Circular Economy

| General activities | Specific activities |
|---|---|
| Innovative product use and manufacturing | R0. Refuse concerning making the product redundant by providing another way of offering the function performed by the product |
| | R1. Rethink or, in other words, making product use more intensive |
| | R2. Reduce, meaning an increase in efficiency in product manufacturing while consuming fewer resources |
| Extend lifespan of the product and its parts | R3. Re-use meaning re-use of the same product by another consumer |
| | R4. Repair or, in other words, making it possible for the product to be used for its original function again. |
| | R5. Refurbish, meaning restoring an old product and bringing it up to date |
| | R6. Remanufacture or, in other words, the use of the product or its parts in the product with the same function |
| | R7. Repurpose or, in other words, the use of the discarded product for a new function |
| Useful application of materials | R8. Recycling of materials of similar quality |
| | R9. Recover, meaning recovery of energy from incineration of the product |

Source: Elaborated based on Potting et al. (2017).

*Table 16.3* Assessment of the importance of risks for elements of the Circular Economy model

| Dimension of the CE | Consumers | Fashion | Technological knowledge | Flow of raw materials | Specificity of products | Cannibalisation of products | Operational risks | Partner restrictions | Capital tied up | Lack of regulation |
|---|---|---|---|---|---|---|---|---|---|---|
| R0 | : | : | : | . | : | : | × | × | × | × |
| R1 | : | : | : | × | : | . | : | : | . | × |
| R2 | : | × | : | × | : | . | : | : | : | : |
| R3 | : | : | : | . | : | : | × | × | : | × |
| R4 | . | : | : | : | : | : | . | : | . | : |
| R5 | : | . | : | . | : | : | . | : | . | × |
| R6 | × | × | : | . | : | × | × | : | . | × |
| R7 | . | × | : | . | : | × | × | : | . | : |
| R8 | : | × | : | : | : | . | : | : | × | : |
| R9 | . | × | . | : | : | × | × | × | × | : |

Legend: Scale of assessment of importance of the risk for the given area: × – not significant, · – minor, :: – average, ··· – major.

frequently associated with the Circular Economy of recycling, which is dependent on 8 out 10 areas of resilience, with the exception of fashion and capital tied up, as being of a different nature.

For the next area R7, technological knowledge and product specificity are of key importance, because they are decisive in determining the possibility of adapting the product to new functions. In the case of R6, similar areas of risk play a dominant role, as do those with a weaker impact. The next dimension of the Circular Economy, R5, is vulnerable to threat from all the areas of risk mentioned in Table 16.3 with a particular emphasis on product specificity and the cannibalisation of products along with a lesser impact of the remaining areas. This is justified by the fact that the recovery of the product engages more elements of the business model. For R5, nearly all areas of resilience are of importance except for legal regulations. The potential impact of cannibalisation of products is of particular importance, which results from the importance of repairs of used products prolonging the lifespan of the product and the 4R model. The remaining areas of risk may have a disruptive effect on this activity to a lesser degree. For the last element related to the extension of the lifespan of the product or, in other words, R3, it is external factors that are of main significance, especially consumers, fashion, cannibalisation of products, and a lack of regulation, which have an impact on whether other people are going to make use of used products or not. In this respect, areas relating to the interior of the organisation, such as operational risk and partner restrictions, are not important.

In the group of elements dedicated to smart production and product use, R2 is the most susceptible to risk from the producer perspective. The following areas have a potential impact on more efficient production with more economic use of resources: technological knowledge, specificity of products, partner restrictions, and a lack of regulation. However, it is also affected to a lesser degree by external factors such as fashion or the flow of raw materials. For R1 – making product use more intensive – the following areas are key to maintaining durability: Consumers, specificity of products, and operational risk. The remaining areas of resilience are of less importance. However, for R0, which corresponds to a creative way of fulfilling a given function, the areas which are of most importance are both external to the organisation (consumer, fashion) and internal to it due to the importance of the product and the technological knowledge related to it.

The set of results of the assessment provided in Table 16.4 make it possible to indicate which of the areas of the Circular Economy are most vulnerable to the threat-related performance of function. This will be shown by a greater total number for major importance of the risk. The results show that elements related to an innovative product and manufacturing (R0, R1, R2) and then to extension of the product lifespan (R3, R4, R5) are the most at risk of a loss of resilience. However, elements of the Circular Economy concerning the useful application of the materials are the least at risk. On this basis, it is possible to draw the conclusion that the more complicated the model of the Circular Economy is, the greater the consequences for the system are as a result of the occurrence of risks related to that model. The aforementioned observation is confirmed by the example of the

*Table 16.4* Number of occurrences of risks for individual elements of the Circular Economy model

| Dimension of the CE | Major importance | Moderate importance | Minor importance | Not of significant importance | Total |
|---|---|---|---|---|---|
| | .. | .. | . | × | |
| R0 | 5 | 0 | 1 | 4 | 10 |
| R1 | 3 | 2 | 3 | 2 | 10 |
| R2 | 4 | 2 | 2 | 2 | 10 |
| R3 | 4 | 2 | 2 | 2 | 10 |
| R4 | 1 | 5 | 3 | 1 | 10 |
| R5 | 2 | 5 | 3 | 0 | 10 |
| R6 | 2 | 1 | 2 | 5 | 10 |
| R7 | 2 | 2 | 4 | 2 | 10 |
| R8 | 4 | 3 | 2 | 1 | 10 |
| R9 | 0 | 4 | 0 | 6 | 10 |

*Table 16.5* Number of occurrences of nature of risk

| Dimension of the CE | Major importance | Moderate importance | Minor importance | Not of significant importance | Total |
|---|---|---|---|---|---|
| | ... | .. | . | × | |
| Consumers | 3 | 2 | 2 | 2 | 10 |
| Fashion | 2 | 2 | 1 | 5 | 10 |
| Technological knowledge | 4 | 4 | 1 | 0 | 10 |
| Flow of raw materials | 1 | 2 | 4 | 2 | 10 |
| Specificity of products | 7 | 3 | 0 | 0 | 10 |
| Cannibalisation of products | 4 | 0 | 3 | 3 | 10 |
| Operational risks | 1 | 3 | 2 | 4 | 10 |
| Partner restrictions | 2 | 3 | 2 | 3 | 10 |
| Capital tied up | 0 | 2 | 5 | 3 | 10 |
| Lack of regulation | 3 | 3 | 1 | 3 | 10 |

Swedish company Unicykel, which produces bicycles, a study of which showed that there was a higher risk associated with the scaling up of a circular business model compared to that of a linear one (Linder & Williander, 2017).

The set of results in Table 16.5 indicate that the most important risks for the Circular Economy are those of specificity of products and a lack of technological knowledge. The others in the increasing order of importance are a lack of regulation, cannibalisation of products, and consumers. It is worth drawing attention to the fact that fashion, like operational risk, affects selected elements of the Circular Economy model. Their limited impact overall does not mean that they are not able to disrupt the functioning of the whole system. A narrower scope of impact is evidence of their specialist nature.

# Conclusion

This chapter has drawn attention to the issue of resilience of the Circular Economy. It was examined through the prism of the 9R model and areas of risk resulting from the specifics of the Circular Economy. In the text, attention was drawn to those elements which are particularly vulnerable to threats or to those risks which are important for the resilience of that system based on the Circular Economy. In the context of this chapter, this is related to problems with maintaining the sustainability of the circular business model, which, with reference to the economic sphere expressed as a closed loop in the biophysical sense, is impossible (Kovacic, Strand, & Völker, 2019).

A critical examination of circular business models enables one to focus not just on the positive aspects, but also on threats. These result above all from the changes which have to be made to the product design, the revenue model, and socio-institutional change (Potting et al., 2017). These types of changes in economic activity require appropriate financial resources. For this reason, not every company will be ready to make such changes, and equally not every consumer will be willing to accept circular products.

A good example of disruptions to the economy, including circular business models, is the occurrence of the COVID-19 pandemic. The effects that it has had, as well as the socio-economic restrictions related to it have had an impact on the production not only of sustainable products, but also on those produced in a traditional way. It was possible to see changes in customer behaviours and disruptions to global supply chains. Paradoxically, what caused problems for companies may change as a result of the transition to the Circular Economy. This is a question of relying on shorter supply chains and on the reuse of products without being dependent on imports (Ibn-Mohammed et al., 2021). The shortages of certain products which occurred drew manufacturers' attention to the need to look for new methods of production based on locally available resources (Wuyts, Marin, Brusselaers, & Vrancken, 2020), including in food production (Giudice, Caferra, & Morone, 2020). Therefore, despite potential weaknesses, the Circular Economy, due to its specific features, may be useful, thus demonstrating the predominance of resilience over the effects of the potential risks.

The conclusions from the analysis conducted here make it possible to formulate several guidelines for practitioners intending to implement a circular business model or operating based on such models.

First, the more advanced the implementation of the Circular Economy is, the more one should expect the occurrence of undesirable events to be in areas such as not only having to adapt the product to circularity and maintain it with respect to adapting it to consumers' needs, but also the technological capacity of the producer and business partners. In particular, attention should be paid to the need to maintain the appropriate technological knowledge to conduct circular activities.

Second, most at threat is the performance of activities focussed on innovative product use and manufacturing or, in other words, increasing product functionality and improving efficiency within the organisation.

Third, due to requirements to have expert technological knowledge, it is more appropriate to make the transition to circular business models for original equipment manufacturers (OEM), who can provide additional services such as warranty services (Lüdeke-Freund et al., 2019). However, studies for OEM suppliers indicate limited options for creating profitable closed-loop supply chains (van Loon & Van Wassenhove, 2018). Expert knowledge is correlated with the degree of complexity of the product and the need to make changes to it and prepare it for activity under the 9R model.

The analyses conducted allow for a conclusion to be drawn about barriers similar to the implementation of circular business models to those identified in publications on the topic of sustainable innovation (Guldmann & Huulgaard, 2020). This thus broadens the scope within which solutions useful in strengthening the resilience of the Circular Economy may be sought.

## References

Bianchini, A., Rossi, J., & Pellegrini, M. (2019). Overcoming the main barriers of circular economy implementation through a new visualization tool for circular business models. *Sustainability, 11*(23), 6614. doi:10.3390/su11236614

Bocken, N., de Pauw, I., Bakker, C., & van der Grinten, B. (2016). Product design and business model strategies for a circular economy. *Journal of Industrial and Production Engineering, 33*(5), 308–320. doi:10.1080/21681015.2016.1172124

Calvo-Porral, C., & Lévy-Mangin, J. P. (2020). The circular economy business model: Examining consumers' acceptance of recycled goods. *Administrative Sciences, 10*(2), 28. doi:10.3390/admsci10020028

Chamberlin, L., & Boks, C. (2018). Marketing approaches for a circular economy: Using design frameworks to interpret online communications. *Sustainability, 10*(6).

Ellen MacArthur Foundation. (2015). *Delivering the circular economy: A toolkit for policymakers*. Cowes: Ellen MacArthur Foundation.

Ethirajan, M., Arasu, M. T., Kandasamy, J., Kek, V., Nadeem, S. P., & Kumar, A. (2021). Analysing the risks of adopting circular economy initiatives in manufacturing supply chains. *Business Strategy and the Environment, 30*(1), 204–236. doi:10.1002/bse.2617

European Environment Agency. (2017). *Circular by design: Products in the circular economy*. (Vol. 6). Publications Office of the European Union. Retrieved from https://data.europa.eu/doi/10.2800/860754

Eurostat. (2021). *Recycling rate of municipal waste*. Eurostat. Retrieved from https://ec.europa.eu/eurostat/databrowser/view/cei_wm011/default/table?lang=en

Folke, C. (2006). Resilience: The emergence of a perspective for social – ecological systems analyses. *Global Environmental Change, 16*(3), 253–267. doi:10.1016/j.gloenvcha.2006.04.002

García-Quevedo, J., Jové-Llopis, E., & Martínez-Ros, E. (2020). Barriers to the circular economy in European small and medium-sized firms. *Business Strategy and the Environment, 29*(6), 2450–2464. doi:10.1002/bse.2513

Giudice, F., Caferra, R., & Morone, P. (2020). COVID-19, the food system and the circular economy: Challenges and opportunities. *Sustainability, 12*(19), 7939. doi:10.3390/su12197939

Govindan, K., & Hasanagic, M. (2018). A systematic review on drivers, barriers, and practices towards circular economy: A supply chain perspective. *International Journal of Production Research, 56*(1–2), 278–311. doi:10.1080/00207543.2017.1402141

Guiltinan, J. (2009). Creative destruction and destructive creations: Environmental ethics and planned obsolescence. *Journal of Business Ethics, 89*(S1), 19–28. doi:10.1007/s10551-008-9907-9

Guldmann, E., & Huulgaard, R. D. (2020). Barriers to circular business model innovation: A multiple-case study. *Journal of Cleaner Production, 243*, 118160. doi:10.1016/j.jclepro.2019.118160

Hazen, B. T., Mollenkopf, D. A., & Wang, Y. (2017). Remanufacturing for the circular economy: An examination of consumer switching behavior. *Business Strategy and the Environment, 26*(4), 451–464. doi:10.1002/bse.1929

Hosseini, S., Barker, K., & Ramirez-Marquez, J. E. (2016). A review of definitions and measures of system resilience. *Reliability Engineering & System Safety, 145*, 47–61. doi:10.1016/j.ress.2015.08.006

Ibn-Mohammed, T., Mustapha, K. B., Godsell, J., Adamu, Z., Babatunde, K. A., Akintade, D. D., . . . Koh, S. C. L. (2021). A critical analysis of the impacts of COVID-19 on the global economy and ecosystems and opportunities for circular economy strategies. *Resources, Conservation and Recycling, 164*, 105169. doi:10.1016/j.resconrec.2020.105169

King, A., Miemczyk, J., & Dufton, D. (2006). Photocopier remanufacturing at Xerox UK. In D. Brissaud, S. Tichkiewitch, & P. Zwolinski (Eds.), *Innovation in life cycle engineering and sustainable development* (pp. 173–183). Cham: Springer.

Kirchherr, J., Hekkert, M., Bour, R., Huijbrechtse-Truijens, A., Kostense-Smit, E., & Muller, J. (2017). *Breaking the barriers to the circular economy*. Deloitte, Utrecht University. Retrieved from https://dspace.library.uu.nl/handle/1874/356517

Kirchherr, J., Piscicelli, L., Bour, R., Kostense-Smit, E., Muller, J., Huibrechtse-Truijens, A., & Hekkert, M. (2018). Barriers to the circular economy: Evidence from the European Union (EU). *Ecological Economics, 150*, 264–272. doi:10.1016/j.ecolecon.2018.04.028

Kirchherr, J., Reike, D., & Hekkert, M. (2017). Conceptualizing the circular economy: An analysis of 114 definitions. *Resources, Conservation and Recycling, 127*, 221–232. doi:10.1016/j.resconrec.2017.09.005

Kjaer, L. L., Pigosso, D. C. A., Niero, M., Bech, N. M., & McAloone, T. C. (2019). Product/service-systems for a circular economy: The route to decoupling economic growth from resource consumption? *Journal of Industrial Ecology, 23*(1), 22–35. doi:10.1111/jiec.12747

Kovacic, Z., Strand, R., & Völker, T. (2019). *The circular economy in Europe: Critical perspectives on policies and imaginaries* (1st ed.). London: Routledge. doi:10.4324/9780429061028

Labaka, L., Hernantes, J., Laugé, A., & Sarriegi, J. M. (2012). Resilience: Approach, definition and building policies. In N. Aschenbruck, P. Martini, M. Meier, & J. Tölle (Eds.), *Future security* (Vol. 318, pp. 509–512). Berlin and Heidelberg: Springer. doi:10.1007/978-3-642-33161-9_74

Lewandowski, M. (2016). Designing the business models for circular economy – towards the conceptual framework. *Sustainability, 8*(1), 43. doi:10.3390/su8010043

Linder, M., & Williander, M. (2017). Circular business model innovation: Inherent uncertainties. *Business Strategy and the Environment, 26*(2), 182–196. doi:10.1002/bse.1906

Lüdeke-Freund, F., Gold, S., & Bocken, N. M. P. (2019). A review and typology of circular economy business model patterns. *Journal of Industrial Ecology, 23*(1), 36–61. doi:10.1111/jiec.12763

Mo, H., Wen, Z., & Chen, J. (2009). China's recyclable resources recycling system and policy: A case study in Suzhou. *Resources, Conservation and Recycling, 53*(7), 409–419. doi:10.1016/j.resconrec.2009.03.002

Niinimäki, K. (2017). Fashion in a circular economy. In C. E. Henninger, P. J. Alevizou, H. Goworek, & D. Ryding (Eds.), *Sustainability in fashion* (pp. 151–169). Cham: Springer International Publishing. doi:10.1007/978-3-319-51253-2_8

Östlin, J., Sundin, E., & Björkman, M. (2009). Product life-cycle implications for remanufacturing strategies. *Journal of Cleaner Production, 17*(11), 999–1009. doi:10.1016/j.jclepro.2009.02.021

Potting, J., Hekkert, M., Worrell, E., & Hanemaaijer, A. (2017). *Circular economy: Measuring innovation in product chains* (Policy Report). The Hague: PBL Netherlands Environmental Assessment Agen.

RECAL. (n.d.). *Recykling*. Retrieved February 19, 2021, from https://recal.pl/recykling/

Reike, D., Vermeulen, W. J. V., & Witjes, S. (2018). The circular economy: New or refurbished as CE 3.0? – Exploring controversies in the conceptualization of the circular economy through a focus on history and resource value retention options. *Resources, Conservation and Recycling, 135*, 246–264. doi:10.1016/j.resconrec.2017.08.027

Singh, P., & Giacosa, E. (2019). Cognitive biases of consumers as barriers in transition towards circular economy. *Management Decision, 57*(4), 921–936. Scopus. doi:10.1108/MD-08-2018-0951

Sumter, D., de Koning, J., Bakker, C., & Balkenende, R. (2020). Circular economy competencies for design. *Sustainability, 12*.

Urbinati, A., Chiaroni, D., & Chiesa, V. (2017). Towards a new taxonomy of circular economy business models. *Journal of Cleaner Production, 168*, 487–498. doi:10.1016/j.jclepro.2017.09.047

Urbinati, A., Chiaroni, D., & Toletti, G. (2019). Managing the introduction of circular products: Evidence from the beverage industry. *Sustainability, 11*(13), 3650. doi:10.3390/su11133650

Van Loon, P., Diener, D., & Harris, S. (2021). Circular products and business models and environmental impact reductions: Current knowledge and knowledge gaps. *Journal of Cleaner Production, 288*, 125627. doi:10.1016/j.jclepro.2020.125627

Van Loon, P., & Van Wassenhove, L. N. (2018). Assessing the economic and environmental impact of remanufacturing: A decision support tool for OEM suppliers. *International Journal of Production Research, 56*(4), 1662–1674. doi:10.1080/00207543.2017.1367107

Wijkman, A., & Skånberg, K. (2015). *The circular economy and benefits for society: Jobs and climate clear winners in an economy based on renewable energy and resource efficiency*. Club of Rome. Retrieved from https://clubofrome.org/wp-content/uploads/2020/03/The-Circular-Economy-and-Benefits-for-Society.pdf

Wuyts, W., Marin, J., Brusselaers, J., & Vrancken, K. (2020). Circular economy as a COVID-19 cure? *Resources, Conservation and Recycling, 162*, 105016. doi:10.1016/j.resconrec.2020.105016

Zwiers, J., Jaeger-Erben, M., & Hofmann, F. (2020). Circular literacy: A knowledge-based approach to the circular economy. *Culture and Organization, 26*(2), 121–141. doi:10.1080/14759551.2019.1709065

# Biographies

## Editors' biographies

**Magdalena Wojnarowska** – PhD in Economics in the field of Commodity Science, Assistant Professor at the Department of Technology and Ecology of Products at the Cracow University of Economics, Poland; manager of scientific research and development works; manager and contractor of research projects funded by the Polish Ministry of Science and Higher Education, as well as Poland's National Science Centre and National Centre for Research and Development, and the European H2020 programme; and author of numerous publications in the field of sustainable production and sustainable consumption, LCA, and eco-design. Her research focuses on the following issues: the Circular Economy, the bioeconomy, environmental management tools, environmental labelling, sustainable production, and sustainable consumption.

**Marek Ćwiklicki** – Prof. Dr. Habil., Prof. at the Cracow University of Economics. He is Habilitated Doctor of Economics and holds a PhD in Economics in the field of Management Science, both obtained from the Cracow University of Economics in Poland. As Head of the Department of Public Management, he researches, writes, and lectures on organisation theory, business research methodology, and public management. He is also Member of the Advisory Board of the Research Competence Centre for Corporate Social Responsibility at the FOM University of Applied Sciences in Germany [FOM KompetenzCentrum für Corporate Social Responsibility].

**Carlo Ingrao** – Assistant Professor in "Commodity Science" at the Department of Economics of University of Foggia, Italy; National qualification for associate professorship in "Commodity Science"; PhD in "Civil Infrastructure for the Territory" at the Kore University of Enna (Italy) in the field of energy efficiency and sustainability of buildings and building materials; PhD in "Geotechnical Engineering" at the University of Catania (Italy) addressing relevant environmental issues connected with the disposal of municipal solid waste; MSc degree in "Engineering for the Environment and the Territory" at the University of Catania on the application of Life Cycle Assessment (LCA) to a precast reinforced concrete shed for non-perishable goods storage; post-doc work experience: Several collaboration contracts at the Universities of Catania and Turin (Italy); research grant at the University of Foggia for collaboration with the research activity for the 7FP European project "Scientific and Technological Advancement in Research on Agroenergy" (STAR*Agroenergy), and at the Department of Engineering of University of Messina (Italy), on the application of LCA to material resource recovery. Research interests and activities include the development of life-cycle energy and environmental assessments in the fields of commodity science, industrial and environmental engineering, buildings, renewable energy, agriculture, and food production and packaging. He is an author of around 80 publications in international journals, as well as of book chapters and conference proceedings.

## Contributors' biographies

**Urszula Balon** – PhD (Eng.), Associate Professor, Department of Quality Management, Cracow University of Economics, Poland. Her research interests include quality management systems, costs of quality, customer satisfaction research, and market research into customer behaviour. She is the author of publications in the area of costs of quality and market research into customer behaviour. She has participated in national and international conferences.

**Riccardo Beltramo** – Full Professor at the University of Turin, Department of Management in the area of Commodity Sciences since 2005, holding a professorship in "Environmental Management Systems and Certification". Since 1985, he has been involved in the field of Industrial Ecology and conducts research into integrated management systems, applied to the manufacturing and service sectors, as well as industrial areas. His research interests range from responsible tourism, especially in mountain areas, to applications of the Internet of Things (IoT). He currently teaches Operations Management, Industrial Ecology, Tourism Eco-management, and Environmental Management Systems at the School of Management and Economics. He has been Chair of the Degree Program in Business Administration in complex learning and President and Director of the NatRisk Interdepartmental Centre on natural risks, a network for theoretical, experimental, and applied research and for the dissemination of information in the field of prediction, prevention, and management of the risk of natural disasters in mountain and hilly areas. As a creative thinker in the field of IoT, he has invented the Scatol8® System, a remote sensing network to monitor, display, and elaborate environmental and management variables, giving rise to the academic spin-off "The Scatol8 for Sustainability srl", http://scatol8.net.

**Barbara Campisi** – Associate Professor, PhD in Commodity Science at the University of Trieste, Italy; Member – in her capacity as former President – of the University Quality Assurance Committee. Her research interests span both quality and improvement-oriented approaches, environmental sustainability, voluntary reporting and certification.

**Giulio Mario Cappelletti** obtained his PhD in "Commodity Science" before going on to be awarded a 2-year post-doctoral scholarship in "Commodity Science". Since 2000, he has been Researcher and Aggregate Professor in Commodity Sciences at the Department of Economics, Management and Territory of the University of Foggia. His research interests concern the application and study of the LCA (Life Cycle Assessment) methodology, the environmental aspects of the agri-food sector, the recovery of agri-food by-products, as well as the quality and typicality of agri-food products. He is the author of over 90 publications published in proceedings of international conferences and national and international journals, and he has participated in 17 research projects on agri-food issues, 2 of which were at European level, with 8 at national and 7 at local level.

**Agnieszka Cholewa-Wójcik**, PhD in Management and Quality Sciences; Associate Professor at the Institute of Quality Sciences and Product Management at the Cracow University of Economics, Poland. Her research interests includes analysis of changes in packaging materials features during storage and usage, the role of packaging in quality assurance of products, and packaging design-concerning needs and requirements of consumers. An author of many expert opinions and opinions for state authorities, she cooperates with national and international companies in the packaging and logistics industry.

**Stefano Duglio,** Associate Professor at the University of Turin, Department of Management. He teaches "Operations Management" and "Environmental Management Systems". His research interests include environmental management for both the private and public sectors, with a specific focus on the tourism industry and the hospitality sector in mountain areas. He is co-founder of the academic spin-off "The Scatol8 for Sustainability Srl".

**Janina Filek** – Prof. Dr. Hab. at the Department of Philosophy at the Cracow University of Economics; Member of the CSR Working Group for the Polish Ministry of Development from 2014 to 2017 and a member of the Working Group on University Social Responsibility for the Polish Ministry of Funds and Regional Policy from 2018 to 2021; from 2016 to 2019, Vice-Rector of Cracow University of Economics for Communication and Cooperation, and, from 2019 to 2020, Vice-Rector for Education and Students; from 2016 to 2020, Member of the Communication and Social Responsibility Committee of the Conference of Rectors of Academic Schools in Poland (CRASP) and currently serving as an expert for this Committee. He is an author of many works on CSR and USR.

**Gianluigi Gallenti** – Full Professor in Agricultural Economics at the Department of Economics, Business, Mathematics and Statistics (DEAMS) at the University of Trieste, Italy; former Dean of the Faculty of Economics and Head of Department; Deputy Rector for Sustainability. Currently, he teaches Agri-Food Economics, Agri-Business Management, International Agri-Food Markets and Policy. He is the author of numerous publications in national and international scientific books and journals. His area of research concerns agricultural policies, agricultural risk management, sustainable agri-food production, sustainable agri-food consumer behaviour, and a sustainable agri-food system.

**Jakub Głowacki** – PhD (Economics); Assistant Professor, Department of Public Economics, Cracow University of Economics (Poland); Member of the Board of the Małopolska Social Economy Fund (since 2009). His research interests include new technologies in the public sector, public policy analysis, local government, water policy, and renewable energy sources.

**Marek Jabłoński** – PhD; Associate Professor. Currently, he is the Head of Department of Process Management, Cracow University of Economics (Poland). His research interests include modern management concepts, human resource management, and technology management. He has served as an expert on numerous scientific projects and also provides expert services in the field of design and improvement of remuneration and bonus systems in enterprises and public institutions.

**Olga Janikowska** – PhD in Political Sciences; after doing an internship at the Wissenschaftszentrum Berlin für Sozialforschung (WZB) and the University of California, she graduated with a PhD in Political Sciences. She has many years of experience as Lecturer at Polish and international institutions, including as Assistant Professor at the Department of Strategic Research at Mineral and Energy Economic Research Institute of the Polish Academy of Sciences. She is the author of several dozen scientific articles and five scientific books and editor of one scientific book. She has been a participant in international projects in education, corporate social responsibility, Sustainable Development, industrial symbiosis, and deliberation. She has also been a proponent of the concept of global justice and deliberative democracy.

**Bartłomiej Kabaja** – PhD Eng. in Economics; Assistant Professor at the Department of Product Packaging at the Cracow University of Economics, Poland; Post-graduate course: Waste and hazardous substances management, Cracow University of Technology. An author of several scientific articles and chapters in monographs, he has participated in many scientific conferences in Poland and abroad. A member of the Polish Logistics Association, the Association of Polish Engineers and Technicians, and the Polish Association of Commodity Science, he has received the Rector's Award (Cracow University of Economics) for individual scientific achievement on two occasions.

**Piotr Kafel** – PhD, Associate Professor at the Cracow University of Economics, Department of Quality Management. His research interests include quality and environmental management systems, food safety systems, and management system integration, as well as the implementation and certification of voluntary food quality systems (e.g. processing of organic farming products). He also has ties to an accredited product certification body, with which he has been cooperating since 2005. He is responsible for activities related to the maintenance and continuous improvement of the quality management system in accordance with the requirements of the ISO/IEC 17065 standard and conducting third-party audits, for example in terms of reducing packaging weight at source.

**Agnieszka Kawecka**, PhD in Economics in the field of Commodity Science; Assistant Professor at the Institute of Quality Sciences and Product Management at the Cracow University of Economics, Poland. Her basic areas of research are focussed on packaging science, environmental performance of packaging materials and packaging, safety of interactions between packaging and packed products, and safety management systems in manufacturing packaging facilities. An author of many expert opinions and opinions for state authorities and companies, she cooperates with national and regional economic organisations and associating entrepreneurs.

**Piotr Kopyciński** – PhD (Economics); Assistant Professor, Department of Public Economics, Cracow University of Economics (Poland). He was Director of the Małopolska School of Public Administration, Cracow University of Economics (2014–2016). His research interests include public policy (particularly innovation policy and territorial policy), modes of governance (public governance, the neo-Weberian state), and reforms in Central and Eastern Europe.

**Joanna Kulczycka**, Professor at the AGH University of Science and Technology – Head of the Department of Strategic Research at Mineral and Energy Economic Research Institute of the Polish Academy of Sciences, and Deputy Dean for Cooperation and Development of the Faculty of Management at the AGH University of Science and Technology, Poland; President of the Waste Management and Recycling Cluster (Key National Clusters in Poland) and Director of the Office of the Highway to Technology and Innovation Institute [*Instytut Autostrada Technologii i Innowacji – IATI*] in Krakow; author of over 200 publications (scientific articles, books, and chapters of books); member of many committees and advisory bodies at EU and national level, including the UNEP Resource Panel, the European Circular Economy Stakeholder Platform (2017–2020), and the Committee for the Sustainable Management of Raw Materials of the Polish Academy of Sciences (2016–2024). Joanna has been the organiser of many conferences and workshops in the area of the Circular Economy and the use of Life Cycle Assessment methods for the analysis of the impact of processes and products on the environment.

**Erik Roos Lindgreen** – PhD candidate, Department of Economics, University of Messina, Italy. He is an Early Stage Researcher on the CRESTING project, funded by the EU Horizon2020 programme. His research revolves around the success factors for the implementation of sustainability assessment methods at company level and designing best practices for CE and sustainability assessment. He holds a BA in the Liberal Arts and Sciences, an MS in Environment & Resource Management, and an MS in Earth Sciences.

**Patrizia de Luca** – PhD in Business and Management; Full Professor in Marketing and Management at the University of Trieste, Italy. Currently, she is Deputy Rector for Communication and Brand Strategy. Her research interests lie in marketing innovation and consumer behaviour from different perspectives, such as sustainable, digital, and experiential marketing.

**Mateusz Malinowski** – PhD in Technical Sciences (Environmental Engineering), Assistant Professor, Department of Bioprocess Engineering, Energetics and Automation, University of Agriculture in Krakow (Poland). His research interests include municipal solid waste management, with particular emphasis on biological treatment methods of bio-waste, as well as the application of biochar in waste treatment.

**Łukasz Mamica** – PhD; Associate Professor. Currently, he is Head of the Department of Public Economics, Cracow University of Economics. He is the coordinator of Research Area Industrial Policy and Development at European Association for Evolutionary Political Economy (EAEPE). He has co-authored all the regional innovation strategies for the Małopolska region and has served as an expert on several projects funded by OECD, the European Union, Polish government departments, and regional administration units.

**Małgorzata Miśniakiewicz** – PhD in Economics; Assistant Professor in the Department of Food Product Quality at the Cracow University of Economics where she has been a faculty member since 2000. She holds a PhD degree in Economics in the field of Commodity Science from the CUE. Her research interests are centred on innovativeness in food sector, consumer preferences, and sustainable consumption. She specifically focuses on food product development and risk management in FPD. Dr Miśniakiewicz is an experienced academic who adopts a very practice-oriented approach to her research. She works for business as a coach and consultant in the area of FPD and has developed many joint projects with food industry partners from Poland and other EU countries.

**Magdalena Muradin** – PhD (Economics), Assistant Professor, Department of Quality Management, Poznań University of Economics and Business (Poland). As Academic Lecturer, Dr Muradin teaches in a number of areas of management. Her research area includes environmental Life Cycle Assessment of products and circular business models. She also works at the Mineral and Energy Economy Research Institute of the Polish Academy of Sciences as an LCA specialist. She is involved as an expert in European, national, and regional projects in LCA and the Circular Economy.

**Tomasz Nitkiewicz** – PhD (management); Associate Professor, Department of Business Informatics and Ecosystems, Częstochowa University of Technology, Poland; Visiting Professor at the Cracow University of Economics, Poland. His research interests include different applications of Life Cycle Assessment with a focus on reverse logistics and end-of-life scenarios, environmental management, sustainability assessment, and its impacts on business competitiveness. He is Founder and the current head of the Life Cycle Modeling Centre at Częstochowa University of Technology. He is also engaged in initiatives in developing education in management and engineering towards meeting the new challenges of Industry 4.0 and the Circular Economy (MSIE4.0 project).

**Agnieszka Nowaczek**, MSc; Assistant in the Division of Strategic Research at Mineral and Energy Economy Research Institute of the Polish Academy of Sciences; Graduate in Sociology and Environmental Law; currently completing doctoral studies at the AGH University of Science and Technology in Krakow. She has the experience in managing and participating in national and international projects, including in the area of Circular Economy, Sustainable Development, waste management, and recovery of raw materials. She is the author and co-author of several scientific publications in this field.

**Paweł Nowicki** – PhD, Assistant Professor at the Cracow University of Economics, Department of Quality Management. His research interests include food quality and safety assurance and management systems, quality and environmental management systems, risk analysis in quality management systems and food safety assurances systems, standardised management systems in the Circular Economy, and circular economics. An author of numerous publications in the field of standardised management systems, risk management, and the Circular Economy, he has participated in several dozen international and national scientific conferences.

**Katelin Opferkuch** – PhD candidate, Department of Science and Technology, Universidade Aberta and Center for Environmental and Sustainability Research (CENSE), NOVA University, Portugal. She is an Early Stage Researcher on the CRESTING project, funded by the EU Horizon2020 programme. Her research combines the Circular Economy, non-financial reporting, and sustainability assessments to determine the value of Circular Economy reporting and communicating strategies for companies. She holds a BS in Environmental Health and Health & Safety and an MA in Sustainable Development.

**Linda O'Riordan** – PhD; Professor at the FOM University of Applied Sciences, Essen, Germany. In her role as a reflective practitioner, she is currently Director of a Research Competence Centre for Corporate Social Responsibility at the FOM University of Applied Sciences in Germany. Her academic activities include lecturing on business studies and international management at leading universities. Her research work focusing on sustainable approaches for business in society has been published in internationally renowned research publications. She is the author, editor, and reviewer of various academic books and peer-reviewed journals. Her latest highly acclaimed book is *Managing Sustainable Stakeholder Relationships: Corporate Approaches to Responsible Management*. Before becoming an academic, she gained business and consultancy experience from working in industry. Some of her former employers include Accenture, UCB-Schwarz Pharma, and the Government of Ireland (Irish Food Board/Bord Bia).

**Kamila Pilch,** MA in Sociology (specialisation: economy and market research); Research Assistant in the Department for Management of Public Organisations at the Cracow University of Economics. She has completed her postgraduate studies in Public Relations at the Jagiellonian University. She has experience in conducting qualitative and quantitative research projects. Her research interests include the methodology of marketing and social research and the analysis of public policies.

**Anna Prusak** – PhD; Associate Professor, Department of Process Management, Cracow University of Economics, Poland. Her research interests include multicriteria decision support methods, pair wise comparisons methods, and project management. She received a distinction from the Polish Academy of Sciences for scientific achievements in application of the AHP in the field of agricultural sciences. She is the author and co-author of numerous publications. She has participated in over a hundred scientific and R&D projects, both at national and international levels and also cooperates with the industry. She has managed and continues to manage projects financed by the Polish National Science Center, the Polish Ministry of Science and Higher Education, and the Polish National Center for Research and Development.

**Andrea Raggi** – Full Professor, Department of Economic Studies, University G. d'Annunzio of Chieti-Pescara, Italy. He has conducted research and taught for more than three decades in environmental management and industrial ecology and chaired the interdisciplinary PhD programme on Business, Institutions, Markets. His research perspectives include environmental management systems, life cycle sustainability assessment, industrial symbiosis, and Circular Economy. His interdisciplinary research projects also involve engagement with industry stakeholders both at a national and international level. He has authored or co-authored more than 230 publications.

**Alberto Simboli** – PhD (Economics); Associate Professor, Department of Economic Studies, University G. d'Annunzio of Chieti-Pescara, Italy. He teaches and conducts research in commodity science and technology, the organisation and management of production systems, and supply chains and environmental studies, mainly focussed on industrial symbiosis and Life Cycle Assessment. He is involved in international and national research groups, with focus on local and regional territory, stakeholders, and companies. He is also a member of the Quality Committee and Third Mission Deputy of the Department of Economic Studies.

**Mariusz Sołtysik** – PhD in Economics in the field of Management; Assistant Professor at the Department of Management Process, College of Management and Quality Sciences, Cracow University of Economics. Post-graduate course: Manager of scientific research and development works; Lecturer in postgraduate studies in Innovative Project Management. Sołtysik has been an author of numerous publications in renowned foreign journals in the field of innovative project management, Sustainable Development of enterprises, and designing organisational strategies.

**Walter J.V. Vermeulen** – Associate Professor, Copernicus Institute for Sustainable Development, Utrecht University, the Netherlands. He has long-standing experience in analysing implementation progress of company sustainability practices in different countries. His work is focused on new forms of private governance in supply chains in Europe and beyond and on Circular Economy practices. He has also published articles on the essence of Sustainable Development and its social dimension and the need for transdisciplinary research. He was President of the International Sustainable Development Research Society.

**Enrica Vesce** – Associate Professor at the University of Turin, Italy. She is a member of Department of Management Commodity Science Research Group, and, since October 2020, she has been the director of the Business Administration Degree Course in complex learning. She has been involved in many national and international research projects in her main fields of research: eco-labelling of products and services, tools and models for environmental analysis, industrial ecology, ecologically-equipped industrial areas, and food products. She teaches courses in "Operations Management, "Tools and technologies for an industrial ecology", and "Food science technology".

**Anna M. Walker** – PhD candidate, Department of Economic Studies, University G. d'Annunzio of Chieti-Pescara, Italy. She is an Early Stage Researcher on the CRESTING project, funded by the EU Horizon2020 programme, on advancing a sustainable Circular Economy. Her research focusses on identifying and applying suitable approaches to assess the sustainability of circular supply chains and industrial symbiosis. She holds a BA in International Affairs and an MA in Standardization, Social Regulation & Sustainable Development.

# Index

Printed in the United States
by Baker & Taylor Publisher Services

Printed in the United States
by Baker & Taylor Publisher Services